Advances in Understanding of Unit Operations in Non-ferrous Extractive Metallurgy 2021

Advances in Understanding of Unit Operations in Non-ferrous Extractive Metallurgy 2021

Editors

Srecko Stopic
Bernd Friedrich

MDPI • Basel • Beijing • Wuhan • Barcelona • Belgrade • Manchester • Tokyo • Cluj • Tianjin

Editors
Srecko Stopic
IME Process Metallurgy and
Metal Recycling, RWTH
Aachen University,
52056 Aachen, Germany

Bernd Friedrich
IME Process Metallurgy and
Metal Recycling, RWTH
Aachen University,
52056 Aachen, Germany

Editorial Office
MDPI
St. Alban-Anlage 66
4052 Basel, Switzerland

This is a reprint of articles from the Special Issue published online in the open access journal *Metals* (ISSN 2075-4701) (available at: https://www.mdpi.com/journal/metals/special_issues/cl_non_ferrous_extractive_metallurgy).

For citation purposes, cite each article independently as indicated on the article page online and as indicated below:

LastName, A.A.; LastName, B.B.; LastName, C.C. Article Title. *Journal Name* **Year**, *Volume Number*, Page Range.

ISBN 978-3-0365-4573-8 (Hbk)
ISBN 978-3-0365-4574-5 (PDF)

© 2022 by the authors. Articles in this book are Open Access and distributed under the Creative Commons Attribution (CC BY) license, which allows users to download, copy and build upon published articles, as long as the author and publisher are properly credited, which ensures maximum dissemination and a wider impact of our publications.

The book as a whole is distributed by MDPI under the terms and conditions of the Creative Commons license CC BY-NC-ND.

Contents

About the Editors vii

Srecko Stopic and Bernd Friedrich
Advances in Understanding of Unit Operations in Non-Ferrous Extractive Metallurgy 2021
Reprinted from: *Metals* **2022**, *12*, 554, doi:10.3390/met12040554 1

Kristina Božinović, Nada Štrbac, Aleksandra Mitovski, Miroslav Sokić, Duško Minić, Branislav Marković and Jovica Stojanović
Thermal Decomposition and Kinetics of Pentlandite-Bearing Ore Oxidation in the Air Atmosphere
Reprinted from: *Metals* **2021**, *11*, 1364, doi:10.3390/met11091364 7

Malte Drobe, Frank Haubrich, Mariano Gajardo and Herwig Marbler
Processing Tests, Adjusted Cost Models and the Economies of Reprocessing Copper Mine Tailings in Chile
Reprinted from: *Metals* **2021**, *11*, 103, doi:10.3390/met11010103 21

Elif Emil Kaya, Ozan Kaya, Srecko Stopic, Sebahattin Gürmen and Bernd Friedrich
NdFeB Magnets Recycling Process: An Alternative Method to Produce Mixed Rare Earth Oxide from Scrap NdFeB Magnets
Reprinted from: *Metals* **2021**, *11*, 716, doi:10.3390/met11050716 43

Jonas Mitterecker, Milica Košević, Srecko Stopic, Bernd Friedrich, Vladimir Panić, Jasmina Stevanović and Marija Mihailović
Electrochemical Investigation of Lateritic Ore Leaching Solutions for Ni and Co Ions Extraction
Reprinted from: *Metals* **2022**, *12*, 325, doi:10.3390/met12020325 57

Vesna Marjanovic, Aleksandra Peric-Grujic, Mirjana Ristic, Aleksandar Marinkovic, Radmila Markovic, Antonije Onjia and Marija Sljivic-Ivanovic
Selenate Adsorption from Water Using the Hydrous Iron Oxide-Impregnated Hybrid Polymer
Reprinted from: *Metals* **2020**, *10*, 1630, doi:10.3390/met10121630 69

Hugo Lucas, Srecko Stopic, Buhle Xakalashe, Sehliselo Ndlovu and Bernd Friedrich
Synergism Red Mud-Acid Mine Drainage as a Sustainable Solution for Neutralizing and Immobilizing Hazardous Elements
Reprinted from: *Metals* **2021**, *11*, 620, doi:10.3390/met11040620 85

Yiqian Ma, Srecko Stopic, Xuewen Wang, Kerstin Forsberg and Bernd Friedrich
Basic Sulfate Precipitation of Zirconium from Sulfuric Acid Leach Solution
Reprinted from: *Metals* **2020**, *10*, 1099, doi:10.3390/met10081099 101

Radmila Markovic, Vesna Krstic, Bernd Friedrich, Srecko Stopic, Jasmina Stevanovic, Zoran Stevanovic and Vesna Marjanovic
Electrorefining Process of the Non-Commercial Copper Anodes
Reprinted from: *Metals* **2021**, *11*, 1187, doi:10.3390/met11081187 115

Andrey Yasinskiy, Sai Krishna Padamata, Ilya Moiseenko, Srecko Stopic, Dominic Feldhaus, Bernd Friedrich and Peter Polyakov
Aluminium Recycling in Single- and Multiple-Capillary Laboratory Electrolysis Cells
Reprinted from: *Metals* **2021**, *11*, 1053, doi:10.3390/met11071053 133

Münevver Köroğlu, Burçak Ebin, Srecko Stopic, Sebahattin Gürmen and Bernd Friedrich
One Step Production of Silver-Copper (AgCu) Nanoparticles
Reprinted from: *Metals* **2021**, *11*, 1466, doi:10.3390/met11091466 143

Gözde Alkan, Milica Košević, Marija Mihailović, Srecko Stopic, Bernd Friedrich, Jasmina Stevanović and Vladimir Panić
Characterization of Defined Pt Particles Prepared by Ultrasonic Spray Pyrolysis for One-Step Synthesis of Supported ORR Composite Catalysts
Reprinted from: *Metals* **2022**, *12*, 290, doi:10.3390/met12020290 155

Srecko Stopic, Felix Wenz, Tatjana-Volkov Husovic and Bernd Friedrich
Synthesis of Silica Particles Using Ultrasonic Spray Pyrolysis Method
Reprinted from: *Metals* **2021**, *11*, 463, doi:10.3390/met11030463 167

Rebeka Rudolf, Aleš Stambolić and Aleksandra Kocijan
Atomic Layer Deposition of aTiO_2 Layer on Nitinol and Its Corrosion Resistance in a Simulated Body Fluid
Reprinted from: *Metals* **2021**, *11*, 659, doi:10.3390/met11040659 179

Miroslava Varničić, Miroslav M. Pavlović, Sanja Eraković Pantović, Marija Mihailović, Marijana R. Pantović Pavlović, Srećko Stopić and Bernd Friedrich
Spray-Pyrolytic Tunable Structures of Mn Oxides-Based Composites for Electrocatalytic Activity Improvement in Oxygen Reduction
Reprinted from: *Metals* **2022**, *12*, 22, doi:10.3390/met12010022 197

Duygu Yeşiltepe Özcelik, Burçak Ebin, Srecko Stopic, Sebahattin Gürmen and Bernd Friedrich
Mixed Oxides NiO/ZnO/Al_2O_3 Synthesized in a Single Step via Ultrasonic Spray Pyrolysis (USP) Method
Reprinted from: *Metals* **2022**, *12*, 73, doi:10.3390/met12010073 213

Tatjana Volkov-Husović, Ivana Ivanić, Stjepan Kožuh, Sanja Stevanović, Milica Vlahović, Sanja Martinović, Srecko Stopic and Mirko Gojić
Microstructural and Cavitation Erosion Behavior of the CuAlNi Shape Memory Alloy
Reprinted from: *Metals* **2021**, *11*, 997, doi:10.3390/met11070997 225

Srecko Stopic and Bernd Friedrich
Advances in Understanding of the Application of Unit Operations in Metallurgy of Rare Earth Elements
Reprinted from: *Metals* **2021**, *11*, 978, doi:10.3390/met11060978 237

About the Editors

Srecko Stopic
Education:
1984   High school graduation in Uzice/Serbia
1986–1991 Study of non-ferrous metallurgy at the Faculty of Technology and Metallurgy, University of Belgrade, Serbia
          degree: Diplom-Engineer
1991–1994 Magister and doctoral study at the Faculty of Technology and Metallurgy, University of Belgrade, Serbia (former Yugoslavia)
1997   Dr.-Ing.-examination, dissertation on "mechanism and kinetics of reduction of nickel chloride"
Professional career:
1991–1999 Scientific assistant at Department of nonferrous metallurgy of the Faculty of Technology and Metallurgy of the University of Belgrade
1999–2001 Assistant professor at non-ferrous metallurgy at the Faculty of Technology and Metallurgy of the University in Belgrade
2002–2003 Research Fellowship by Alexander von Humboldt Foundation at the IME Process Metallurgy and Metal Recycling, RWTH Aachen University
since 04.2003 scientific IME Process Metallurgy and Metal Recycling, chair and Institute of the RWTH Aachen University
30.04.2014 Professional thesis on Synthesis of metallic nanosized particles by ultrasonic spray pyrolysis (Privat Dozent at the RWTH Aachen University)
06.2020 Visiting Professor, Technical Faculty, Cacak, University in Kragujevac, Serbia

Bernd Friedrich
Education:
1977   High school graduation in Aachen/Germany 1978–1983 Study of non-ferrous metallurgy at RWTH Aachen University,
          degree: Diplom-Engineer
1984–1987 Scientist at IME Process Metallurgy and Metal-Recycling, RWTH Aachen University
1988   Dr.-Ing.-examination, dissertation on "electrolytic refining of recycling-tin using three dimensional-electrodes"
Professional career:
1988–1992 Head of R&D-Institute at GfE, Nuremberg/Germany (refractory metals, ferro-alloys, advanced materials, metal-recycling and residue-utilisation)
1992–1995 Head of profit center "hydride-technology and advanced materials" at GfE, Nuremberg/Germany
1995–1999 Plant manager NiCad/NiMH at Varta Batterie AG in Hagen/Germany and Ceska Lipa/Czech. Republic, head of R&D-center "innovative rechargeable battery systems R&D-coordinator 3C-Alliance (Varta-Toshiba-Duracel)
since 07.1999 Director of IME Process Metallurgy and Metal Recycling, chair and Institute of the RWTH Aachen University

Editorial

Advances in Understanding of Unit Operations in Non-Ferrous Extractive Metallurgy 2021

Srecko Stopic * and Bernd Friedrich

IME Process Metallurgy and Metal Recycling, RWTH Aachen University, 52056 Aachen, Germany; bfriedrich@ime-aachen.de
* Correspondence: sstopic@ime-aachen.de; Tel.: +49-17-6782-61674

Citation: Stopic, S.; Friedrich, B. Advances in Understanding of Unit Operations in Non-Ferrous Extractive Metallurgy 2021. *Metals* **2022**, *12*, 554. https://doi.org/10.3390/met12040554

Received: 11 March 2022
Accepted: 14 March 2022
Published: 25 March 2022

Publisher's Note: MDPI stays neutral with regard to jurisdictional claims in published maps and institutional affiliations.

Copyright: © 2022 by the authors. Licensee MDPI, Basel, Switzerland. This article is an open access article distributed under the terms and conditions of the Creative Commons Attribution (CC BY) license (https:// creativecommons.org/licenses/by/ 4.0/).

The high demand for critical materials, such as rare earth elements, indium, gallium, and scandium, raises the need for an advance in understanding the unit operations in non-ferrous extractive metallurgy. Unit metallurgical operation processes are usually separated into three categories: (1) hydrometallurgy (leaching, mixing, neutralization, precipitation, cementation, crystallization), (2) pyrometallurgy (roasting, smelting), and (3) electrometallurgy (aqueous electrolysis and molten salt electrolysis). In hydrometallurgy, the aimed metal is first transferred from ores and concentrates to a solution using a selective dissolution (leaching; dry digestion) under an atmospheric pressure below 100 °C and under a high pressure (40–50 bar) and high temperature (below 270 °C) in an autoclave and tube reactor. The purification of the obtained solution was performed using neutralization agents such as sodium hydroxide and calcium carbonate or more selective precipitation agents such as sodium carbonate and oxalic acid. The separation of metals is possible using a liquid/liquid process (solvent extraction in mixer-settler) and solid–liquid (filtration in filter-press under high pressure). Crystallization is the process by which a metallic compound is converted from a liquid into a solid crystalline state via a supersaturated solution. The final step is metal production using electrochemical methods (aqueous electrolysis for basic metals such as copper, zinc, silver and molten salt electrolysis for rare earth elements and aluminum). Advanced processes, such as ultrasonic spray pyrolysis and microwave-assisted leaching, can be combined with reduction processes in order to produce metallic powders. Some preparation for the leaching process is performed using a roasting process in a rotary furnace, where the sulphidic ore was first oxidized in an oxidic form, which is suitable for the metal transfer to water solution. During the smelting process, the target metal is further refined at high temperatures and reduced to its pure form. The pyrometallurgical treatment of the ore was performed in an electric furnace and combined with a refining during distillation. Unit Operations in Non-ferrous Extractive metallurgy can be successfully used for the recovery of non-ferrous metals from secondary materials.

The first Special Issue "Advances in Understanding of Unit Operations in Non-ferrous Extractive Metallurgy 2021" contains 17 papers divided in six groups:

1. Pyrometallurgical treatment (Thermal Decomposition and Kinetics of Pentlandite-Bearing Ore Oxidation in the Air Atmosphere [1])

The roasting of sulfide ores and concentrates is one of the most important steps in pyrometallurgical metal production from primary raw materials, due to the necessity of excess sulfur removal, present in the virgin material. Pentlandite is one of the main sources for nickel pyrometallurgical production. Raw pentlandite-bearing ore from the Levack mine (Ontario, Canada) was subjected to oxidative roasting in the air atmosphere. Thermochemical calculations, a phase analysis and construction of Kellogg diagrams for Ni-S-O and Fe-S-O systems at 298 K, 773 K, 923 K and 1073 K were used for proposing the theoretical reaction mechanism. A thermal analysis (TG/DTA—Thermogravimetric and Differential Thermal Analyses) was conducted in temperature range 298–1273 K, under a heating rate of 15 °C min^{-1}. A kinetic analysis was conducted according to the non-isothermal method of Daniels and Borchardt. Calculated activation energies of 113 kJ mol^{-1}, 146 kJ mol^{-1} and 356 kJ mol^{-1} for three oxidation stages imply that in every examined stage of the oxidation process, temperature is a dominant factor determining the reaction rate.

2. Dissolution of primary and secondary materials (1) Leaching, Processing Tests, Adjusted Cost Models and the Economies of Reprocessing Copper Mine Tailings in Chile; (2) NdFeB Magnets Recycling Process: An Alternative Method to Produce Mixed Rare Earth Oxide from Scrap NdFeB Magnets; and (3) Electrochemical Investigation of Lateritic Ore Leaching Solutions for Ni and Co Ions Extraction

To increase resource efficiency, mining residues—especially tailings—have come into the focus of research, companies, and politics. Tailings still contain varying amounts of unextracted elements of value and minerals that were not of economic interest during production. As for primary mineral deposits, only a small share of tailings offers the possibility for an economic reprocessing. To minimize exploration expenditure, a stepwise process is followed during exploration, to estimate the likelihood of a project to become a mine or in this case a reprocessing facility. During this process, costs are continuously estimated at least in an order of magnitude. Reprocessing flowsheets for copper mine tailings in Chile were developed, and costs and revenues of possible products from reprocessing were examined for a rough economic assessment. Standard cost models with capex and opex for flotation, leaching, and magnetic separation were adopted to the needs of tailings reprocessing. A copper tailing (around 2 Mt) that also contains magnetite was chosen as a case study [2]. A combination of magnetic separation and leaching gave the best economic results for copper and magnetite. The adopted cost models showed positive results at this early stage of investigation (semi-technical scale processing tests)

Spent NdFeB-magnets [3] and nickel lateritic ore [4] were chosen for the leaching process. The recovery of rare earth elements has become essential to satisfy this demand in recent years. In the present study, rare earth elements recovery from NdFeB magnets as new promising process flowsheet is proposed as follows; firstly acid baking process is performed to decompose the NdFeB magnet to increase in the extraction efficiency for Nd, Pr, and Dy. Than iron was removed from the leach liquor during hydrolysis. Finally, a production of REE-oxide from leach liquor was performed using ultrasonic spray pyrolysis method.

To examine the possible pathways of intrinsic electrochemical extraction of the crucial elements Ni and Co, it was necessary to make model solutions of these elements and to subject them to electrochemical examination techniques in order to obtain a benchmark. Beside Ni and Co, the model system for Fe had to be evaluated. The leachate examination results were compared to separate model solutions, as well as to their combinations in concentrations and to pH values comparable to those of the leachate [4].

3. Purification of solution during adsorption, precipitation and neutralization (1). Selenate Adsorption from Water Using the Hydrous Iron Oxide-Impregnated Hybrid Polymer; (2) Synergism Red Mud-Acid Mine Drainage as a Sustainable Solution for Neutralizing and Immobilizing Hazardous Elements; and (3) Basic Sulfate Precipitation of Zirconium from Sulfuric Acid Leach Solution

Hybrid adsorbent, based on the cross-linked copolymer impregnated with hydrous iron oxide, was applied for the first time for Se(VI) adsorption from water [5]. The influence of the initial solution pH, selenate concentration and contact time to adsorption capacity was investigated in detail. Adsorbent regeneration was explored using a full factorial experimental design in order to optimize the volume, initial pH value and concentration of the applied NaCl solution as a reagent. In addition to the experiments with synthetic solutions, the adsorbent performances in drinking water samples were explored, showing the purification efficiency up to 25%, depending on the initial Se(VI) concentration and water pH. Determined sorption capacity of the cross-linked copolymer impregnated with hydrous iron oxide and its ability for regeneration, candidate this material for further research, as a promising anionic species sorbent.

Acid mine drainage (AMD) and red mud (RM) are frequently available in the metallurgical and mining industry. Treating AMD solutions require the generation of enough alkalinity to neutralize the acidity excess. RM, recognized as a waste generating high alkalinity solution when it is in contact with water, was chosen to treat AMD from South Africa at room temperature. A German and a Greek RM have been evaluated as a potential low-cost material to neutralize and immobilize harmful chemical ions from AMD. Results showed that heavy metals and other hazardous elements such as As, Se, Cd, and Zn had been immobilized in the mineral phase [6]. According to European environmental standards, S and Cr, mainly present in RM, were the only two elements not immobilized below the concentration established for inert waste.

H_2SO_4 was ensured to be the best candidate for Zr leaching from the eudialyte. The resulting sulfuric leach solution consisted of Zr(IV), Nb(V), Hf(IV), Al(III), and Fe(III). It was found that ordinary metal hydroxide precipitation was not feasible for obtaining a relatively pure product due to the co-precipitation of Al(III) and Fe(III). A basic zirconium sulfate precipitation method was investigated to recover Zr from a sulfuric acid leach solution of a eudialyte residue after rare earth elements extraction. Nb precipitated preferentially by adjusting the pH of the solution to around 1.0. The precipitate contained 33.77% Zr and 0.59% Hf with low concentrations of Fe and Al. It was found that a high-quality product of ZrO_2 could be obtained from the basic sulfate precipitate [7]

4   Electrochemical methods for metal refining and winning (1) Electrorefining Process of the Non-Commercial Copper Anodes and (2) Aluminum Recycling in Single- and Multiple-Capillary Laboratory Electrolysis Cells

The electrorefining process of the non-commercial Cu anodes was tested on the enlarged laboratory equipment over 72 h [8]. Cu anodes with Ni content of 5 or 10 wt.% and total content of Pb, Sn, and Sb of about 1.5 wt.% were used for the tests. The real waste solution of sulfuric acid character was a working electrolyte of different temperatures (T1 = 63 ± 2 °C and T2 = 73 ± 2 °C). The current density of 250 A/m^2 was the same as in the commercial process. Tests were confirmed that those anodes can be used in the commercial copper electrorefining process based on the fact that the elements from anodes were dissolved, the total anode passivation did not occur, and copper is deposited onto cathodes.

The single- and multiple-capillary cells were designed and used to study the kinetics of aluminum reduction in LiF–AlF$_3$ and equimolar NaCl–KCl with 10 wt.% AlF$_3$ addition at 720–850 °C [9]. The cathodic process on the vertical liquid aluminum electrode in NaCl–KCl (+10 wt.% AlF$_3$) in the 2.5 mm length capillary had mixed kinetics with signs of both diffusion and chemical reaction control. The apparent mass transport coefficient changed from 5.6×10^{-3} cm s^{-1} to 13.1×10^{-3} cm s^{-1} in the mentioned temperature range. The dependence between the mass transport coefficient and temperature follows an Arrhenius-type behavior with an activation energy equal to 60.5 kJ mol^{-1}.

5   Synthesis of metallic, oxidic and composite powders using different methods (1) One Step Production of Silver-Copper (AgCu) Nanoparticles; (2) Synthesis and Characterization of a Metal Catalyst prepared by Ultrasonic Spray Pyrolysis as Pre-Definition Step for Titanium oxide-supported Platinum (3) Synthesis of Silica Particles Using Ultrasonic Spray Pyrolysis Method; (4) Atomic Layer Deposition of a TiO$_2$ Layer on Nitinol and Its Corrosion Resistance in a Simulated Body Fluid; (5) Spray-Pyrolytic Tunable Structures of Mn Oxides-Based Composites for Electrocatalytic Activity Improvement in Oxygen Reduction, and (6) Mixed Oxides NiO/ZnO/Al$_2$O$_3$ Synthesized in a Single Step via Ultrasonic Spray Pyrolysis (USP) Method

Synthesis of metallic, oxidic and composite methods using different methods was performed by ultrasonic spray pyrolysis method and atomic layer deposition. AgCu nanoparticles were prepared through hydrogen-reduction-assisted Ultrasonic Spray Pyrolysis (USP) and the Hydrogen Reduction (HR) method [10]. The changes in the morphology and crystal structure of nanoparticles were studied using different concentrated precursors. The average particle size decreased from 364 nm to 224 nm by reducing the initial solution concentration from 0.05 M to 0.4 M. These results indicate that the increase in concentration also increases the grain size. Antibacterial properties of nanoparticles against Escherichia coli were investigated. The obtained results indicate that the produced particles show antibacterial activity (100%).

Polygonal Pt nanoparticles were synthesized using ultrasonic spray pyrolysis (USP) at different precursor concentrations [11]. Physicochemical analysis of the synthesized Pt particles involved thermogravimetric, microscopic, electron diffractive, and light absorptive/refractive characteristics. Electrochemical properties and activity in the oxygen reduction reaction (ORR) of the prepared material were compared to commercial Pt black. Registered electrochemical behavior is correlated to the structural properties of synthesized powders by impedance characteristics in ORR. The reported results confirmed that Pt nanoparticles of a characteristic and uniform size and shape, suitable for incorporation on the surfaces of interactive hosts as catalyst supports, were synthesized. It is found that USP-synthesized Pt involves larger particles than Pt black, with the size being slightly dependent on precursor concentration. Among ORR-active planes, the least active (111) structurally defined the synthesized particles. These two morphological and structural characteristics caused the USP-Pt to be made of lower Pt-intrinsic capacitive and redox currents, as well as of lower ORR activity. Although being of lower activity, USP-Pt is less sensitive to the rate of ORR current perturbations at higher overpotentials. This issue is assigned to less-compact catalyst layers and uniform particle size distribution, and consequently, of activity throughout the catalyst layer with respect to Pt black.

Silica has sparked strong interest in hydrometallurgy, catalysis, the cement industry, and paper coating. The synthesis of silica particles was performed at 900 °C using the ultrasonic spray pyrolysis (USP) method [12]. Ideally, spherical particles are obtained in one horizontal reactor from an aerosol. The controlled synthesis of submicron particles of silica was reached by changing the concentration of precursor solution. The experimentally obtained particles were compared with theoretically calculated values of silica particles. The obtained silica by ultrasonic spray pyrolysis had an amorphous structure. In comparison to other methods such as sol–gel, acidic treatment, thermal decomposition, stirred bead milling, and high-pressure carbonation, the advantage of the ultrasonic spray method for preparation of nanosized silica controlled morphology is the simplicity of setting up individual process segments and changing their configuration, one-step continuous synthesis, and the possibility of synthesizing nanoparticles from various precursors.

Nitinol is a group of nearly equiatomic alloys composed of nickel and titanium, which was developed in the 1970s. Its properties, such as superelasticity and shape memory effect, have enabled its use, especially for biomedical purposes. Due to the fact that Nitinol exhibits good corrosion resistance in a chloride environment, an unusual combination of strength and ductility, a high tendency for self-passivation, high fatigue strength, low Young's modulus and excellent biocompatibility, its use is still increasing [13]. In this research, Atomic Layer Deposition (ALD) experiments were performed on a continuous vertical cast (CVC) NiTi rod (made in-house) and on commercial Nitinol as the control material, which was already in the rolled state. The ALD deposition of the TiO_2 layer was accomplished in a Beneq TFS 200 system at 250 °C. The pulsing times for $TiCl_4$ and H_2O were 250 ms and 180 ms, followed by appropriate purge cycles with nitrogen (3 s after the $TiCl_4$ and 2 s after the H_2O pulses).

Hybrid nanomaterials based on manganese, cobalt, and lanthanum oxides of different morphology and phase compositions were prepared using a facile single-step ultrasonic spray pyrolysis (USP) process and tested as electrocatalysts for oxygen reduction reaction (ORR). Electrochemical performance was characterized by cyclic voltammetry and linear sweep voltammetry in a rotating disk electrode assembly [14]. All synthesized materials were found electrocatalytically active for ORR in alkaline media. Two different manganese oxide states were incorporated into a Co_3O_4 matrix, δ-MnO_2 at 500 and 600 °C and manganese (II,III) oxide-Mn_3O_4 at 800 °C. The difference in crystalline structure revealed flower-like nanosheets for birnessite-MnO_2 and well-defined spherical nanoparticles for material based on Mn_3O_4. Electrochemical responses indicate that the ORR mechanism follows a preceding step of MnO_2 reduction to MnOOH. The calculated number of electrons exchanged for the hybrid materials demonstrate a four-electron oxygen reduction pathway and high electrocatalytic activity towards ORR.

Mixed oxides have received remarkable attention due to the many opportunities to adjust their interesting structural, electrical, catalytic properties, leading to a better, more useful performance compared to the basic metal oxides. Mixed oxides $NiO/ZnO/Al_2O_3$ were synthesized in a single step via USP method using nitrate salts, and the temperature effects of the process were investigated (400, 600, 800 °C) [15]. The synthesized samples were characterized by means of scanning electron microscopy, energy-dispersive spectroscopy, X-ray diffraction and Raman spectroscopy analyses. The results showed Al_2O_3, NiO–Al_2O_3 and ZnO–Al_2O_3 systems with spinel phases. Furthermore, the Raman peaks supported the coexistence of oxide phases, which strongly impact the overall properties of nanocomposite.

6   Characterization and behavior of the produced materials (Microstructural and Cavitation Erosion Behavior of the CuAlNi Shape Memory Alloy)

Microstructural and cavitation erosion testing was carried out on Cu-12.8Al-4.1Ni (wt.%) shape memory alloy (SMA) samples produced by continuous casting followed by heat treatment consisting of solution annealing at 885 °C for 60 min and, later, water quenching [16]. Cavitation resistance testing was applied using a standard ultrasonic vibratory cavitation set up with stationary specimen. Surface changes during the cavitation were monitored by metallographic analysis using an optical microscope (OM), atomic force microscope (AFM), and scanning electron microscope (SEM) as well as by weight measurements. The results revealed a martensite microstructure after both casting and quenching. Microhardness value was higher after water quenching than in the as-cast state. After 420 min of cavitation exposure, a negligible mass loss was noticed for both samples showing excellent cavitation resistance.

Finally, the combined process for the winning of rare earth elements and their oxides from primary and secondary materials was presented in the review paper: Advances in Understanding of the Application of Unit Operations in Metallurgy of Rare Earth Elements, prepared by B. Friedrich and S. Stopic [17].

Author Contributions: Conceptualization, S.S., methodology and writing—review and editing, S.S. and B.F.; supervision B.F. All authors have read and agreed to the published version of the manuscript.

Funding: This research received no external funding.

Conflicts of Interest: The authors declare no conflict of interest.

References

1. Božinović, K.; Štrbac, N.; Mitovski, A.; Sokić, M.; Minić, D.; Marković, B.; Stojanović, J. Thermal Decomposition and Kinetics of Pentlandite-Bearing Ore Oxidation in the Air Atmosphere. *Metals* **2021**, *11*, 1364. [CrossRef]
2. Drobe, M.; Haubrich, F.; Gajardo, M.; Marbler, H. Processing Tests, Adjusted Cost Models and the Economies of Reprocessing Copper Mine Tailings in Chile. *Metals* **2021**, *11*, 103. [CrossRef]
3. Kaya, E.; Kaya, O.; Stopic, S.; Gürmen, S.; Friedrich, B. NdFeB Magnets Recycling Process: An Alternative Method to Produce Mixed Rare Earth Oxide from Scrap NdFeB Magnets. *Metals* **2021**, *11*, 716. [CrossRef]
4. Mitterecker, J.; Košević, M.; Stopic, S.; Friedrich, B.; Panić, V.; Stevanović, J.; Mihailović, M. Electrochemical investigation of lateritic ore leaching solutions for Ni and Co ions extraction. *Metals* **2022**, *12*, 325. [CrossRef]
5. Marjanović, V.; Perić-Grujić, A.; Ristić, M.; Marinković, A.; Marković, R.; Onjia, A.; Šljivić-Ivanović, M. Selenate Adsorption from Water Using the Hydrous Iron Oxide-Impregnated Hybrid Polymer. *Metals* **2020**, *10*, 1630. [CrossRef]
6. Lucas, H.; Xakalashe, B.; Stopic, S.; Ndlovu, N.; Friedrich, B. Synergism Red Mud-Acid Mine Drainage as a Sustainable Solution for Neutralizing and Immobilizing Hazardous Elements. *Metals* **2021**, *11*, 620. [CrossRef]

7. Ma, Y.; Stopic, S.; Wang, H.; Forsberg, K.; Friedrich, B. Basic Sulfate Precipitation of Zirconium from Sulfuric Acid Leach Solution. *Metals* **2020**, *10*, 1099. [CrossRef]
8. Markovic, R.; Krstic, V.; Friedrich, B.; Stopic, S.; Stevanovic, J.; Stevanovic, Z.; Marjanovic, V. Electrorefining Process of the Non-Commercial Copper Anodes. *Metals* **2021**, *11*, 1187. [CrossRef]
9. Yasinskiy, A.; Padamata, S.K.; Moiseenko, I.; Stopic, S.; Feldhaus, D.M.; Friedrich, B.; Polyakov, P. Aluminium Recycling in Single- and Multiple-Capillary Laboratory Electrolysis Cells. *Metals* **2021**, *11*, 1053. [CrossRef]
10. Köroğlu, M.; Ebin, B.; Stopic, S.; Gürmen, S.; Friedrich, B. One Step Production of Silver-Copper (AgCu) Nanoparticles. *Metals* **2021**, *11*, 1466. [CrossRef]
11. Alkan, G.; Košević, M.; Mihailović, M.; Stopic, S.; Friedrich, B.; Stevanović, J.; Panić, V. Characterization of Defined Pt Particles Prepared by Ultrasonic Spray Pyrolysis for One-Step Synthesis of Supported ORR Composite Catalysts. *Metals* **2022**, *12*, 290. [CrossRef]
12. Stopic, S.; Wenz, F.; Husovic, T.-V.; Friedrich, B. Synthesis of Silica Particles Using Ultrasonic Spray Pyrolysis Method. *Metals* **2021**, *11*, 463. [CrossRef]
13. Rudolf, R.; Stambolic, A.; Kocijan, A. Atomic Layer Deposition of a TiO_2 Layer on Nitinol and Its Corrosion Resistance in a Simulated Body Fluid. *Metals* **2021**, *11*, 659. [CrossRef]
14. Varničić, M.; Pavlović, M.M.; Pantović, S.E.; Mihailović, M.; Pavlović, M.R.P.; Stopić, S.; Friedrich, B. Spray-Pyrolytic Tunable Structures of Mn Oxides-Based Composites for Electrocatalytic Activity Improvement in Oxygen Reduction. *Metals* **2022**, *12*, 22. [CrossRef]
15. Özcelik, D.; Ebin, B.; Stopic, S.; Gürmen, S.; Friedrich, B. Mixed Oxides $NiO/ZnO/Al_2O_3$ Synthesized in a Single Step via Ultrasonic Spray Pyrolysis (USP) Method. *Metals* **2022**, *12*, 73. [CrossRef]
16. Volkov-Husović, T.; Ivanić, I.; Kožuh, S.; Stevanović, S.; Vlahović, M.; Martinović, S.; Stopic, S.; Gojić, M. Microstructural and Cavitation Erosion Behavior of the CuAlNi Shape Memory Alloy. *Metals* **2021**, *11*, 997. [CrossRef]
17. Stopic, S.; Friedrich, B. Advances in Understanding of the Application of Unit Operations in Metallurgy of Rare Earth Elements. *Metals* **2021**, *11*, 978. [CrossRef]

Thermal Decomposition and Kinetics of Pentlandite-Bearing Ore Oxidation in the Air Atmosphere

Kristina Božinović [1], Nada Štrbac [1], Aleksandra Mitovski [1,*], Miroslav Sokić [2], Duško Minić [3], Branislav Marković [2] and Jovica Stojanović [2]

1. Technical Faculty in Bor, University of Belgrade, 19210 Bor, Serbia; kbozinovic@tfbor.bg.ac.rs (K.B.); nstrbac@tf.bor.ac.rs (N.Š.)
2. Institute for Technology of Nuclear and Other Mineral Raw Materials, 11000 Belgrade, Serbia; m.sokic@itnms.ac.rs (M.S.); b.markovic@itnms.ac.rs (B.M.); j.stojanovic@itnms.ac.rs (J.S.)
3. Faculty of Technical Science, University in Priština, 40000 Kosovska Mitrovica, Serbia; dusko.minic@pr.ac.rs
* Correspondence: amitovski@tfbor.bg.ac.rs

Abstract: The roasting of sulfide ores and concentrates is one of the most important steps in pyrometallurgical metal production from primary raw materials, due to the necessity of excess sulfur removal, present in the virgin material. Pentlandite is one of the main sources for nickel pyrometallurgical production. The knowledge of its reaction mechanism, products distribution during oxidation and reaction kinetics is important for optimizing the production process. Raw pentlandite-bearing ore from the Levack mine (Ontario, Canada) was subjected to oxidative roasting in the air atmosphere. A chemical analysis of the initial sample was conducted according to EDXRF (Energy-Dispersive X-ray Fluorescence) and AAS (Atomic Adsorption Spectrometry) results. The characterization of the initial sample and oxidation products was conducted by an XRD (X-ray Diffraction) and SEM/EDS (Scanning Electron Microscopy with Energy Dispersive Spectrometry) analysis. Thermodynamic calculations, a phase analysis and construction of Kellogg diagrams for Ni-S-O and Fe-S-O systems at 298 K, 773 K, 923 K and 1073 K were used for proposing the theoretical reaction mechanism. A thermal analysis (TG/DTA—Thermogravimetric and Differential Thermal Analyses) was conducted in temperature range 298–1273 K, under a heating rate of 15° min^{-1}. A kinetic analysis was conducted according to the non-isothermal method of Daniels and Borchardt, under a heating rate of 15° min^{-1}. Calculated activation energies of 113 kJ mol^{-1}, 146 kJ mol^{-1} and 356 kJ mol^{-1} for three oxidation stages imply that in every examined stage of the oxidation process, temperature is a dominant factor determining the reaction rate.

Keywords: pentlandite; oxidation; reaction mechanism; phase analysis

1. Introduction

Nickel has a long history of being used for coin manufacturing. Interest in nickel production from its ores, concentrates and secondary resources increased rapidly in recent decades with the expansion of its commercial use in new products. Nickel resistance to high temperatures, heat and good corrosion stability, allows its application in durable materials without replacement for a long time. Nickel is used in materials and alloys which can resist aggressive environments such as jet engines, offshore installations and power generation plants. Nickel is also applied as an alloying element in cast irons, austenitic stainless steels and non-ferrous alloys. Nickel steel is widely used for armor plating. Other nickel alloys are used in boat propeller shafts and turbine blades. Nickel is also used in batteries, including rechargeable nickel–cadmium batteries and nickel–metal hydride batteries used in hybrid vehicles. It is expected to form an ever-larger proportion of future batteries [1–7]; namely, pentlandite, complex nickel sulfide, may also be used as is, in the electrochemical reaction of hydrogen evolution [8–10].

Nickel can be extracted from its ores by both hydrometallurgical and pyrometallurgical routes. Almost 70% of ore deposits are processed by hydrometallurgical extraction. Although this route is more convenient, pyrometallurgical processing of nickel, as well as copper [11,12], from its sulfide ores is commonly used worldwide. It is known that nearly all of nickel in its sulfide ores occurs in the form of binary sulfide mineral pentlandite ((Ni,Fe)$_9$S$_8$). Pentlandite is the most common and abundant mineral used for nickel extraction, accounting for over 60% of worldwide nickel production, mostly from sulfide copper–nickel ore deposits [13–15]. Pentlandite is usually associated with iron sulfides–pyrite (FeS$_2$) and pyrrhotite (Fe$_7$S$_8$) and copper-iron sulfide chalcopyrite (CuFeS$_2$). Nickel sulfides are intergrown within the iron sulfides, with nickel dissolved in the crystal lattice. In metallurgical practice, this can cause technical difficulties in nickel concentrate production [16].

In the pyrometallurgical production of nickel from its sulfide ores and concentrates, one of the most important processing steps is controlling the iron sulfide oxidation process. Reactions which occur during the nickel sulfides oxidation have a highly exothermal character. The amount of sulfur removed during roasting is the main parameter for controlling the matte grade in the following smelting stage. Thus, the partial roasting of nickel sulfide ores and concentrates, in order to decrease the sulfur amount in nickel calcine, has been practiced for a few decades. Sulfur removal is conducted by roasting in fluidized bed reactors, in multiple hearth roasters or by smelting in a flash smelting furnace. Strong off-gases are generated and are suitable for SO$_2$ capture and fixation. Certain changes were recently introduced into the process, which indicate to the most possible method of iron sulfide removal already in the ore milling stage [17].

The mechanism of sulfide minerals oxidation in the air atmosphere at elevated temperatures is quite complex, since it depends on sulfide species present in minerals, roasting conditions and the interactions between solid and gaseous phases in the investigated Me-S-O systems. A thermogravimetric (TG) and differential thermal analysis (DTA) under laboratory conditions, although having much milder oxidation conditions than the industrial roasting process, provides useful information on the reaction mechanism and kinetics of gas–solid reactions, for the possibility of a more precise control of experimental conditions and monitoring variable experimental parameters.

A literature review on reaction mechanisms of complex sulfide oxidation processes refers to two main possible processes: (1) forming a protective oxide layer on the particle surface, which inhibits the diffusion of reactive gas to unreacted inner core, and (2) predominant diffusion of one cation species in binary sulfides towards an oxygen rich interface, which leads to its preferential oxidation. Older reports on the oxidation of pentlandite, primarily involved heating it in an inert atmosphere [18–20].

Mkhonto et al. [21] in an ab initio study of oxygen adsorption on a nickel-rich pentlandite mineral surface confirmed that iron atoms are more reactive than nickel atoms, which implies preferential iron oxidation. Thornhill and Pidgeon [22] reported forming dense outer oxide shells below 923 K, during pentlandite particles roasting. They also confirmed selective oxidation when roasting chalcopyrite and pentlandite, with a preferential oxidation of iron species in both particles and the increase in the copper content in chalcopyrite and nickel content in pentlandite. This is confirmed by the findings of Ellingham [23], whose diagrams show that change in Gibbs free energy values for iron oxides are lower in comparison to nickel oxide, while the trend is opposite in the case of sulfides. Ashcroft [24] assumed that iron acts such as a catalyzer in the formation of copper, zinc and nickel sulfates during the treatment of copper and nickel ores and concentrates. Tanabe et al. [25] determined that the outer and inner layers during the oxidation of dense pentlandite in a mixed O$_2$–N$_2$ atmosphere consisted of duplex oxide layers of Fe$_2$O$_3$ and Fe$_3$O$_4$, with the rapid increase in the Fe$_3$O$_4$ layer at the initial period of oxidation. Consequently, there can be observed the iron depletion in the unreacted sulfide core in a pentlandite sample heated in air to 923 K. The oxidation behavior of pyrrhotite in air and oxygen was investigated by Kennedy and Sturman [26]. Dunn and Kelly [27] reported the oxidation of synthetic millerite prior to the investigation of natural mineral pentlandite oxidation at elevated temperatures in a dynamic oxygen atmosphere [28]. Results showed that in temperature interval 733–973 K, pentlandite decomposed into Fe_2O_3, $NiSO_4$, NiO, NiS and $NiFe_2O_4$. Iron sulfate decomposed above 913 K. Above 1073 K, only oxides NiO and Fe_2O_3 and spinel-phase trevorite $NiFe_2O_4$ were detected. Zhu et al. [29] reported the existence of nickel and sulfur-rich phases as intermediate phases in unreacted cores during the oxidation of synthetic pentlandite at 973 K in isothermal conditions. Hematite, trevorite and nickel oxide were observed as the final oxidation products. An SEM/EDS analysis showed the formation of gaps and holes between the oxide shell and sulfide core. Reaction rate was rapid in the first two hours, then it slowed down. Xia et al. [14] also investigated a sintetic pentlandite oxidation behavior at the temperature range of 530–600 °C by microscopic and kinetic methods. The proposed reaction mechanism involved the transformation of pentlandite to a monosulfide solid solution (mss), (Fe, Ni)S, following to α-NiS in the second step, oxidation from Fe_3O_4 to Fe_2O_3 in the third step, and the oxidation of α-NiS to NiO as the final step. Corresponding activation energies (140 kJ mol^{-1}, 151 kJ mol^{-1} and 127 kJ mol^{-1}) were calculated according to the Avrami/Arrhenius analysis. A thermogravimetric analysis of oxidation of a floated nickel sulfide concentrate mixture in an oxygen and air atmosphere, under various heating rates, was performed by Dunn and Jayaweera [30]. Two mass gains were observed, which correspond to metal sulfates formation. The thermoanalytical results showed a significant dependence on the heating rate. The air atmosphere reactions occur to a lesser extent comparing to the results of oxidation in the oxygen atmosphere. Pandher and Utigard [31] studied roasting of three nickel concentrates, consisting of pentlandite, pyrrhotite, silica and chalcopyrite by TGA analysis in an inert or oxidizing atmosphere in order to determine the reaction mechanism. Two broad peaks on the TG curve were observed, instead of one gradual bell-shaped peak while heating the nickel concentrates in inert atmospheres. According to mass changes, it was assumed that sulfates form in a temperature range from 773 K to 973 K. Yu and Utigard [32] reported TG/DTA results on the oxidation of nickel concentrate from an ambient temperature up to 1273 K. A reaction mechanism was proposed, where the preferential oxidation of iron sulfide occurred between 623 K and 973 K. At 1086 K, a nickel sulfide core melted and decomposed, forming NiO and $NiSO_4$. At 1215 K sulfate decomposed and all remaining nickel sulfide transformed into oxide due to the absence of a protective sulfate layer. Results of a kinetic analysis indicate that the diffusion of oxygen controls the reaction rate. Although some of the results agreed on some level with Dunn and Kelly [28], some discrepancies can be noticed regarding oxidation products and following the mechanism of pentlandite oxidation.

In recent years, pentlandite has been broadly researched due to its newly discovered usage. On the other hand, the mechanism of the pentlandite oxidation process itself has not been fully understood, considering the complexity of natural sulfide minerals. Nearly all of nickel in its sulfide ores occurs in the form of pentlandite. As a contribution to a better understanding of the pentlandite-bearing ore oxidation process in the air atmosphere at elevated temperatures, the results of characterization, thermodynamic, thermal and kinetic analysis are presented in this paper.

2. Materials and Methods

Pentlandite-bearing ore, from the Levack mine (Ontario, Canada) was used for the experimental investigation. Initial samples were prepared by crushing and milling a solid piece of pentlandite-bearing ore and then used in powder form. Sampling was performed by standard procedure of bulk mineral sampling division, first by quartering and then by chess field method, in order to ensure a homogeneous distribution and representative portion of all present mineral fractions in the initial sample.

Qualitative chemical analysis of the initial sample was conducted according to EDXRF (Energy-Dispersive X-ray Fluorescence) analysis, on an analyzer (Canberra Packard, Schwadorf, Austria) with radioisotopes for Cd-109 excitation (22.1 keV). The weight of the sample was

0.5 g. Quantitative chemical analysis was carried out using atomic adsorption spectrometry (AAS) (AAS PinAAcle 900 T, PerkinElmer, Überlingen, Germany). Examined samples (in three replicates) were previously dissolved by the nitric–perchloric acid digestion method described by Hseu [33], and the contents of selected inorganics were determined from the solution. A part of Si that remained undissolved in the form of a precipitate was further determined by the hydrochloric acid dehydration (gravimetric) technique [34].

Characterization of the initial sample and the oxidation products was conducted by X-ray diffraction method on X-ray diffractometer, model PW-1710 (PHILIPS, Eindhoven, The Netherlands), with a curved graphite monochromator and a scintillation counter. The intensities of diffracted CuKα X-rays (λ = 1.54178Å) were measured at room temperature at intervals of 0.02 °2θ and a time of 1 s in the range from 4 to 65 °2θ. The X-ray tube was loaded with a voltage of 40 kV and a current of 30 mA, while the slots for directing the primary and diffracted beam were 1° and 0.1 mm. The samples investigated by XRD method were isothermally roasted for 30 min in previously preheated furnace in the air atmosphere at 673 K, 773 K, 873 K, 973 K and 1073 K. Oxidation products were taken out of the furnace and cooled in the air atmosphere at room temperature.

The initial sample and the oxidation products were analyzed by scanning electron microscope with energy dispersive spectrometry (SEM/EDS) using a VEGA3 Scanning Electron Microscope (TESCAN, Brno, Czech Republic). In order to better understand the reaction mechanism of the oxidation process of the pentlandite-bearing ore, a thermal analysis of the sample was also performed. TG/DTA analysis was performed with a single heating rate of 15° min^{-1}, in temperature range 298–1273 K, using a STA 409 EP (Netzsch, Selb, Germany). Thermal analysis and X-ray diffraction experiments were performed on 100 mg initial samples.

Proposing theoretical reaction mechanism included thermodynamic calculation, phase analysis and construction of Kellogg's diagrams for Ni-S-O and Fe-S-O systems at various temperatures: 298 K, 773 K, 923 K and 1073 K. Based on the constructed Kellogg's diagrams, with values of the oxygen partial pressures which correspond to industrial conditions, the possible reaction paths in these two systems at 773 K, 923 K and 1073 K were proposed. Thermodynamically possible reactions in the Ni-S-O and Fe-S-O systems were proposed and the changes in Gibbs free energy at 298 K, 773 K, 923 K and 1073 K were calculated.

Kinetic analysis of pentlandite oxidation process in the air atmosphere was conducted using non-isothermal method of Daniels and Borchardt [35,36], at heating rate of 15° min^{-1}.

3. Results and Discussion

3.1. Results of Chemical Analysis

A qualitative chemical analysis of the examined sample was performed by EDXRF analysis and the presence of the following elements was detected in the sample: iron, nickel, copper, sulfur, as well as other accompanying components of tailings.

The content (quantitative analysis) of the elements that EDXRF indicated and which were the subject of our research was determined by the AAS method and is shown in Table 1.

Table 1. Chemical composition of the initial pentlandite sample.

Element	Fe	S	Ni	Si	Cu	Ca	Mn	Pb
Mass %	41.63	25.90	5.12	4.8	0.4	2.74	0.034	0.016

3.2. X-ray Diffraction Results

The characterization of the initial pentlandite sample at 298 K and oxidation products at 673 K, 773 K, 873 K, 973 K and 1073 K was conducted by an X-ray diffraction method on a polycrystalline powder sample. The results of the X-ray analysis of the initial sample and oxidation products at different temperatures are given in Figure 1. In the initial ore sample, at room temperature, the following sulfide minerals were observed: pyrrhotite (FeS), pentlandite ((Fe,Ni)$_9$S$_8$), pyrite (FeS$_2$), magnetite (Fe$_3$O$_4$), chalcopyrite (CuFeS$_2$), gangue minerals quartz (SiO$_2$) and feldspar ((K,Na)Si$_3$O$_8$). The most abundant were pyrrhotite and pentlandite, to a lesser extent pyrite and magnetite, while quartz, feldspar and chalcopyrite were significantly less prevalent. According to Figure 1, the decomposition of pentlandite began below 673 K, which was why there was NiS present at 673 K. The intensification of the pentlandite oxidation process occurs below 873 K, that is why the characterization of the oxidation products at 873 K showed no pentlandite. It can be observed that NiS was present even at 1073 K. Since copper minerals were significantly less present, chalcopyrite oxidation products were not shown on the XRD diffractograms of oxidation products at higher temperatures.

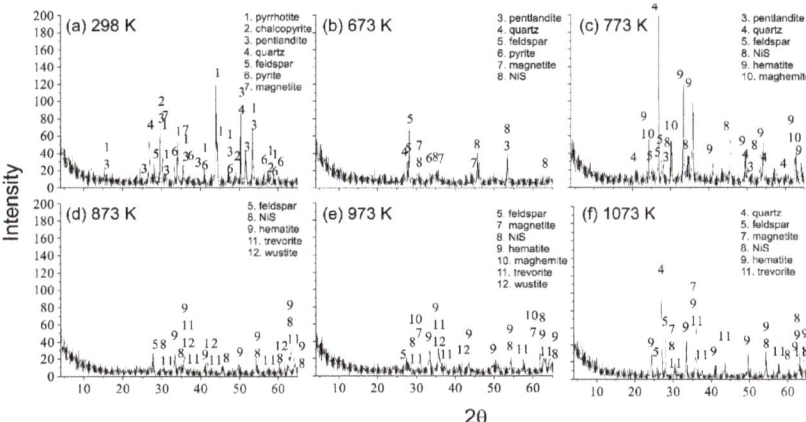

Figure 1. The results of XRD analysis for the (**a**) initial sample, (**b**) oxidation product at 673 K, (**c**) 773 K, (**d**) 873 K, (**e**) 973 K and (**f**) 1073 K.

3.3. SEM/EDS Analysis

The SEM/EDS analysis was performed on the initial sample and on solid residues after 30 min roasting at 673 K, 773 K, 873 K, 973 K and 1073 K, in order to compare the obtained results with the results of the XRD analysis. The SEM microphotographs are given in Figure 2, while marked areas were used for the EDS analysis. Based on the EDS analysis results (atomic percentages of elements on the sample surface), phase compositions of the initial sample and the oxidation products were calculated and given in Table 2.

The SEM/EDS analysis showed that nickel-bearing sulfides were intergrown with quartz or gangue minerals at lower temperatures. With the temperature increase, to above 873 K, still present NiS was found incorporated within the iron oxides hematite and wustite. The occurrence of the spinel phase trevorite (NiFe$_2$O$_4$) was also observed at higher temperatures.

3.4. The Results of Thermodynamic Analysis

The determination of thermodynamically stable phases in the Me-S-O system (Me = Ni, Fe) required a construction of Kellogg's (PAD = Predominance Area Diagrams) (Figures 3 and 4) diagrams, as a function:

$$\log \text{pSO}_{2(g)} = f(\log \text{pO}_{2(g)}) \quad (1)$$

$$\log \text{pO}_{2(g)} = f(T) \quad (2)$$

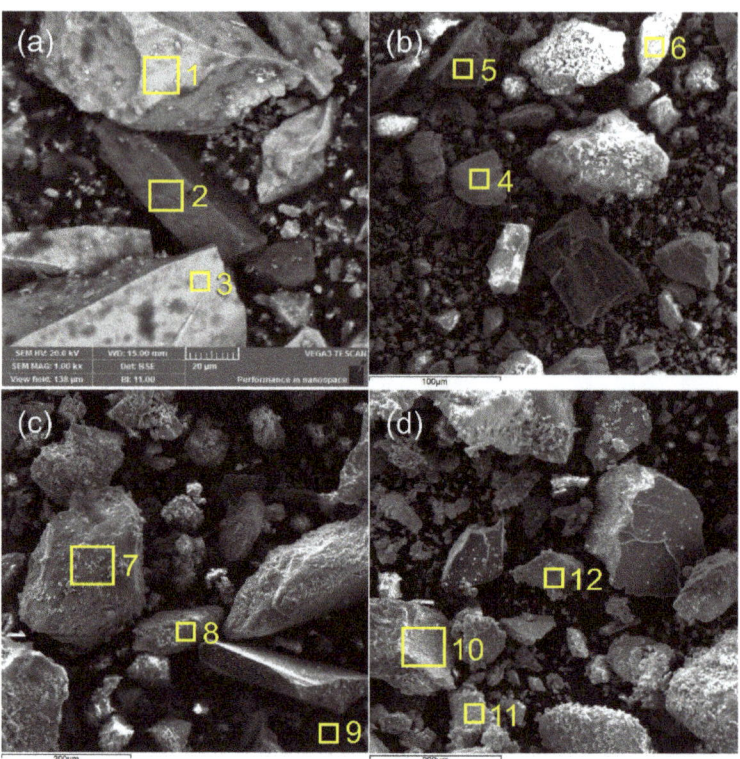

Figure 2. Microphotographs with EDS analysis of the pentlandite-bearing ore (**a**) 298 K and the initial sample roasted at (**b**) 773 K, (**c**) 873 K and (**d**) 1073 K.

Table 2. Phase composition of the initial sample and the oxidation products according to EDS analysis.

Temperature K	Spectrum	Species
298	1	quartz (SiO_2)
	2	pentlandite (($Fe,Ni)_9S_8$)
	3	pyrite (FeS_2)
773	4	NiS + hematite (Fe_2O_3)
	5	pentlandite (($Fe,Ni)_9S_8$) + quartz (SiO_2)
	6	pentlandite (($Fe,Ni)_9S_8$) + gangue minerals
873	7	NiS + hematite (Fe_2O_3)
	8	NiS + wustite (FeO) + quartz (SiO_2)
	9	trevorite ($NiFe_2O_4$)
1073	10	NiS + hematite (Fe_2O_3) + gangue minerals
	11	trevorite ($NiFe_2O_4$) + hematite (Fe_2O_3)
	12	magnetite (Fe_3O_4) + trevorite ($NiFe_2O_4$)

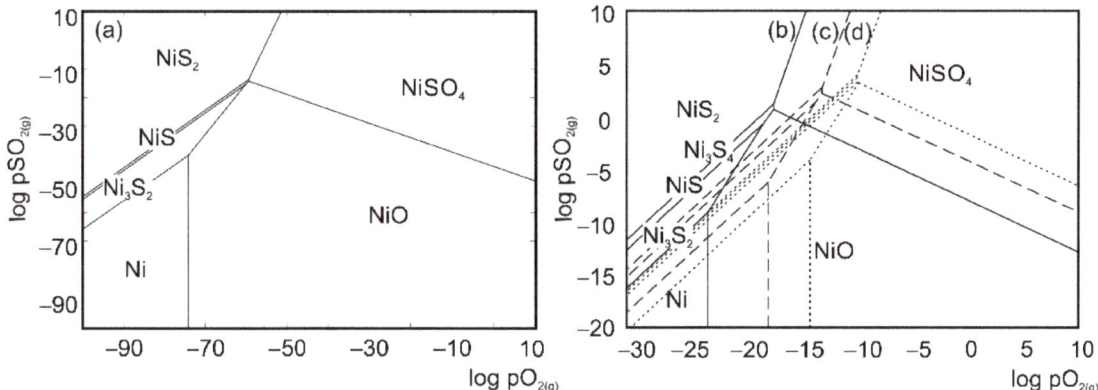

Figure 3. Kellogg diagrams constructed for Ni-S-O system at (**a**) 298 K, (**b**) 773 K, (**c**) 923 K and (**d**) 1073 K [37,38].

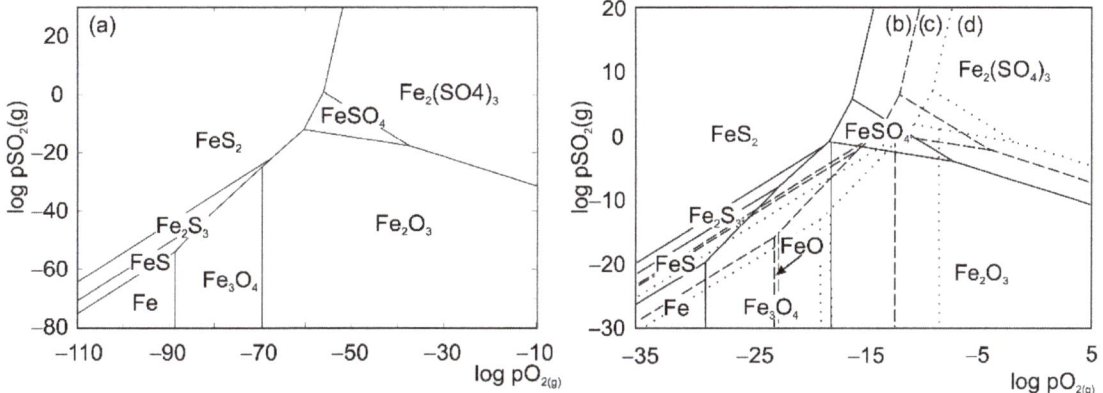

Figure 4. Kellogg diagrams constructed for Fe-S-O system at (**a**) 298 K, (**b**) 773 K, (**c**) 923 K and (**d**) 1073 K [37,38].

For the Kellogg diagrams construction solid phases such as oxides, sulfates, sulfides and metals, they were taken in consideration and diagrams were constructed in four characteristic temperatures (298 K, 773 K, 923 K, and 1073 K). Partial pressures in the HSC Chemistry software (9.0, Outotec) for PAD construction were given in bar units [36].

Regarding the constructed diagrams, the theoretically proposed reaction paths for Ni-S-O and Fe-S-O systems are presented in Table 3.

The theoretically obtained reaction paths were proposed for constant values of SO_2 partial pressure (10^{-4} bar and 1 bar), which correspond to industrial partial pressures of SO_2 gas [38]. Observing the constructed PAD diagrams, the regarded partial pressures marked off an area within which the theoretical mechanism of oxidative roasting was defined. It can be observed that for every investigated temperature and corresponding pressure based on the theoretical mechanism, in the Ni-S-O system, the final stable phase was always $NiSO_4$ and there could not be a direct sulfide transformation to oxide. In the Fe-S-O system, the final theoretical stable phase was always $Fe_2(SO_4)_3$.

Table 3. Theoretically proposed reaction paths for oxidation process in the Ni-S-O and Fe-S-O systems.

System	Theoretical Reaction Path
Ni-S-O	773 K $pSO_{2(g)} = 10^5 Pa \Rightarrow NiS_2 \to Ni_3S_4 \to NiO \to NiSO_4$ $pSO_{2(g)} = 10 Pa \Rightarrow NiS_2 \to Ni_3S_4 \to NiS \to Ni_3S_2 \to NiO \to NiSO_4$ 923 K $pSO_{2(g)} = 10^5 Pa \Rightarrow NiS_2 \to Ni_3S_4 \to NiS \to Ni_3S_2 \to NiO \to NiSO_4$ $pSO_{2(g)} = 10 Pa \Rightarrow NiS_2 \to Ni_3S_4 \to NiS \to Ni_3S_2 \to NiO \to NiSO_4$ 1073 K $pSO_{2(g)} = 10^5 Pa \Rightarrow NiS_2 \to Ni_3S_4 \to NiS \to Ni_3S_2 \to NiO \to NiSO_4$ $pSO_{2(g)} = 10 Pa \Rightarrow NiS_2 \to Ni_3S_4 \to NiS \to Ni_3S_2 \to NiO \to NiSO_4$
Fe-S-O	773 K $pSO_{2(g)} = 10^5 Pa \Rightarrow FeS_2 \to Fe_3O_4 \to Fe_2O_3 \to FeSO_4 \to Fe_2(SO_4)_3$ $pSO_{2(g)} = 10 Pa \Rightarrow FeS_2 \to Fe_2S_3 \to Fe_3O_4 \to Fe_2O_3 \to Fe_2(SO_4)_3$ 923 K $pSO_{2(g)} = 10^5 Pa \Rightarrow FeS_2 \to Fe_2S_3 \to Fe_3O_4 \to Fe_2O_3 \to FeSO_4 \to Fe_2(SO_4)_3$ $pSO_{2(g)} = 10 Pa \Rightarrow FeS_2 \to Fe_2S_3 \to FeS \to Fe_3O_4 \to Fe_2O_3 \to Fe_2(SO_4)_3$ 1073 K $pSO_{2(g)} = 10^5 Pa \Rightarrow FeS_2 \to FeS \to Fe_3O_4 \to Fe_2O_3 \to Fe_2(SO_4)_3$ $pSO_{2(g)} = 10 Pa \Rightarrow FeS_2 \to FeS \to Fe_3O_4 \to Fe_2O_3 \to Fe_2(SO_4)_3$

In addition, the Tpp Predominance Area Diagrams (temperature—partial pressure diagrams) were calculated for both systems as a function:

$$pSO_{2(g)} = f(T) \qquad (3)$$

with a partial pressure of oxygen 0.1 bar. Calculated diagrams (Figure 5) imply that during an oxidation process, Ni and Fe sulfates were expected to form even at lower temperatures.

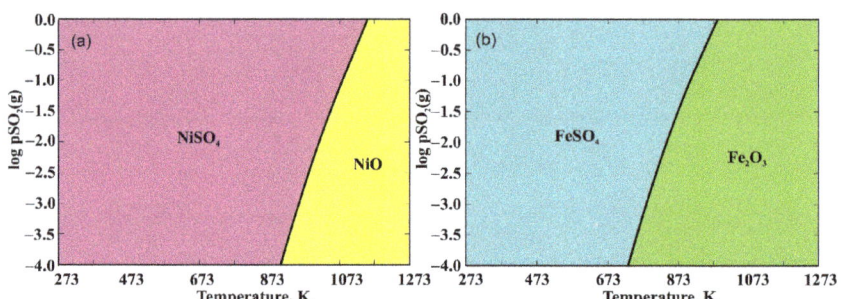

Figure 5. Tpp Predominance Area Diagrams for (**a**) Ni-S-O and (**b**) Fe-S-O system for a constant value of oxygen partial pressure [37].

Based on the theoretical calculations, reactions in the given systems were proposed and changes in Gibbs free energy were calculated. Thermodynamically possible reactions for both systems are given in Table 4.

Table 4. Values of change in Gibbs free energy calculated for theoretically proposed reactions for Ni-S-O and Fe-S-O systems [37].

Reaction	ΔG_T^0 (kJ mol^{-1})		
	773 K	923 K	1073 K
$3NiS_2 + 2O_{2(g)} = Ni_3S_4 + 2SO_{2(g)}$	−543	−552	−562
$Ni_3S_4 + 5.5O_{2(g)} = 3NiO + 4SO_{2(g)}$	−1439	−1404	−1368
$Ni_3S_4 + O_{2(g)} = 3NiS + SO_{2(g)}$	−270	−273	−275
$3NiS + O_{2(g)} = Ni_3S_2 + SO_{2(g)}$	−257	−261	−269
$Ni_3S_2 + 3.5O_{2(g)} = 3NiO + 2SO_{2(g)}$	−912	−870	−824
$2NiO + 2SO_{2(g)} + O_{2(g)} = 2NiSO_4$	−231	−148	−66
$3FeS_2 + 8O_{2(g)} = Fe_3O_4 + 6SO_{2(g)}$	−2254	−2232	−2210
$2FeS_2 + O_{2(g)} = Fe_2S_3 + SO_{2(g)}$	−283	−284	−283
$FeS_2 + O_{2(g)} = FeS + SO_{2(g)}$	−271	−281	−291
$4Fe_3O_4 + O_{2(g)} = 6Fe_2O_3$	−266	−220	−175
$1.5Fe_2S_3 + 6.5O_{2(g)} = Fe_3O_4 + 4.5SO_{2(g)}$	−1829	−1805	−1786
$Fe_2S_3 + O_{2(g)} = 2FeS + SO_{2(g)}$	−259	−278	−299
$3FeS + 5O_{2(g)} = Fe_3O_4 + 3SO_{2(g)}$	−1440	−1388	−1337
$2Fe_2O_3 + 4SO_{2(g)} + O_{2(g)} = 4FeSO_4$	−201	−72	55
$Fe_2O_3 + 3SO_{2(g)} + 1.5O_{2(g)} = Fe_2(SO_4)_3$	−230	−108	14
$2FeSO_4 + SO_{2(g)} + O_{2(g)} = Fe_2(SO_4)_3$	−129	−71	−13

3.5. The Results of Thermal Analysis

For the investigated ore sample, the thermogravimetric and differential thermal analyses were conducted in air atmosphere, with a heating rate of 15° min^{-1}. The temperature range was 298–1273 K. The comparative results of the TG/DTA analysis are given in Figure 6.

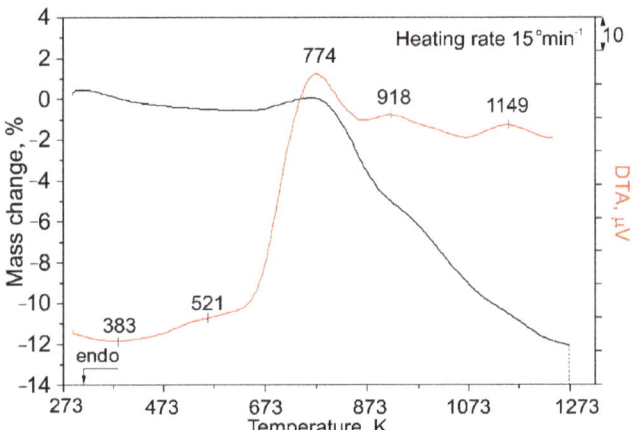

Figure 6. Comparative results of TG/DTA analysis for the investigated pentlandite-bearing ore sample.

Obtained TG/DTA heating curves were analyzed in comparison to the results of XRD and SEM/EDS analyses, as well as the calculated thermodynamically possible reaction paths (Table 3) and theoretically proposed reactions (Table 4). Based on the above, the actual oxidation mechanism was proposed.

Heating the investigated ore sample at the temperature range of 473–573 K gave a slight exothermic peak on the DTA curve at 521 K and a small mass decrease on the TG curve, which correspond to the beginning of the pyrite oxidation, according to Equation (4):

$$FeS_2 + O_2 \rightarrow FeS + SO_2 \tag{4}$$

Although it is expected that pyrite oxidation took place at higher temperatures [20], the formation of pyrrhotite from oxidation in the air atmosphere could occur at lower temperatures due to the easy oxidation of sulfur by oxygen to SO_2 [39].

Based on the available literature [40], and the results of the XRD analysis, it was assumed that formed, and already existing, FeS was further oxidized up to 673 K, according to Equation (5):

$$3FeS + 5O_2 \rightarrow Fe_3O_4 + 3SO_2 \tag{5}$$

Over 773 K, there can be noticed three different exothermic peaks on the DTA curve with appropriate changes of mass on the TG curve. Regarding the first DTA peak (774 K), the area beneath spread from 623 K up to almost 873 K. In the beginning of the suggested temperature range, there can be seen a mass increase on the TG curve, until the first DTA peak, after which the mass of the investigated sample began sharply decreasing. It was assumed that these results suggested sulfate formation by Equation (6):

$$2Fe_2O_3 + 4SO_2 + O_2 \rightarrow 4FeSO_4 \tag{6}$$

even though, because of their instability, they were not spotted according to the XRD analysis. The assumption was that prolonged roasting was a necessity in order to achieve thermodynamic equilibrium, for sulfate detection on the XRD diffractograms. The formation of sulfates would imply a mass increase, which was not the case for the whole investigated temperature range. A mass loss, which occurred on the TG curve, could be explained by the decomposition of pentlandite and an intense SO_2 gas removal, which in this temperature range "masked" the sulfate formation. Based on the XRD analysis results, the presence of the high-temperature NiS phase could be detected even at 673 K, pinpointing the decomposition of pentlandite. Based on the results of the XRD and EDS analyses, and prior investigations [20], the existence of pentlandite could be determined even at 773 K. This implies that pentlandite decomposition took place in a wide temperature range giving an impoverished pentlandite [25], FeS_2 and NiS. Further heating brought a second exothermic peak at 918 K on the DTA curve, with a corresponding mass loss on the TG curve. Based on the same research [20], and the obtained results, this peak could be explained as a sulfate decomposition, according to Equation (7):

$$FeSO_4 \rightarrow FeO + SO_2 + 0.5O_2 \tag{7}$$

In this temperature interval, the decomposition of impoverished pentlandite led to the formation of $NiFe_2O_4$ spinel, trevorite [41,42], which was confirmed together with Fe_2O_3 in the results of the XRD analysis at 873 K [25], given by Equation (8):

$$(Fe, Ni)_{8 \pm x} S_{6 \pm y} + O_2 \rightarrow NiFe_2O_4 + Fe_2O_3 + NiO + SO_2 \tag{8}$$

It was assumed that the reaction between the formed NiO and existing Fe_2O_3 took place, leading to the subsequent formation of Ni-Fe spinel. The mass loss on the TG curve, corresponding to this exothermic peak, was explained with SO_2 emission. The higher temperature brought to another exothermic peak on the DTA curve with a corresponding mass loss on the TG curve. This change at 1149 K was assumed to be the oxidation of the high-temperature NiS phase, which was present even at 1073 K, with a constant spinel formation and SO_2 emission, explaining this mass loss.

Accessible literature data [27] and the theoretically proposed reaction mechanism suggest that NiS oxidation and sulfate formation occurred at temperature range of 733–988

K. One of the possible reasons for the difficulty of sulfate detection in the oxidation products was the non-equilibrium conditions (roasting time of 30 min) [20]. Another possible reason for the absence of sulfates could be explained with the preferential oxidation of iron, followed by the formation of protective coatings of the iron oxidation products on the mineral particle surface. This prevented oxygen diffusion to the reaction zone while allowing the existence of NiS in the inner core of the particle [20,25,27,31].

3.6. The Results of Kinetic Analysis

The kinetic analysis of the pentlandite-bearing ore oxidation process was performed using the method of non-isothermal kinetics of Daniels and Borchardt under a constant heating rate. The heating of the sample was performed in an air atmosphere, in a temperature range of 298–1273 K under a heating rate of 15° min^{-1}. Based on the results of the DTA analysis (Figure 6), three characteristic stages were defined. Three exothermal peaks (774 K, 918 K and 1149 K) were used for the kinetic parameters calculation. Based on the given method, activation energies were calculated for the observed oxidation stages and the results are given in Figure 7 and Table 5.

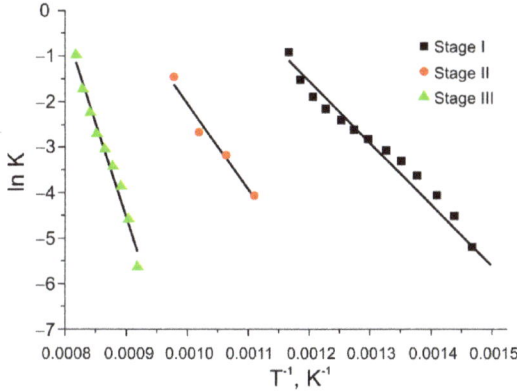

Figure 7. Arrhenius diagram of the pentlandite-bearing ore oxidation process.

Table 5. Calculated values of the activation energies.

Temperature Range (K)	Stage	Ea (kJ Mol^{-1})
667–856	I	113
901–1023	II	156
1089–1223	III	346

Calculated activation energies indicated that all three stages of the oxidation process occurred in a kinetic field [43]. These values were implying that in every examined stage of the oxidation process, temperature was a dominant parameter, determining the reaction rate. With the temperature increase, the values of Ea increased too, which meant that the oxidation process was even more "introduced" into the kinetic field. The temperature and the reaction surface had a more emphasized influence on the speed of the oxidation process. This also signified that a major part of the oxidation process took place on the surface boundary between the solid and gas phases.

4. Conclusions

The oxidation process of pentlandite-bearing ore was investigated in this paper. The characterization of the initial sample and the oxidation products was performed in a temperature range of 298–1073 K, at different elevated temperatures. Additionally, thermodynamic, thermal and kinetic analyses results of the oxidation process were obtained.

The results obtained via the XRD and SEM/EDS analyses were in a good mutual agreement, confirming that the initial sample consisted of pentlandite, pyrrhotite, chalcopyrite, pyrite, magnetite and gangue minerals (quartz and feldspar). The oxidation products hematite, magnetite and spinel phase trevorite were detected at elevated temperatures. NiS was observed up to 1073 K.

The thermodynamic analysis of the investigated pentlandite-bearing sample involved the construction of Kellogg diagrams for the Ni-S-O and Fe-S-O systems. These diagrams were constructed at several temperatures: 298 K, 773 K, 923 K and 1073 K. Based on the constructed diagrams, the theoretical reaction paths were suggested at 773 K, 923 K and 1073 K with defined partial pressures of SO_2 and O_2 gas, corresponding to industrial conditions. Thermodynamically possible reactions were proposed and values of change in Gibbs free energies were calculated at given temperatures.

The thermal analysis of the pentlandite-bearing ore oxidation process involved heating the initial sample in temperature range 298–1273 K, in an air atmosphere, with a heating rate of 15° min^{-1}. Obtained TG/DTA heating curves were analyzed and the results were compared to the XRD and SEM/EDS analyses and theoretical reaction paths, leading to proposing the actual reaction mechanism of the investigated oxidation process.

The kinetic analysis of the pentlandite-bearing ore oxidation process, was conducted by applying a non-isothermal kinetics calculation, according to the Daniels and Borchardt method. The kinetic analysis of the investigated oxidation process was based on the DTA heating curve with a constant heating rate of 15° min^{-1}. Three different exothermal peaks were observed on the DTA heating curve, corresponding to three oxidation stages. Calculated activation energies were 113 kJ mol^{-1}, 146 kJ mol^{-1} and 356 kJ mol^{-1}, respectively. These values implied that all three stages of the oxidation process occurred in a kinetic field, i.e., temperature was a dominant parameter, determining the reaction rate. A major part of the oxidation process took place on the surface boundary between the solid and gas phases.

The proposed reaction mechanism and kinetic analysis results presented in this paper could be considered for a better optimization of the oxidative roasting of nickel-bearing ores and concentrates in industrial conditions. Obtained results were in a good agreement with the available literature and represent a contribution to a better understanding of the complex sulfide ores oxidation process.

Author Contributions: N.Š., D.M. and M.S. designed and directed the project; D.M. provided the study materials; D.M., B.M., A.M., J.S. and K.B. performed the experiments and performed the analytic calculations; K.B. and A.M. prepared the original draft. All authors have read and agreed to the published version of the manuscript.

Funding: The research presented in this paper was conducted with the financial support of the Ministry of Education, Science and Technological Development of the Republic of Serbia, within the funding of the scientific research work at the University of Belgrade, Technical Faculty in Bor, according to the contract with registration number 451-03-9/2021-14/200131 and the Institute for Technology of Nuclear and Other Mineral Raw Materials, according to the contract with registration number 451-03-9/2021-14/200023.

Institutional Review Board Statement: Not applicable.

Informed Consent Statement: Not applicable.

Data Availability Statement: The data presented in this study are available on request from the corresponding author.

Acknowledgments: Authors would also like to express their gratitude to Vaso Manojlović from the Faculty of Technology and Metallurgy of the University of Belgrade, for his technical support and immerse help. The contribution of Gordana Mundrić, for proofreading the manuscript, is also greatly appreciated.

Conflicts of Interest: The authors declare no conflict of interest.

References

1. Kuck, P.H. *Nickel Mineral Commodity Summaries*; United States Geological Survey: Washington, DC, USA, 2010; Available online: https://s3-us-west-2.amazonaws.com/prd-wret/assets/palladium/production/mineral-pubs/nickel/mcs-2010-nicke.pdf (accessed on 12 February 2020).
2. Kuck, P.H. *Nickel Mineral Commodity Summaries*; United States Geological Survey: Washington, DC, USA, 2015; Available online: https://s3-us-west-2.amazonaws.com/prd-wret/assets/palladium/production/mineral-pubs/nickel/mcs-2015-nicke.pdf (accessed on 12 February 2020).
3. McRae, M.E. *Nickel Mineral Commodity Summaries*; United States Geological Survey: Washington, DC, USA, 2020. Available online: https://pubs.usgs.gov/periodicals/mcs2020/mcs2020-nickel.pdf (accessed on 12 February 2020).
4. Betteridge, W. *Nickel and Its Alloys*; Ellis Horwood: Chichester, UK; Halsted Press: New York, NY, USA, 1984.
5. Vracar, R.Z. *Theory and Practice of Extraction of Non-Ferrous Metals/Teorija i Praksa Dobijanja Obojenih Metala*; Serbian Association of Metallurgy Engineers: Belgrade, Serbia, 2010.
6. Berndt, D. *Maintenance-Free Batteries: Lead-Acid, Nickel/Cadmium, Nickel/Metal Hydride: A Handbook of Battery Technology*, 2nd ed.; Research Studies Press: Baldock, UK, 1997.
7. Linden, D.; Reddy, T.B. *Handbook of Batteries*, 3rd ed.; McGraw-Hill: New York, NY, USA, 2002.
8. Gong, M.; Zhou, W.; Tsai, M.C.; Zhou, J.; Guan, M.; Lin, M.C.; Zhang, B.; Hu, Y.; Wang, D.Y.; Yang, J.; et al. Nanoscale nickel oxide/nickel heterostructures for active hydrogen evolution electrocatalysis. *Nat. Commun.* **2014**, *5*, 4695. [CrossRef]
9. Tasker, S.Z.; Standley, E.A.; Jamison, T.F. Recent advances in homogeneous nickel catalysis. *Nature* **2014**, *509*, 299–309. [CrossRef]
10. Konkena, B.; Junge Puring, K.; Sinev, I.; Piontek, S.; Khavryuchenko, O.; Dürholt, J.P.; Schmid, R.; Tüysüz, H.; Muhler, M.; Schuhmann, W.; et al. Pentlandite rocks as sustainable and stable efficient electrocatalysts for hydrogen generation. *Nat. Commun.* **2016**, *7*, 12269. [CrossRef]
11. Sokić, M.; Ilić, I.; Živković, D.; Vučković, N. Investigation of mechanism and kinetics of chalcopyrite concentrate oxidation process. *Metalurgija* **2008**, *47*, 109–113.
12. Mitovski, A.; Štrbac, N.; Mihajlović, I.; Sokić, M.; Stojanović, J. Thermodynamic and kinetic analysis of the polymetallic copper concentrate oxidation process. *J. Therm. Anal. Calorim.* **2014**, *118*, 1277–1285. [CrossRef]
13. Crundwell, F.K.; Moats, M.S.; Ramachandran, V.; Robinson, T.G.; Davenport, W.G. *Extractive Metallurgy of Nickel, Cobalt and Platinum-Group Metal*, 1st ed.; Elsevier: Amsterdam, The Netherlands, 2011.
14. Xia, F.; Pring, A.; Brugger, J. Understanding the mechanism and kinetics of pentlandite oxidation in extractive pyrometallurgy of nickel. *Miner. Eng.* **2012**, *27–28*, 11–19. [CrossRef]
15. Goryachev, A.A.; Chernousenko, E.V.; Potapov, S.S.; Tsvetov, N.S.; Makarov, D.V. A Study of the Feasibility of Using Ammonium Sulfate in Copper—Nickel Ore Processing. *Metals* **2021**, *11*, 422. [CrossRef]
16. Janjić, S.; Ristić, P. *Mineralogy/Mineralogija*; Scientific Book: Belgrade, Serbia, 1995.
17. Diaz, C.M.; Landolt, C.A.; Vahed, A.; Warner, A.E.M.; Taylor, W. A review of nickel pyrometallurgical operations. *JOM* **1988**, *40*, 28–33. [CrossRef]
18. Naldrett, A.J.; Craig, J.R.; Kullerud, G. The central portion of the Fe-Ni-S system and its bearing on pentlandite solution in iron-nickel sulfide ores. *Econ. Geol.* **1967**, *62*, 826–847. [CrossRef]
19. Mishra, K.C.; Fleet, M.E. The chemical composition of synthetic and natural pentlandite assemblages. *Econ. Geol.* **1973**, *68*, 518–539. [CrossRef]
20. Dunn, J.G. The oxidation of sulfide minerals. *Thermochim. Acta* **1997**, *300*, 127–139. [CrossRef]
21. Mkhonto, P.P.; Chauke, H.R.; Ngoepe, P.E. Ab initio studies of O2 adsorption on (110) Nickel-rich pentlandite ($Fe_4Ni_5S_8$) mineral surface. *Minerals* **2015**, *5*, 665–678. [CrossRef]
22. Thornhill, P.G.; Pidgeon, L.M. Micrographic study of sulfide roasting. *JOM* **1957**, *9*, 989–995. [CrossRef]
23. Ellingham, H.J.T. Reducibility of oxides and sulphides in metallurgical processes. *J. Soc. Chem. Ind.* **1944**, *63*, 125–133.
24. Ashcroft, E.A. Process for the Treatment of Ores or Materials Containing Copper and/or Nickel. U.S. Patent 1,851,885, 29 March 1932.
25. Tanabe, T.; Kawaguchi, K.; Asaki, Z.; Kondo, Y. Oxidation kinetics of dense pentlandite. *Trans. JIM* **1987**, *28*, 977–985. [CrossRef]
26. Kennedy, T.; Sturman, B.T. The oxidation of iron (II) sulphide. *J. Therm. Anal. Calorim.* **1975**, *8*, 329–337. [CrossRef]
27. Dunn, J.G.; Kelly, C.E. A TG/MS and DTA study of the oxidation of nickel sulphide. *J. Therm. Anal. Calorim.* **1977**, *12*, 43–52. [CrossRef]
28. Dunn, J.G.; Kelly, C.E. A TG/DTA/MS study of the oxidation of pentlandite. *J. Therm. Anal. Calorim.* **1980**, *18*, 147–154. [CrossRef]
29. Zhu, H.; Chen, J.; Deng, J.; Yu, R.; Xing, X. Oxidation behavior and mechanism of pentlandite at 973 K (700 °C) in air. *Metall. Mater. Trans. B* **2012**, *43*, 494–502. [CrossRef]
30. Dunn, J.G.; Jayweera, S.A. Effect of heating rate on the TG curve during the oxidation of nickel sulfide concentrates. *Thermochim. Acta* **1983**, *61*, 313–317. [CrossRef]
31. Pandher, R.; Utigard, T. Roasting of Nickel Concentrates. *Metall. Mater. Trans. B* **2010**, *41*, 780–789. [CrossRef]
32. Yu, D.; Utigard, T.A. TG/DTA study on the oxidation of nickel concentrate. *Thermochim. Acta* **2012**, *533*, 56–65. [CrossRef]
33. Hseu, Z.Y. Evaluating heavy metal contents in nine composts using four digestion methods. *Bioresour. Technol.* **2004**, *95*, 53–59. [CrossRef] [PubMed]

34. Shultz, J.I.; Bell, R.K.; Rains, T.C.; Menis, O. *Methods of Analysis of NBS Clay Standards*; National Bureau of Standards Special Publication 260-37; National Bureau of Standards: Washington, DC, USA, 1972; pp. 3–4.
35. Borchard, H.J.; Daniels, F. The Application of Differential Thermal Analysis to the Study of Reaction Kinetics. *J. Am. Chem. Soc.* **1957**, *79*, 41–46. [CrossRef]
36. Živković, D.; Živković, Ž. *Problems in the Theory of Metallurgical Processes/Zbirka Zadataka iz Teorije Metlurških Procesa*; University of Belgrade, Technical Faculty in Bor: Belgrade, Serbia, 2001.
37. Roine, A. *HSC Chemistry® v 9.0*; Research Oy Center, Outotec: Pori, Finland, 2016.
38. Shamsuddin, M. Roasting of Sulfide Minerals. In *Physical Chemistry of Metallurgical Processes*; John Wiley & Sons: Hoboken, NJ, USA, 2016; pp. 39–71.
39. Zhang, Y.; Li, Q.; Liu, X.; Xu, B.; Yang, Y.; Jiang, T. A Thermodynamic Analysis on the Roasting of Pyrite. *Minerals* **2019**, *9*, 220. [CrossRef]
40. Smirnov, V.I.; Tihonov, A.I. *Incineration of Copper Ores and Concentrates/Obţig Mednih Rud i Koncentratov*; Metallurgy: Moscow, Russia, 1966.
41. Mayangsari, W.; Prasetyo, A.B. Phase Transformation of Limonite Nickel Ores with Na_2SO_4 Addition in Selective Reduction Process. *IOP Conf. Ser. Mater. Sci. Eng.* **2017**, *202*, 012016. [CrossRef]
42. Dunn, J.G.; Howes, V.L. The oxidation of violarite. *Thermochim. Acta* **1996**, *282–283*, 305–316. [CrossRef]
43. Živković, Ž. *Theory of Metallurgical Processes/Teorija Mealurških Proces*; University of Belgrade, Technical Faculty in Bor: Belgrade, Serbia, 1991.

Article

Processing Tests, Adjusted Cost Models and the Economies of Reprocessing Copper Mine Tailings in Chile

Malte Drobe [1,*], Frank Haubrich [2], Mariano Gajardo [3] and Herwig Marbler [4]

1 Federal Institute for Geosciences and Natural Resources (BGR), Stilleweg 2, 30655 Hannover, Germany
2 G.E.O.S. Ingenieurgesellschaft mbH, Schwarze Kiefern 2, 09633 Halsbruecke, Germany; f.haubrich@geosfreiberg.de
3 Servicio Nacional de Geología y Minería Av. Santa María, Providencia 0104, Chile; mariano.gajardo@sernageomin.cl
4 German Mineral Resources Agency (DERA) at the Federal Institute for Geosciences and Natural Resources (BGR), Wilhelmstrasse 25-30, 13593 Berlin, Germany; herwig.marbler@bgr.de
* Correspondence: malte.drobe@bgr.de; Tel.: +49-511-643-3189

Citation: Drobe, M.; Haubrich, F.; Gajardo, M.; Marbler, H. Processing Tests, Adjusted Cost Models and the Economies of Reprocessing Copper Mine Tailings in Chile. *Metals* **2021**, *11*, 103. https://doi.org/10.3390/met11010103

Received: 4 December 2020
Accepted: 20 December 2020
Published: 6 January 2021

Publisher's Note: MDPI stays neutral with regard to jurisdictional claims in published maps and institutional affiliations.

Copyright: © 2021 by the authors. Licensee MDPI, Basel, Switzerland. This article is an open access article distributed under the terms and conditions of the Creative Commons Attribution (CC BY) license (https://creativecommons.org/licenses/by/4.0/).

Abstract: To increase resource efficiency, mining residues–especially tailings–have come into the focus of research, companies, and politics. Tailings still contain varying amounts of unextracted elements of value and minerals that were not of economic interest during production. As for primary mineral deposits, only a small share of tailings offers the possibility for an economic reprocessing. To minimize exploration expenditure, a stepwise process is followed during exploration, to estimate the likelihood of a project to become a mine or in this case a reprocessing facility. During this process, costs are continuously estimated at least in an order of magnitude. Reprocessing flowsheets for copper mine tailings in Chile were developed and costs and revenues of possible products from reprocessing were examined for a rough economic assessment. Standard cost models with capex and opex for flotation, leaching, and magnetic separation were adopted to the needs of tailings reprocessing. A copper tailing (around 2 M t) that also contains magnetite was chosen as a case study. A combination of magnetic separation and leaching gave the best economic results for copper and magnetite. The adopted cost models showed positive results at this early stage of investigation (semi-technical scale processing tests).

Keywords: tailings reprocessing; early stage cost estimation; magnetic separation; leaching; flotation

1. Introduction

Copper is by value one of the main mineral commodities in the world [1]. Due to the nature of mineral deposits, not only ore minerals are extracted, but also barren minerals that have to be deposited as tailings [2]. According to [3], more than 500 M t of tailings were produced in Chile, in 2019 alone. Over the past 30 years, more than 7 billion tons of tailings have been stockpiled or discarded in Chile, assuming a copper grade of 2% (most probably lower, leading to even more tailings). In many cases, the tailings still contain some copper and possibly other metals of value with potential for recovery [4–6]. The ore processing technology has improved significantly over the last decades, making it possible to mine lower grade ore bodies [7] and reducing the metal grade that enters into a tailings storage facility (TSF) from 0.75% at the beginning of the 20th century to around 0.1% in the late 1990s [8]. This makes old tailings or tailings originating from processing facilities with a non-homogeneous ore feed or with a non-homogeneous copper mineralogy (both factors are harmful for a good recovery) optimal targets for a possible economic reprocessing. Due to weathering, reprocessing is often more cost-intensive compared to primary ore processing, as described by [9] and references therein, for Russia. As tailings have already been floated once, re-flotation in some cases is possible [10], but can result in low recoveries because of non-floatable minerals or very fine grain sizes. The small grain size can at least in part be overcome by using a tailored size distribution of the tailings material [11]. Besides froth flotation, leaching is the second most common method to process copper ore. Leaching can be carried out via conventional sulfuric acid leach or with support of bacteria. An overview of the share of (bio)leaching in copper processing is given by [12]. Bioleaching, especially, is in the focus of research, as it can be an alternative to conventional ore processing, when the share of floatable minerals is too low [12–17]. A review about the various possible reuses of mine tailings is given by [18] and references therein. They conclude that: "Various methods have been suggested for recycling but there is a lack of well-documented cost and performance data under a variety of operating conditions." Nevertheless, a business case is the prerequisite for reprocessing mine waste, unless the environmental harm is so severe that a mining company is forced to intervene or a public agency is funding the cleanup. Different recycling flowsheets need to be designed for every single case study and economic analyses have to be undertaken, to assess the recycling potential. This is especially the case, as the grade of valuable elements in the tailings is not of value, but the recoverable grade–as for any orebody. In the case of tailings, which are influenced by weathering over years or decades, the mineralogy can change due to the climate and make the recovery more challenging [9,19]. Especially a mixture of sulfide and oxide minerals is difficult to process, as a combination of sulfide and oxide flotation or a combination of leaching and flotation have to be applied. Reference [8] showed that there is a substantial amount of copper in tailings, waste rock, and slag, and proposed that this copper would soon be recoverable using new technology. Nevertheless, reprocessing copper tailings remains relatively sparse. Only 30,000 t of copper per year were produced from projects that use tailings as primary source [20]. Assuming that not all tailings reprocessing projects are known and incorporated into the mentioned database, the amount is low compared to the more than 20 M t of copper that is produced by mines every year. Reference [21] describes various possibilities how to reuse mine waste as mineral resource, but also mentions that it is only rarely done due to economic reasons. Actually, it is difficult to give hard facts for capital expenditure (capex) and operating expenditure (opex), when applying new technologies, for which no empirical values or rules of thumb exist. Even for established techniques like flotation or leaching, it is difficult to give cost ranges in terms of opex and capex for tailings reprocessing, as standard cost models have to be adjusted and real life examples are rare. In order not to spend too much time and money on uneconomic projects, decisions whether to proceed with exploration or to stop the efforts should be taken as early as possible [22]. As a mining (or reprocessing) company usually has more than one project, these projects should be compared with regard to their economic feasibility and the focus should of course lie on the project with the best economics. To support the economic analysis, the objective of this study is to demonstrate and optimize the technical feasibility of tailings recycling at an early stage and to evaluate the economics of the recycling process. For the economic evaluation, several assumptions considering investment costs, operating costs, and revenues have to be made and are discussed in this paper. On the revenue side, the assumptions relate to achievable product standards and current price levels for the products. On the investment and operating cost side, a reasonable plant capacity has to be chosen and, based on the production scale, mineral processing cost models, generated by costmine© (Infomine), have to be adopted and modified to the specific situation of tailings. Although there is later data from costmine©, we have taken 2017 data, as the costs are better broken down into equipment, supplies, labor, administration, etc., which could be adopted individually to the tailings reprocessing. Finally, these cost data were combined with the results from the processing tests for a rough economic assessment.

The objective of this paper is to discuss different schemes for tailings reprocessing and to apply a simple economic evaluation tool based on comparative cost models for a first economic assessment. Reference [4] focused on historical data of porphyry copper deposits in Chile and demonstrated the potential of economic reprocessing. This study intends to go one step further and demonstrate the feasibility based on processing tests to calculate possible revenues and compare these revenues with the capex and opex in an early stage, when not too much money has been spent [22,23]. The particular advantage of Chile is a data repository with superficial grade and tonnage of several hundred tailings [24], which makes it possible to find comparative examples to the investigated site.

2. Materials and Methods

2.1. Tailings Material and Sampling

There are several hundred tailings storage facilities (TSFs) in Chile. For an initial screening to identify the most promising in terms of economic reprocessing, several criteria were evaluated, including age, tonnage, metal grades, type of ore feed (homogeneous or heterogeneous), possible by-products, possible pollutants, ownership, and infrastructure (amongst others). The most challenging in the selection of a TSF is the legal aspect or the simple question: "Do I get the permission to sample on the TSF?" As the project is a collaboration of BGR and a state organization (SERNAGEOMIN), we had access to the tailings of the state owned company ENAMI (Empresa Nacionál de Minería, Santiago de Chile, Chile). As this company buys ore from different small and medium scale enterprises, the ore feed is heterogeneous, which is an advantage for reprocessing, as the flotation cannot be optimized for all mines. Therefore, we first sampled tailings of ENAMI in Ovalle and of some companies close to Ovalle and afterwards tailings of ENAMI in Taltal and of some companies close to Taltal. Due to confidentiality, we cannot name the TSFs. After finishing the sampling, it turned out that the TSF of ENAMI in Taltal (Figure 1) was the most promising in terms of copper grade and possible by-products, in this case magnetite. In order to generate a pilot experiment, the project focused on this relatively small sized TSF of 2 M t. The tailings site contained as main mineral species magnetite (around 20 wt.% of the tailings) and could also be considered copper rich (0.56%). During the construction phase especially of small TSFs, the tailings are size-classified by hydrocyclones, located at the crest of the embankment. The sand-like underflow of the hydrocyclone is used to raise the embankment and the silt-like overflow is deposited on the beach of the enclosed area [25,26]. As a consequence, inside the tailings bodies, there is a vertical and horizontal gradation of the material, according to the deposition velocity of the tailings particles: the further away from the embankment of the TSF, the finer the particle sizes [5,27]. The coarse-sized fraction was usually of black to grey colour and the fine-grained clayish material was either black, reddish or brownish (Figure 2 top and centre). Samples with a high content of clayish material tend to have higher copper grades, compared to black magnetite rich sandy layers. In some samples that are taken from inside of the settling zone of the TSF, layers (a few cm to 20 cm) of yellowish silt could be observed (Figure 2 bottom). These samples had higher contents of arsenic, up to 250 ppm, compared to the average grade of around 90 ppm. The yellowish silt layers are associated with episodic gold processing, in which gold bearing arsenopyrite-ore was milled and floated or density sorted. The average copper grade of all Taltal samples (130) is 0.56% Cu (0.25–1.1% standard deviation 0.19%) and the average iron content is 26% Fe (13–40% Fe, standard deviation 7%). The copper grade is substantially higher compared to the average copper grade in tailings from the 1930s of 0.33% [4] and far above the average of SERNAGEOMIN's cadaster of mine tailings (sampling only the first meter), which gives copper grades of 0.27% for inactive and abandoned tailings <10 M t [24]. The investigated TSF was in operation between 1966 and 2015. As the processed sulfidic ore came from different mines, the flotation process could not be optimized. Additionally there is oxide and sulfide ore present in the area of Taltal, meaning that the flotation feed most probably had a certain copper oxide mineral content. These factors lead to a high residual copper grade in the tailings. The sampling carried out using a hand held jackhammer that drove a percussion probe into the tailings. One of the advantages of this handy, low cost equipment is the high mobility to change the position on a tailings and the short assembly time at new locations. A road access on the surface of the dried TSF sites was not necessary, as the dismantled lightweight device could be carried easily. As a hand-held device was used, sometimes a high degree of physical strength was needed to penetrate the tailings crusts and hardpan layers. It was especially laborious to recover the sample cores that were enclosed in the drilling probe after every meter of drilling (the sampling device was one meter long). Due to physical limitations, it was at best possible to drill to a depth of 11 m, but as a compromise of working efficiency, 7 m was taken as the target depth. The operational depth limits also lead to the conclusion that only tailings of small size and shallow depth containing up to a few million tons of material can be investigated using this technology. However, in deeper tailings sites, at least a first overview about the contents of the TSF could be obtained. The magnetite and copper-rich TSF near Taltal was investigated with 18 perforations (Figure 1) with a depth of up to 11 m below the tailings surface. In total, 119 samples (each representing 1 vertical m) were taken for subsequent processing tests. The material consisted of an alternating strata of a fine- and coarse-grained material. The fine-grained tailings fraction looked similar to clay and showed plastic behavior. The coarse-grained tailings fraction showed no plastic behaviour and could be classified as silty sand. Sieving tests showed that the fine-grained material had a grain size of d80 < 60 µm and the coarse-grained material showed d80 < 120 µm (Supplementary Materials Figure S1).

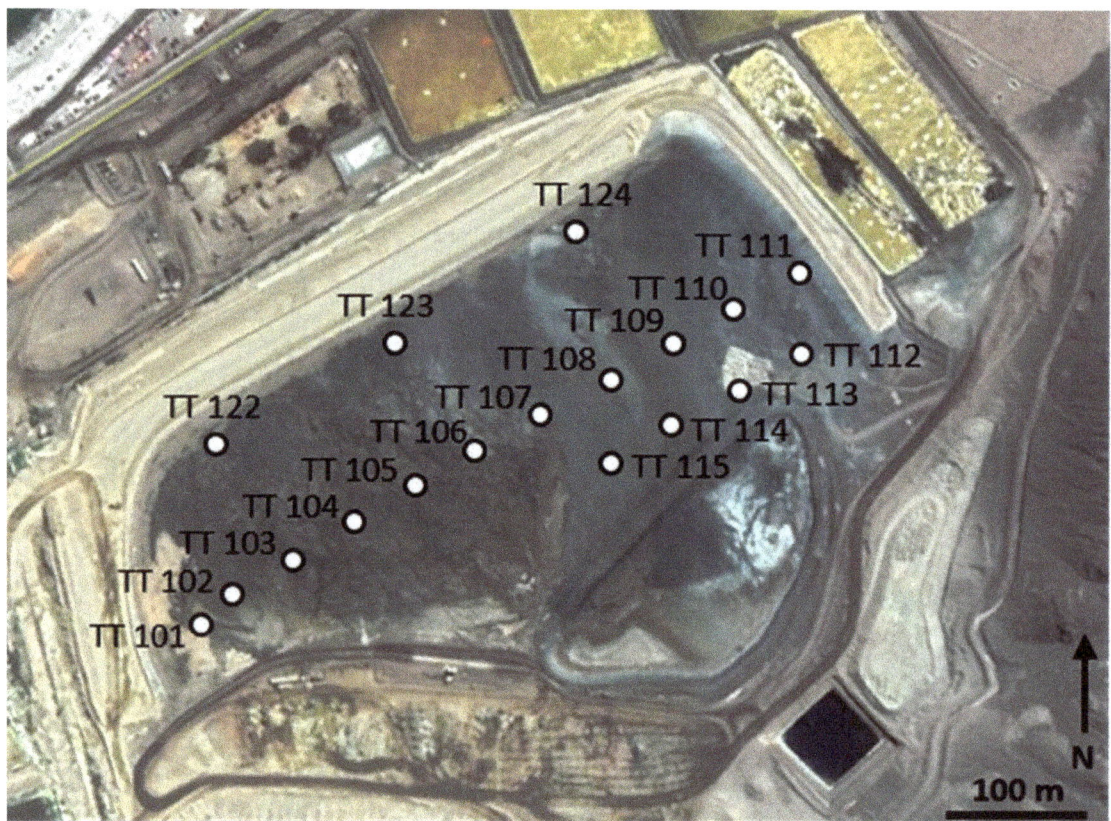

Figure 1. TSF near Taltal (Google Earth) with drilling locations. The tailings have their maximum thickness (15 m) in the northern area (TT 122 to TT 124), where samples were taken up to a depth of 11 m and is thinning to the SSW. There are no drilling spots SE and SW of TT 115, as the tailings pond became very shallow (below 4 m).

Figure 2. Probe with sample material from the Taltal tailings. **Top**: reddish copper rich clay. **Center**: dark magnetite rich sandy material. **Bottom**: yellowish layer probably from episodic gold processing (direct information of the owner of the tailings). Length of probe 1 m.

To give a better picture about Cu-grade and grade variation, the Cu-grade is shown by depth and per perforation in Figure 3. There is an increase in copper grade with depth from around 0.45% for the first two meters, up to 0.80% at 5 m depth. Three holes (TT122, TT1223, and TT124) were drilled deeper, at the crest of the embankment. Due to the construction of the TSF, this part is coarser-grained, as hydrocyclones classified the material, using the coarser material to raise the TSF. As only the material from these three drilling locations with relatively coarse material is included into the calculation of the average Cu grade from 7–11 m depth, the average grade is only 0.5% between 9 and 11 m (Figure 4). Although there is a tendency for higher copper grades with depth, the strongest correlation for Cu and Fe grade is given by grain size. The coarse-grained material has an Fe grade of 29% and the fine grained material of only 19%. On the other hand, the Cu grade is 0.7% for the fine (37 samples) material and only 0.5% for the coarse material (82 samples).

2.2. Geochemical and Mineralogical Analysis

The sampled material was first analyzed by X-ray fluorescence analysis (XRF) using Philips PW1480 and PW2400 with Cr and Rh excitation at the Federal Institute for Geosciences and Natural Resources (BGR), Hanover, Germany. The mineralogy was examined using X-ray phase analysis. XRD pattern were recorded using a PANalytical X'Pert PRO MPD Θ-Θ diffractometer (Co-Kα radiation generated at 40 kV and 40 mA), equipped with a variable divergence slit (20 mm irradiated length), primary and secondary soller, diffracted beam monochromator, point detector, and a sample changer (sample diameter 28 mm) at BGR, Hannover, Germany. The samples were investigated from 2° to 75° 2Θ with a step size of 0.03° 2Θ and a measuring time of 3 s per step. For specimen preparation the back loading technique was used. Rietveld refinement of the experimental XRD data was conducted using the software BGMN.

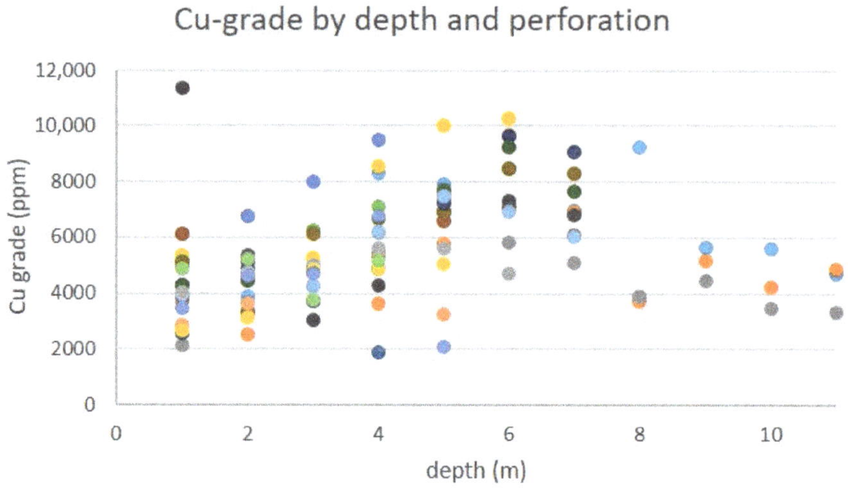

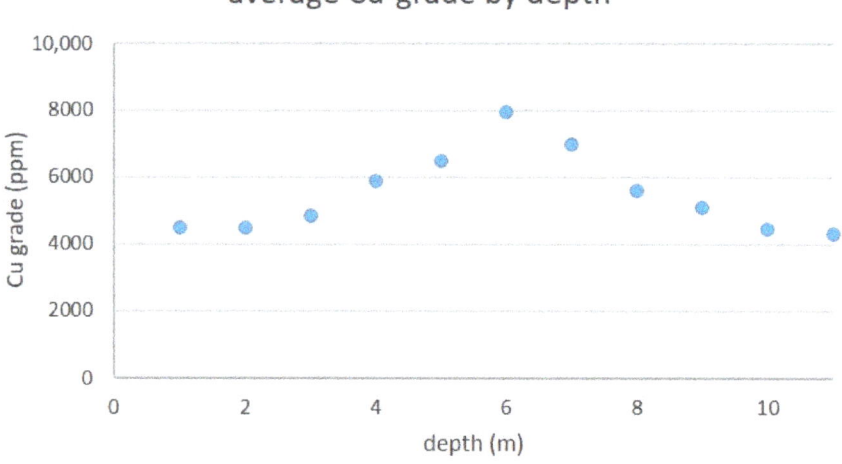

Figure 3. Copper grade for all samples (every sample represents 1 m of drilling). The target depth was 7 m, except for the crest of the embankment, where up to 11 m depth was drilled and sampled.

Figure 4. Average Cu grade by depth for all available samples (18 from 0–4 m and only 3 samples from 7–11 m).

2.3. Magnetic Separation

Magnetic separation of the iron fraction was investigated in two ways. Magnetic separation using a hand magnet (0.1 T) was applied on the sample material before the leaching tests (each 8.5 kg) were carried out. A hand magnet can separate magnetic from non-magnetic material, but tends to carry some non-magnetic minerals into the magnetic fraction. The second approach was to use wet magnetic separation in combination with flotation. In this case, a magnetic separator (Sala at the RWTH Aachen, Germany) 20 cm × 11.6 cm with a magnetic field strength of 0.1 T in a metallurgical laboratory was utilized (internal report, RWTH Aachen, 2018). In general, the magnetic material (mostly magnetite) could be separated from the non-magnetic material, which contained among others the majority of the copper minerals.

2.4. Flotation

In order to produce a copper (pre)concentrate, flotation tests were carried out in the laboratories of the AMD of the RWTH in Aachen. A Denver D-12 flotation cell with a volume of 2.5 L was used. The feed was usually 0.5 kg per test, one time 1 kg and one time 1.4 kg. The solid content was 17% for 0.5 kg, 29% for 1 kg, and 37.5% for 1.4 kg with an aeration of 3 L/s. For activation, a planetary ball mill (PM 400 from Retsch, at the RWTH Aachen, Germany) was used. A list of chemicals used can be found below in Table 1.

Table 1. Reagents used for flotation tests.

Reagents		Type	Comments
PAX	potassium amyl xanthate	promoter	collector for sulfide and oxide Cu-Minerals after sulfidization
AERO 238	sodium di-sec-butyl phosphorodithioate	promoter	collector for sulfide and oxide Cu-Minerals after sulfidization
AERO 404	dithiophosphate and mercaptobenzothiazole	promotor	collector for sulfide and oxide Cu-Minerals after sulfidization
AERO OX-100	hydroxamic acid	promoter	collector for oxide Cu-minerals without sulfidization
$CuSO_4$	copper (II)sulfate	activator	activator for sulfides
CaO	lime	pH-regulation	pH-regulation
MIBC	methylisobutylcarbinol	frother	frother
$NaHS \cdot H_2O$	sodium hydrosulfide hydrate	modifier	sulfidizer
$Na_2S \cdot 9H_2O$	sodium sulfide	modifier	sulfidizer

Ten flotation tests were carried out for the processing scheme flotation followed by magnetic separation (see below). The activation time in the mill was 30 s. Afterwards, various combinations of the above-listed chemicals and pH values were tested. For every test, usually 2–3 chemicals in different dosage were added and flotation was carried out for 3 min. Afterwards, 3–5 more flotation steps were carried out for every test and 1–2 chemicals were added and/or pH adjustments was done, in order to produce the optimum pre-concentrate. For the flotation tests in the scheme magnetic separation before flotation, only four flotation tests with the most promising results from the above-mentioned scheme were carried out, using 0.5 kg of tailings for each test.

2.5. Conventional Sulfuric Acid Chemical Leaching

Conventional leaching tests were carried out, in order to study the recovery and kinetics of acidic copper leaching. The leaching experiments were carried out by G.E.O.S GmbH in Halsbrücke, Germany, on behalf of BGR. The solid to liquid ratio was 1:5 and dilute sulfuric acid was used (concentration of 49 g H_2SO_4/L H_2O, corresponding to 5% sulfuric acid). For the combination of leaching and magnetic separation, for each processing test, 8.5 kg of material were used. The material was mixed with 42.5 L of dilute (5%) sulfuric acid for a solid-liquid ratio of 1 kg: 5 L. After 24 h, the leach residues were washed and prepared for wet-magnetic separation. The separation was done using a NdBFe permanent magnet at 0.1 Tesla.

Compared to conventional leaching in copper mining, the acid grade of 5% is elevated. In heap leaching, a lixiviant with a low acid concentration (0.1–0.5%) is usually applied for months or longer periods for economic reasons. In order to obtain a benchmark result for the leaching process, the decision was taken to carry out the leaching test with a stronger lixiviant applied for a time-period of 24 h within an agitated reactor. This should result in the maximum recovery of copper that is technically and economically feasible. The metal grades in the eluate were analyzed using ICP-OES. The recovery was calculated comparing the XRF results of the input sample with the ICP-OES data of the output eluate (leached out copper).

2.6. Methodological Approach Used for the Economic Assessment

For the economic assessment, standard cost models for (I) flotation and (II) agitation cyanide leach for gold from costmine© were used as a basis. For reprocessing of tailings, mining, crushing, and milling are not necessary, reducing opex and capex. For opex and capex, the daily tonnage is of great importance. According to the TSF tonnage, a mill capacity of 500 t/day was chosen. This daily capacity is relatively low, compared to mining projects of similar size, but will keep the capex and thus the project risk low. Assuming a downtime of 20%, the yearly tonnage would be 146,000 t of material, which translates into a project life of 13.7 years.

In the flotation cost model, the opex is subdivided into four categories and the reduction for reprocessing is given in percent after the category: supplies and materials (66% reduction), labor (45% reduction), administration (33% reduction), diverse (33% reduction). The resulting opex for reprocessing is around 50% lower compared to the original cost model. In order not to underestimate the costs, we chose an opex of 60% of the original costs, which is 13.6 USD/t.

The capex in the flotation model is subdivided into 13 categories and the reduction in capex is given in percent after the category. The categories are equipment (55% reduction), installation labor (50% reduction), concrete (70% reduction), piping (20% reduction), structural steel (70% reduction), instrumentation (50% reduction), insulation (50% reduction), electrics (30% reduction), coatings and sealants (80% reduction), mill building (50% reduction), tailings embankment (10% reduction), and working capital (50% reduction). The reduced capex is 9.4 M USD, around 50% of the original capex.

For the agitation cyanide leach model (the only agitation leach model available), the opex in the cost model is subdivided into comminution, agitation cyanide leach, solid-liquid separation, general operation, and administration. The comminution was completely skipped and the costs for agitation cyanide leach were reduced by 50%. The chemicals for copper leaching are cheaper and due to the fast reaction kinetic, fewer or smaller leaching tanks can be used. This justifies a reduction of 50% of the opex for this category. The total opex is reduced by 60% to 16 USD/t.

The capex is subdivided into comminution, agitation cyanide leach, solid-liquid separation, general, engineering and construction management, and working capital. The comminution costs are omitted and the costs for cyanide leach are halved. Far less volume will be needed for tank leaching due to the reaction kinetic, but as stainless steel is needed in an acid environment, costs can only be reduced by 50%. The same reduction is applied to the general costs, engineering and construction management, and working capital, as comminution is not needed and the volume for tank leaching is far smaller for sulfuric acid leaching, compared to cyanide leaching. The capex sums up to 9.3 M USD, 50% of the original capex.

3. Results and Discussion

3.1. Mineralogy

Besides the geochemical data, mineralogical data was used to determine in which minerals Cu and Fe are located. The geochemical data already showed a high Fe content of 26% on average. A first mineralogical analysis using XRD, focusing on the main mineral phases, confirmed these data with a magnetite grade of 22% for the coarse-grained material and 15% for the fine-grained material (Table 2). A more detailed analysis focusing on minor and trace minerals on the homogenized sample material (coarse- and fine-grained) confirmed the main mineral phases and additionally demonstrated the presence of 0.4% of atacamite and 0.2% chalcopyrite (Table 3). These grades would only translate to a copper grade of 0.31% for the TSF. This leads to the conclusion that the copper minerals were underestimated in the mineralogical analysis, or that there are additional phases not detected by XRD. Microscopic analysis gave hints to the presence of chrysocolla (not detected by XRD). Additionally it is possible that soluble copper phases migrated in the tailings and precipitated on grain surfaces or were adsorbed by clay minerals both not detectable by XRD).

Table 2. Main mineral phases of the coarse- and fine-grained sample material.

-	Quartz	Magnetite	Gypsum	Amphibole	Calcite	Hematite	Muscovite	Plagioclase	Chlorite
coarse	13	22	3	10	3	6	8	23	12
fine	15	15	4	8	4	8	15	19	12

Table 3. Trace minerals in the homogeneous sample material (coarse- and fine-grained).

-	Pyrite	Apatite	Atacamite	Halite	Ankerite	Chalcopyrite
homogenized sample material	1.0	0.5	0.4	0.4	0.4	0.2

3.2. Semi Technical Processing Tests

The elements with the highest product value in the Taltal tailings are copper and iron. Therefore, the approach for tailings reprocessing is focused on these two elements. For the reprocessing of the tailings material, different processing tests (5–10 kg each) were performed, including flotation, magnetic separation, and leaching, which were combined to processing flowsheets (Figure 5). As the processed ore came from various mines and was originally processed via flotation, starting in 1966, there should still be floatable minerals like chalcopyrite in the TSF. Additionally, the region of Taltal is known for oxide ores that are still processed in Taltal. Therefore, collector chemicals for oxides and sulfides were tested (Table 1). Conventional sulfuric acid leaching was tested, in order to evaluate the recovery and reaction kinetics. As a leaching facility is already installed in Taltal, this could make an economic copper recovery process easier, as some processing infrastructure is already available. Density sorting was also tested, but the recoveries for copper were poor and are not discussed here. It would also be possible to apply bioleaching on the tailings. This method can result in high recoveries of >90% [17] under special environments (e.g., temperature). As we knew that a large share of the copper was present as oxides or chloride, we chose only to apply conventional chemical leaching, as a high recovery was achievable at predictable relatively low costs [28].

The objective of the processing tests was to investigate value concentration paths in order to achieve saleable products or intermediate products in an economically efficient way. In total, four concentration paths were investigated (Figure 5).

The aim was to produce a clean, saleable magnetite concentrate and to evaluate whether it is possible to produce a copper (pre)concentrate as the base of a saleable product, or to leach the copper minerals with dilute sulfuric acid.

3.2.1. Flotation–Magnetic Separation and Magnetic Separation–Flotation (Figure 5a,b)

As described in Section 3.1, the tailings from Taltal contain significant amounts of copper, as oxides, chlorides, and sulfides, and iron, which is mostly contained in the mineral magnetite. Due to these two value-containing elements, combinations of flotation (oxides and sulfides) followed by magnetic separation and a processing route vice versa were tested (Figure 5a,b). The tests were performed using several kg of material each.

A detailed flowsheet of flotation followed by magnetic separation is given in Figure 6. The tests were carried out according to the description in 2.3. The average recovery was 61% (46–73%) and the average copper grade was 1.8% (1.2–2.2%, excluding Test 10) as shown in Table 4. Recoveries of over 60% for a low grade ore that consists of oxides and sulfides seems quite satisfactory, but the Cu grades of maximum 2.2% for the (pre)concentrates are far too low for a saleable product and would have to be upgraded with an additional method, most probably leaching in an acidic medium, increasing costs and lowering total recovery.

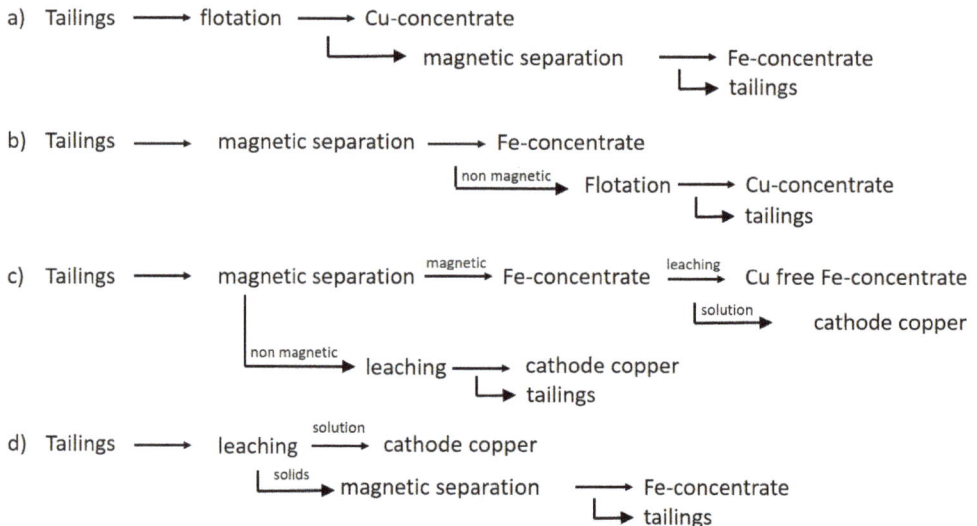

Figure 5. Different processing schemes (**a**–**d**) that were tested for the tailings of Taltal. The schemes (**a**) and (**b**) use flotation in combination with magnetic separation. Schemes (**c**) and (**d**) include leaching and magnetic separation.

Table 4. Results of the performed flotation tests for copper oxides and sulfides. For Test 1, 1 kg of material was used and for Test 10, 1.4 kg was used. The other tests were carried out with 0.5 kg.

Test	Tailings	Pre-Concentrate of Cu		
	Mass (g)	Mass (g)	Grade (%)	Recovery (%)
1	936	182.3	1.2	45.9
2	490	67.5	1.9	56.6
3	490	105.4	1.5	73.1
4	487	80.8	1.3	48.0
5	486	94.5	1.6	65.1
6	485	67.7	2.2	65.1
7	483	75.0	1.9	66.2
8	501	71.9	2.2	65.7
9	487	66.6	2.0	60.2
10	1445	472.4	0.9	64.2

The following magnetic separation (Figure 6) was performed using a magnetic rougher concentration, followed by three magnetic cleaner concentration steps. The result was a concentrate with 60% Fe-content and Fe recovery of 64%. The sulfur content is very low (0.005%), which is a positive fact for selling such a concentrate. Almost 90% of the sulfur is collected in the flotation (pre)concentrate (Table 5).

The second test scheme started with magnetic separation, followed by flotation. The magnetic separation included one rougher, two scavenger steps, and three cleaner separations steps (Figure 7). The recovery of iron into a magnetic concentrate was 74% with an Fe-grade of 60% in the concentrate (Table 6).

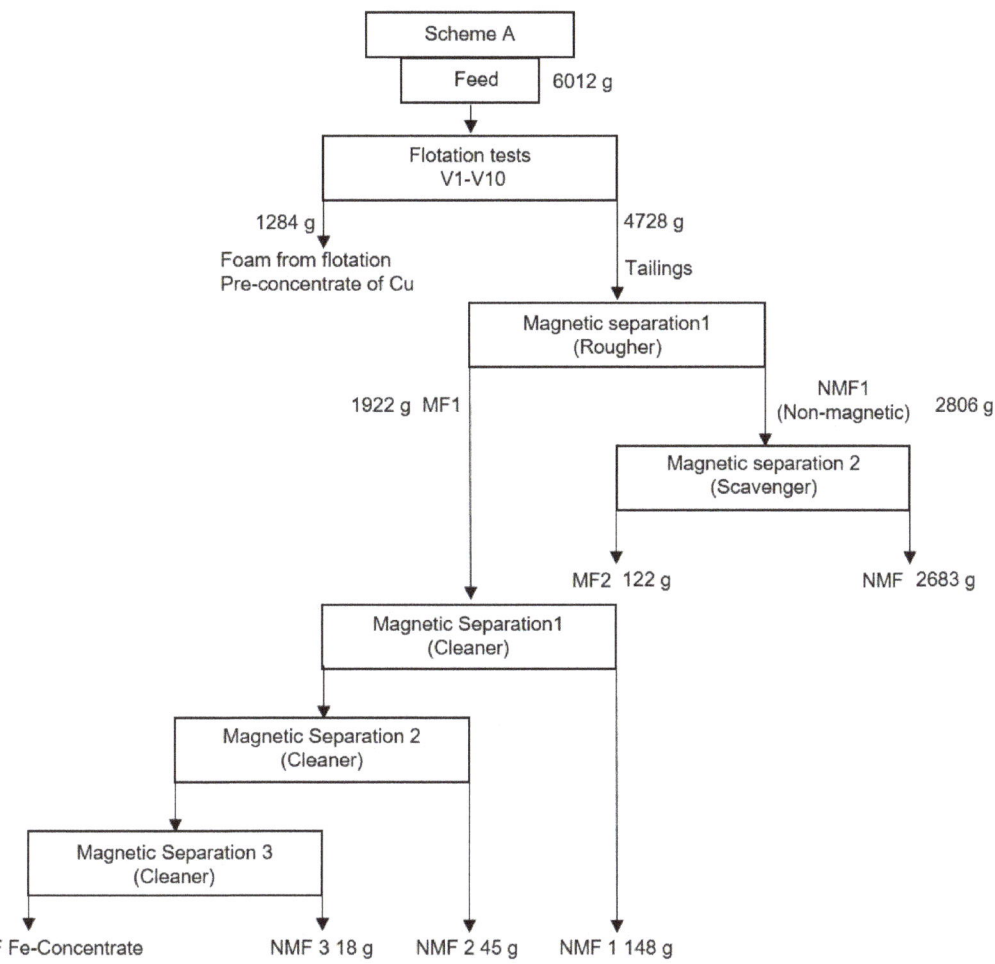

Figure 6. Detailed processing scheme from flotation followed by magnetic separation. MF: magnetic fraction, NMF: non-magnetic fraction.

Table 5. Recoveries and grades for flotation with subsequent magnetic separation.

Product	Mass (g)	Recovery (%)	Cu Grade (%)	Recovery (%)	Fe Grade (%)	Recovery (%)	S Grade (%)	Recovery (%)
Feed	6012	100	0.45	100	26.7	100	0.04	100
Flotation (pre)concentrate	1284	21.4	1.37	65.1	20.6	16.5	0.16	88.1
Fe-concentrate	1711	28.5	0.08	4.8	60.0	64.1	0.01	3.6
tailings	3016	50.2	0.27	30.1	10.3	19.4	0.01	8.3

Four flotation tests with different flotation chemicals were carried out on the non-magnetic material using the most promising setup and combination of chemicals from the previous tests (tests 6–9 from Figure 5). For each test, around 0.5 kg of tailings material was used (Table 7). The copper recovery of the test was 48–57% of the non-magnetic fraction. The corresponding copper grade was in the range of 1.7–2.4% (average 2.0%). The average copper recovery of this test scheme is only 44%, as there are additional copper losses prior to flotation into the magnetic fraction (Table 6). The copper grade in the (pre)concentrates is similarly low, compared to the scheme of flotation with subsequent magnetic separation. Likewise, it could be possible to subject the (pre)concentrate to an acidic leach process, with the aim to produce saleable copper products, as for example copper cathode or copper chemicals, but the overall recovery of copper is very low.

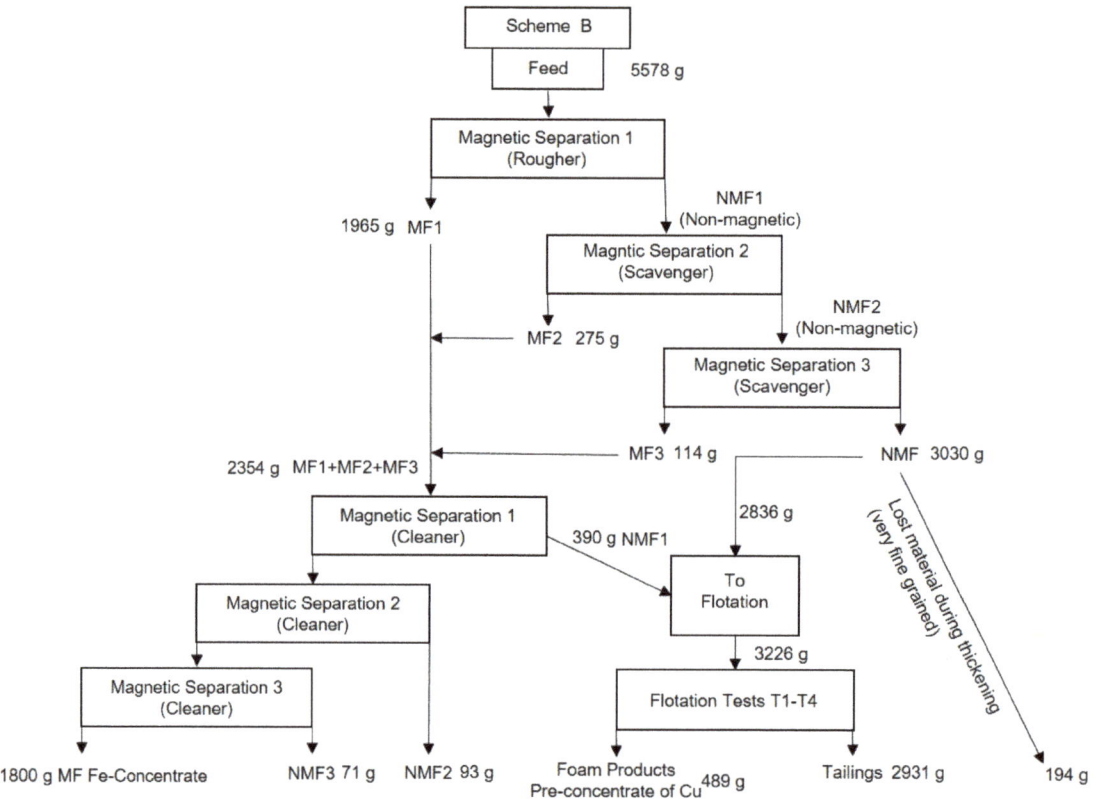

Figure 7. Processing scheme of magnetic separation with subsequent flotation. The magnetic concentrate corresponds to 32% of the mass of the fed tailings. MF: magnetic fraction, NMF: non-magnetic fraction.

Table 6. Recoveries and grades for magnetic separation followed by flotation. The 194 g that were lost during thickening (Figure 7) is not shown in this table, as the geochemical composition is unknown.

Product	Mass (g)	Recovery (%)	Cu Grade (%)	Recovery (%)	Fe Grade (%)	Recovery (%)	S Grade (%)	Recovery (%)
Feed	5384	100	0.42	100	26.9	100	0.0	100
Fe-concentrate	1800	33.4	0.14	11.2	59.8	74.4	0.0	6.6
Flotation (pre)concentrate	489	9.1	2.04	44.2	12.4	4.2	0.2	67.3
tailings	3095	57.5	0.32	44.6	10.0	21.4	0.0	26.1

The advantage of this setup is the higher recovery of Fe into a concentrate, but the copper recovery into a (pre)concentrate is even lower. The relatively high grade of Cu in the Fe concentrate (0.14%) is still far below the restrictions given for several steel grades, e.g., railway steel (0.4%) and some steel used in construction (0.55%) that can be found online [29].

Table 7. Results of the performed flotation tests for copper oxides and sulfides. The tests were carried out with 0.5 kg sample material. As not all the non-magnetic material was used to carry out the tests, the recovered mass (Figure 7 and Table 6) in the Cu pre-concentrates was extrapolated.

Test	Tailings	Pre-Concentrate of Cu		
	Mass (g)	Mass (g)	Grade (%)	Recovery (%)
1	499	62.0	2.2	50.9
2	501	72.1	2.1	52.7
3	501	56.4	2.4	47.6
4	501	95.7	1.7	57.1
total	-	286.2	2.1	52.1

3.2.2. Leaching–Magnetic Separation and Magnetic Separation–Leaching (Figure 5c,d)

The combination of leaching and magnetic separation is schematically shown in Figure 8. Relatively strong sulfuric acid of 5% was used, in order to obtain a maximum value for leaching recovery. Leaching kinetics (Figure 9) are very fast (hours), compared to days in bioleaching [17], at least for the high acid grades applied. From the original (Org) sample material, 70% of Cu could immediately (after 5 min) be leached (Figure 9), as the copper is concentrated in oxide, chloride, or carbonate minerals. After one hour, 77% of the contained copper was already leached. The maximum leaching recovery was 84%. The leaching kinetics and recoveries were similar for the non-ferromagnetic material (NFM) with a recovery of around 80% for both. The gathered Cu bearing solution has a grade of 730 mg Cu/L, probably too low for direct application of solvent extraction.

Magnetic separation of the original, non-leached material resulted in a magnetic fraction (40%) of the material and a non-magnetic fraction (60%). The Fe-grade in the FM fraction was 47%, compared to 26% before the separation. The Fe-grade was higher after leaching (52%), at the expense of SiO_2, CaO, and most other major elements. This Fe-content is relatively low, as a marketable iron concentrate should have 62%. The processing tests of flotation combined with magnetic separation resulted in concentrates with Fe-grades of 60% (Tables 5 and 6). The larger portion of the Fe-concentrate (40% compared to 28–33% in Tables 5 and 6) results in a lower Fe-grade of only 52%, as the separation worked poorer.

When applying magnetic separation after leaching (Figure 5d), the Fe grade in the magnetic fraction was only 43%, not encouraging, as this concentrate is not marketable. The reason for the substantially lower Fe grades might be connected to different equipment used for the magnetic separation, including missing cleaner steps. For the leaching recovery, it is not important if leaching or magnetic separation is applied first. In contrast, it does play a role in the quality of the magnetic separation, whether magnetic separation is done before leaching, or afterwards, as leaching can decrease grainsize, which has a negative effect on magnetic separation.

3.3. Introduction of Adjusted Cost Models

In an industrial scale processing facility, usually lower acid grades of 0.1–0.5% are applied in combination with a longer reaction time. A lower acid grade reduces the grade of undesired elements in solution and therefore reduces the acid consumption, as less material is dissolved (both not tested in this paper). Considering the realized semi-technical tests for reprocessing in this paper, the most promising processing option for the production of a magnetite concentrate and copper intermediates is as follows (Figure 10):

1. Magnetic separation into FM and NFM fraction, including magnetic cleaner steps
2. Separate leaching of the FM and NFM material in stirred tank or horizontal rotary reactor with dilute sulfuric acid (below 5%) in continuous or batch mode (for several hours)
3. Final production of an Fe-concentrate (Cl-, S-, P- and Cu-grades could be an issue)
4. Hydrocyclone and/or settling to separate solids and liquids (Cu-rich solution)
5. Washing of solids to remove rests of copper (dilute Cu-solution)
6. Deposition of the NFM fraction (finer-grained than original tailings material)

7. Further concentration and cleaning of Cu-solution to produce intermediate products, or cathodes by electrowinning

In step 7 above, a further concentration and cleaning of the Cu solution is mentioned as the copper grade of the solution is only 730 mg Cu/L. A combination of the relatively low-grade solution with the higher grade solution of the nearby heap leaching facility in Taltal could be possible. Another option would be to recirculate the solution or to increase the solid-liquid ratio, to increase the copper grade. Afterwards, solvent extraction could be applied. It would also be possible to use a different technique like ion exchange, in order to adsorb the copper on a resin, to produce a high grade copper solution on which electrowinning could directly be applied on.

In the following discussion of the cost models, the presented scheme in Figure 10, as well as flotation are discussed, as flotation is a possible technique for reprocessing many TSFs. Flotation can be applied for economic reprocessing, but also to reduce sulfide grades and minimize acid rock drainage.

For the economic assessment, a revenue of 65 USD/t of magnetite concentrate was assumed. The average price over the last five years was 72 USD/t for 62% Fe iron ore concentrate (cost freight China), but there are still uncertainties about production costs and quality of the concentrate. Therefore, a more conservative revenue is assumed. For copper cathode, the average price in the last five years was around 5500 USD/t, which is used for the economic evaluation. In order to process the tailings deposit of 2,000,000 t, a 500 t/d capacity processing plant is supposed to be a realistic match, in order to minimize the preproduction expenditures. Larger processing facilities would improve the long-term economics of the project [22,30], but also increase the economic project risks. The annual capacity is assumed to be 146,000 t (see Section 2.5). It is generally possible to produce marketable copper concentrates from tailings. This can be seen in the concentrates of Minera Valle Central, although the size of the operation is several orders of magnitude larger compared to the tailings described here. Reference [10] also demonstrates the possibility to produce a concentrate with a copper grade of 19 % from fresh tailings or almost fresh tailings at low cost at a large industrial scale (150,000 t/d). The processing costs would be 5.6 USD/t, taking the data from their publication (Labor, Energy, Chemicals). As the Taltal TSF only contains around 2 M t of material, substantially higher costs have to be assumed.

When reprocessing tailings, a hydraulic transport of the slurried tailings to the plant or transport via truck, an activation grinding, and a desliming classification of the tailings has to be done. According to [10], desliming is possibly not necessary as in certain cases, ultrafine particles can also be floated and even increase the recovery.

3.4. Comparison of Cost Models and NPV

The given capex and opex from costmine© for a flotation processing plant of 500 t/d is 22 USD/t and an investment of around 15.7 Mio USD. As no crushing is needed and if milling only to a very small amount, these costs can be subtracted, leading to processing costs of around 13.2 USD/t and investment costs of 9.4 M USD (Section 2.5 and Table 8). These costs are too high to carry out economic reprocessing in Taltal (Tables 9 and 10). To cover the operating costs of 13.2 USD/t, at a copper price of 5500 USD/t, the recoverable copper grade has to be 0.24%. To refinance the capex over 10 years, additional 170 t of copper (around 940,000 USD) would have to be extracted annually and marketed without deduction, equivalent to additional 0.12% recoverable. This results in a minimum recoverable copper grade of 0.35%, just to cover capex and opex. The opex of the adopted cost model for flotation is more than two times higher, compared to [10], but these authors worked at a processing plant with a capacity of 150,000 t/d, which is the yearly processing capacity, chosen in our cost model. Based on these data, it seems difficult to achieve an economic reprocessing for copper in small tailings dams. In addition to the 0.35% of recoverable copper needed to refinance opex and capex, discounted cash flow (10%) is usually applied, leading to even higher needed recoveries. Furthermore, a net smelter return (NSR) for transport, treatment, and refining the concentrate (deductions of 10–20%) has to be subtracted from the metal value. Nevertheless, the cost estimation can be a basis for rules of thumb, when relatively small tailings have to be reprocessed for environmental reasons, e.g., to reduce acid drainage.

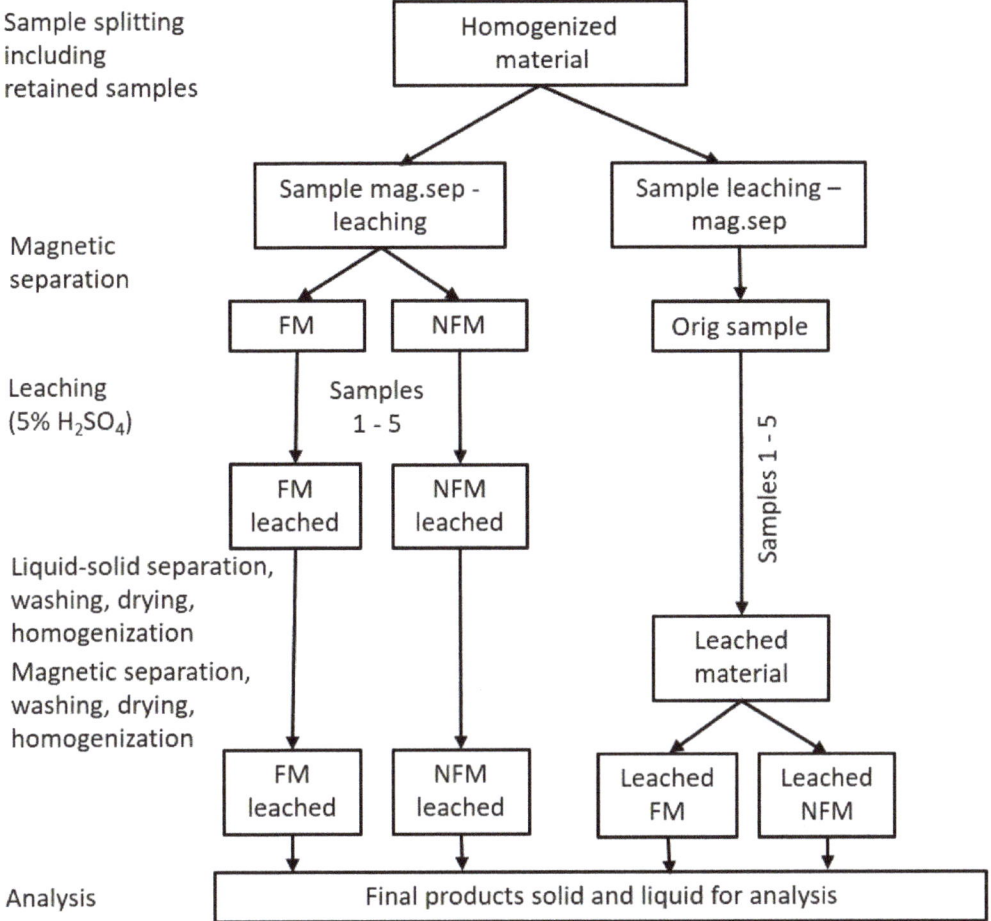

Figure 8. Processing scheme of magnetic separation in combination with leaching. Each test was carried out with around 8.5 kg sample material. FM: ferromagnetic, NFM: non ferromagnetic.

For the required magnetic separation, only a very rough cost estimation was possible, since this processing method is not part of the standard cost models of costmine©. The opex is dominated by freight costs to the next harbor facility. Using the rule-of-thumb-costs for truck transport of 10 US cent per ton and mile [22], additional freight costs of around 15.0 USD/t for the transport of the magnetite concentrate to the next harbor (150 miles to Caldera) have to be added to the plant operation costs. The operation costs consist of the costs for magnetic separation, cleaner separation, thickening, drying, and equipment and were estimated at about 5 USD per processed ton of tailings. The required investment for the plant has been estimated to about 1 M USD. The total costs for the recycling of 2 M t of tailings, including transport to the next harbor, would be in the range of 20 M USD (around 10 USD/t), depending on the amount of concentrate that has to be transported. According to the model costs and the calculated revenue (Tables 9 and 10), magnetic separation would result in revenues that are around two times higher, compared to the costs.

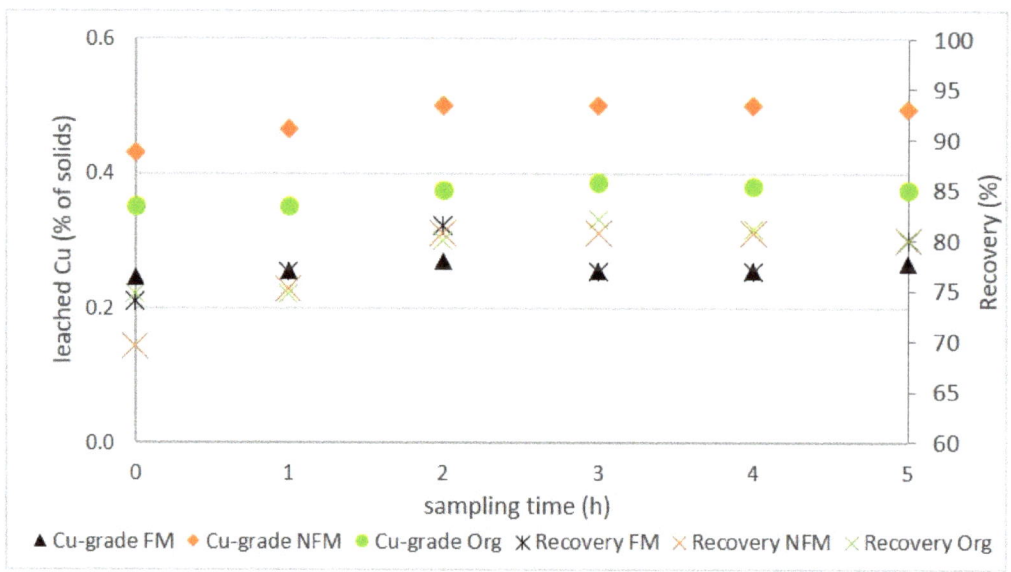

Figure 9. Recoveries (crosses) and leached copper grade (filled symbols) of the original tailings material (Org) and the two fractions with the ferromagnetic material (FM) and the non-magnetic material (NFM). The leached copper is given as % of the solids. The leaching seems to begin at time 0; it is actually 5 min after t 0.

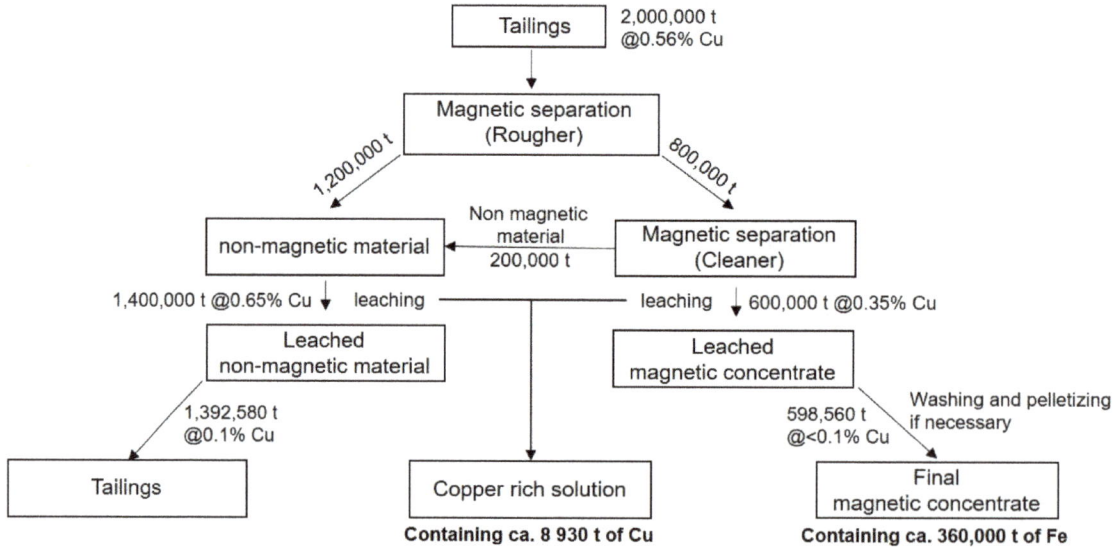

Figure 10. Mass balance for the processing scheme of magnetic separation followed by separate leaching of the magnetic (FM) and non-magnetic fractions (NMF).

Table 8. Original (costmine©) and modified (adjusted to a tailings processing facility) opex in USD/t and capex in M USD for a 500 t/d flotation and magnetic separation processing plant. Costs for magnetic separation are inclusive of freight costs to the port of Caldera.

	Capacity	Original Opex	Adjusted Opex	Original Capex	Adjusted Capex
flotation	500 t/d	22.3 USD/t	13.2 USD/t	15.7 M USD	9.4 M USD
magnetic separation	500 t/d	-	10 USD/t incl. fr.	-	1 M USD
agitated leaching	500 t/d	40 USD/t	16 USD/t	19.4 USD/t	9.3 M USD

Table 9. Data of coarse-grained tailings material and (pre)concentrates of copper and iron, including recoveries, metal value and investment, processing, and transport costs for the scheme of first flotation and second magnetic separation.

Flotation-Magnetic Separation	Mass (M t)	Fe Grade (%)	Fe Recovery (%)	Cu Grade (%)	Flotation Recovery (%)	Cu in Concentrate (t)	Metal Value in Concentrate (M USD)	Capex and Opex (M USD)
tailings mass	2.0	27	-	0.45	-	-	-	-
Cu (pre)concentrate	0.43	-	-	1.37	65	6020	33.1	35.8
Fe concentrate	0.57	60	64	0.08	-	-	37.0	20.2

Table 10. Data of coarse-grained tailings material and (pre)concentrates of copper and iron, including recoveries, metal value and investment, processing, and transport costs for the scheme of first magnetic separation and second flotation.

Magnetic Separation-Flotation	Mass (M t)	Fe Grade (%)	Fe Recovery (%)	Cu Grade (%)	Flotation Recovery (%)	Cu in Concentrate (t)	Metal Value in Concentrate (M USD)	Capex and Opex (M USD)
tailings mass	2.0	27	-	0.45	-	9000	-	-
Fe- concentrate	0.66	60	74	0.14	-	-	42.9	21.7
Cu (pre)concentrate	0.18	-	-	2.04	44	3700	20.4	26.7

The results from the magnetic separation are very encouraging in both processing schemes, although the copper grade and other deleterious elements have to be examined in more detail. This is particularly valid for the chloride grade, which could not be analyzed by XRF. The origin of the chloride is seawater, which is frequently used in the dry coastal areas for processing. Nevertheless, if saleable, the magnetite concentrate would give a very profitable product (Tables 9 and 10). In case of flotation, the concentrates do not pay the costs for processing and investment. Even if it were possible to reduce the costs for investment by purchasing used equipment, flotation does not seem to be an appropriate processing technique, as the concentrate grades are far below any saleable product. The only reason to do flotation would be a desulfurization of the tailings, to prevent acid drainage, as up to 90% of the sulfur can be recovered in the flotation products.

Next to flotation, heap leaching is the second most common processing method for copper ores [12]. Due to the fine-grained tailings material, heap leaching is difficult to apply, as acid tolerant binders to agglomerate the material are limited [31]. Therefore, agitation tank leaching would have to be carried out, which is relatively expensive in terms of capex and opex. Examples for costs can be found in various feasibility studies, for example the Raventhorpe project in Australia, which is slightly larger (two times) compared to the investigated tailings in this paper [32]. The reaction kinetics for leaching are very fast, as almost all of the leachable copper was in solution after 2 h (Figure 9), at least at an admittedly high acid concentration of 5% H_2SO_4. Similar kinetics were also described by [33]. The recovery of approximately 80% for the 2 M t of tailings with an average copper grade of 0.56% (Tables 9 and 10 refer to the coarse-grained material only) leads to around 8900 t of recoverable copper with a metal value of 49.1 M USD.

As a basis for a rough estimation of capex and opex, the agitated cyanide tank-leaching model for Gold from costmine© was chosen and adjusted (Table 8). The new capex is 9.3 M USD (for new equipment)—50% of the model capex—and the opex is 16 USD/t for tailings, 60% less, compared to the model opex. This results in total costs for the processing of about 41.3 M USD. Compared to the calculated revenue of 49.1 M USD, the reprocessing of the tailings would result in a small profit (Table 11). The opex data of 16 USD/t is equivalent to 0.29% of leached copper (at a price of 5500 USD/t). Additionally, 0.12% of Cu has to be recovered to refinance the capex of 9.3 M USD over 10 years. This leads to a total grade of 0.41% that has to be recovered just to cover the costs. The 2 M t would be processed in 13.7 years (146,000 t/a) and 652 t of Cu would be produced per year (8930/13.7), leading to a positive cash flow of 7.8 M USD per year (5500 USD/t Cu) minus investment. A dynamic model using discounted cash flow (10%) results in an NPV of 0.2 M USD. These calculations are prone to changes in copper price and recovery/recoverable grade. A price variation of 10 % for copper (500 USD/t of copper) results in changes of 2.5 M USD in the NPV and thus a negative NPV of −2.3 M USD, or an NPV of 2.7 M USD in a positive scenario. Taking the copper price of December 2020 (7700 USD/t), we would get an NPV of 11.1 M USD, but it must be avoided to use these extremely positive assumptions as prices are cyclical. Similar variations of around 2.8 M USD can be seen when we introduce a scenario with a grade variation of 10% from the original grade of 0.56%. A grade reduction leads to a negative NPV of −2.1 M USD and a higher grade of 0.61% leads to a far more positive result of 3 M USD. A double negative (price and grade) scenario would lead to a devastating NPV of −4.7 M USD and a double positive scenario would lead to relatively promising NPV of 5.7 M USD.

Table 11. Combination of magnetic separation and leaching, which has given the best economic results in the processing tests. Leaching recoveries taken from processing tests. Data from magnetic separation is theoretical, as larger scale processing tests are still to come.

Leaching-Magnetic Separation	Mass (M t)	Fe Grade (%)	Fe Recovery (%)	Cu Grade (%)	Leaching Recovery (%)	Cu Leached (t)	Metal Value in Leachate (M USD)	Capex and Opex (M USD)
magnetic separation	0.6	60	64	-	-	-	39.0	20.7
leaching with dilute sulfuric acid	2.0	-	-	0.56	80	8930	49.1	41.3

The cash flow calculation for the magnetite is clearly positive. The annual cash flow is 1.46 M USD, which sums up to an NPV of 10 M USD using a 10% discount rate, quite high for the amount invested. There are similar changes to the economies, when applying positive or negative scenarios for recovery and prices, but the NPV will always be positive. The biggest risk for magnetite is the production of a marketable concentrate and to enter into an offtake agreement for such low volume.

Mobile processing plants are also discussed in the literature, especially for tailings of relatively small (<100,000 m^3) volume [34] or placer deposits [35]. Other authors [36] supposed such a strategy for a TSF of similar size to Taltal. Nevertheless, it has to be considered that a project life of almost 14 years most probably leads to permanently installed technology, but it should be considered that at least parts of the processing plant are built in a way to use them for other TSFs in the region afterwards, especially for the production of Fe-concentrates.

4. Conclusions

Cost models, including an approximate estimation of the capacity of the processing plant offer the opportunity for a rough economic assessment in an early project stage, before cost-intensive processing tests in a pilot plant scale are conducted. Additionally, this allows comparing projects of different grade and different tonnage. Larger scale processing tests can be carried out on the most promising projects, to check the results of the tests, in order to proceed closer to an industrial scale plant. Early stage cost estimations for tailings reprocessing show that opex can be reduced by 40% and capex by 50% for a flotation plant with a capacity of 500 t/d. The opex is reduced to 13.2 USD/t and the capex is reduced to 9.4 M USD. Used equipment would further reduce costs. To cover opex and capex, without discounted cash flow and neglecting the NSR, a minimum copper grade of 0.35% has to be extracted from the tailings, too much for the investigated example.

In case of a high content of copper oxides in the tailings, leaching with dilute sulfuric acid can be a feasible alternative. The adopted cost-model of cyanide gold leaching resulted in capex of 9.3 M USD and opex of 16 USD/t. This is a reduction of 60% of the opex and 50% of the capex of the model. The advantage of leaching in this case is the high recovery of 80% and the option to produce copper cathode from the pregnant leach solution instead of a (pre)concentrate that needs further treatment. To cover opex and capex, without discounting the cash flow, a copper grade of 0.41% has to be extracted from the tailings. When applying a discount rate of 10%, the extracted grade has to be around 0.44% Cu, to get an NPV of 0.

The disadvantage in leaching is that acids reduce the grain size of the material. This may cause some issues for re-deposition that would have to be investigated in more detail in a pilot-plant study.

In the case of the Taltal tailings, the most economical processing method is the concentration of magnetite. The presented semi-technical scale tests result in a recovery of 30% of the total material (600,000 t) to a magnetic concentrate of around 60% Fe-grade (value around 39 M USD). As opex and capex are low for magnetic separation, the magnetite in the tailings is the most profitable commodity with an NPV (10%) of around 10 M USD.

In Chile, there are around 50 tailings with 25 M t and an Fe_2O_3 grade >30% that have a high potential to produce Fe-concentrates and another around 100 Mio. t close to Copiapó, where similar deposits to Taltal are mined with a potential to produce Fe-concentrates.

Supplementary Materials: The following are available online at https://www.mdpi.com/article/10.3390/met11010103/s1, Figure S1: Grain size distribution of the sample material that was macroscopically divided into "coarse" and "fine" grained.

Author Contributions: Conceptualization, M.D. and H.M.; methodology, M.D.; validation, M.D., M.G., H.M., and F.H.; formal analysis, F.H. and M.D.; investigation, M.D., F.H., and M.G.; writing—original draft preparation, M.D.; writing—review and editing, M.D., F.H., H.M., and M.G.; All authors have read and agreed to the published version of the manuscript.

Funding: This research received no external funding.

Institutional Review Board Statement: Not applicable.

Informed Consent Statement: Not applicable.

Data Availability Statement: The data presented in this study are available in the article.

Acknowledgments: As part of the commodity partnership agreed between Chile and Germany, the Federal Institute for Geosciences and Natural Resources (BGR, Hannover, Germany) has initiated a project together with the Chilean Geological survey SERNAGEOMIN to evaluate the economic potential for reprocessing/recycling of mining residues with the focus on copper and other elements of value. M.D. wants to thank SERNGEOMIN for the excellent cooperation, including sampling, discussion, and logistics. Special thanks go to OR and B. Processing tests were carried out by RWTH Aachen and G.E.O.S. Ingenieurgesellschaft mbH.

Conflicts of Interest: The authors declare no conflict of interest.

References

1. Drobe, M. Vorkommen und Produktion Mineralischer Rohstoffe—Ein Ländervergleich. 2020. Available online: https://www.bgr.bund.de/DE/Themen/Min_rohstoffe/Downloads/studie_Laendervergleich_2017.pdf?__blob=publicationFile&v=7 (accessed on 1 July 2020).
2. Lottermoser, B. *Mine Wastes: Characterization, Treatment, and Environmental Impacts*, 3rd ed.; Springer: Berlin, Germany, 2010.
3. SERNAGEOMIN. Anuario de la Minería de Chile 2019. Available online: https://www.sernageomin.cl/pdf/anuario_2019_act100720.pdf (accessed on 14 July 2020).
4. Alcalde, J.; Kelm, U.; Vergara, D. Historical assessment of metal recovery potential from old mine tailings: A study case for porphyry copper tailings, Chile. *Miner. Eng.* **2018**, *127*, 334–338. [CrossRef]
5. Nikonow, W.; Rammlmair, D.; Furche, M. A multidisciplinary approach considering geochemical reorganization and internal structure of tailings impoundments for metal exploration. *J. Appl. Geochem.* **2019**, *104*, 51–59. [CrossRef]
6. Araya, N.; Kraslawski, A.; Cisternas, L.A. Towards mine tailings valorization: Recovery of critical materials from Chilean mine tailings. *J. Clean. Prod.* **2020**, *263*, 121555. [CrossRef]
7. Giurco, D.; Prior, T.; Mudd, G.; Mason, L.; Behrisch, J. *Peak Minerals in Australia: A Review of Changing Impacts and Benefits*; Department of Civil Engineering, Monash University: Clayton, Australia, 2010; Available online: https://opus.lib.uts.edu.au/bitstream/10453/31155/1/2009003150OK.pdf (accessed on 15 September 2020).
8. Gordon, R.B. Production residues in copper technological cycles. *Resour. Conserv. Recycl.* **2002**, *36*, 87–106. [CrossRef]
9. Evdokimov, S.I.; Evdokimov, V.S. Metal recovery from old tailings. *J. Min. Sci.* **2014**, *50*, 800–808. [CrossRef]
10. Yin, Z.; Sun, W.; Hu, Y.; Zhang, C.; Guan, Q.; Wu, K. Evaluation of the possibility of copper recovery from tailings by flotation through bench-scale, commissioning, and industrial tests. *J. Clean. Prod.* **2018**, *171*, 1039–1048. [CrossRef]
11. Mackay, I.; Videla, A.R.; Brito-Parade, P.R. The link between particle size and froth stability—Implications for reprocessing of flotation tailings. *J. Clean. Prod.* **2020**, *242*, 118436. [CrossRef]
12. Schippers, A.; Hedrich, S.; Vasters, J.; Drobe, M.; Sand, W.; Willscher, S. Biomining: Metal Recovery from Ores with Microorganisms. In *Geobiotechnology I. Advances in Biochemical Engineering/Biotechnology*; Schippers, A., Glombitza, F., Sand, W., Eds.; Springer: Berlin/Heidelberg, Germany, 2013; Volume 141, pp. 1–47. [CrossRef]
13. Xie, Y.; Xu, Y.; Yan, L.; Yang, R. Recovery of nickel, copper and cobalt from low-grade Ni-Cu sulfide tailings. *Hydrometallurgy* **2005**, *80*, 54–58. [CrossRef]
14. Olson, G.J.; Brierley, C.L.; Briggs, A.P.; Calmet, E. Biooxidation of thiocyanate-containing refractory gold tailings from Minacalpa, Peru. *Hydrometallurgy* **2006**, *81*, 159–166. [CrossRef]
15. Schippers, A.; Nagy, A.A.; Kock, D.; Melcher, F.; Gock, E.-D. The use of FISH and real-time PCR to monitor the biooxidation and cyanidation for gold and silver recovery from a mine tailings concentrate (Ticapampa, Peru). *Hydrometallurgy* **2008**, *94*, 77–81. [CrossRef]
16. Marrero, J.; Coto, O.; Goldmann, S.; Graupner, T.; Schippers, A. Recovery of nickel and cobalt from laterite tailings by reductive dissolution under aerobic conditions using Acidithiobacillus species. *Environ. Sci. Technol.* **2015**, *49*, 6674–6682. [CrossRef] [PubMed]
17. Falagán, C.; Grail, B.M.; Johnson, D.B. New approaches for extracting and recovering metals from mine tailings. *Miner. Eng.* **2017**, *106*, 71–78. [CrossRef]
18. Edraki, M.; Baumgartl, T.; Manlapig, E.; Bradshaw, D.; Franks, D.M.; Moran, C.J. Designing mine tailings for better environmental, social and economic outcomes: A review of alternative approaches. *J. Clean. Prod.* **2014**, *84*, 411–420. [CrossRef]
19. Dold, B.; Fondbote, L. Element cycling and secondary mineralogy in porphyry copper tailings as a function of climate, primary mineralogy, and mineral processing. *J. Geochem. Explor.* **2001**, *77*, 3–55. [CrossRef]
20. S&P Global. SNL Database. Commercial Online-Databank. Available online: https://platform.marketintelligence.spglobal.com/web/client#dashboard/metalsAndMining (accessed on 14 July 2020).
21. Lottermoser, B.G. Recycling, reuse and rehabilitation of mine wastes. *Elements* **2011**, *7*, 405–410. [CrossRef]
22. Wellmer, F.W.; Dalheimer, M.; Wagner, M. *Economic Evaluations in Exploration*, 3rd ed.; Springer: Berlin/Heidelberg, Germany, 2008; p. 250. [CrossRef]
23. Wellmer, F.W.; Drobe, M. A quick estimation of the economics of exploration projects—Rules of thumb for mine capacity revisited—The input for estimating capital and operating costs. *Bol. Geol. Min.* **2019**, *130*, 7–26. [CrossRef]
24. SERNAGEOMIN. Datos de Geoquímica de Depósitos de Relaves de Chile (actualización: 13/01/2020). Available online: https://www.sernageomin.cl/datos-publicos-deposito-de-relaves/ (accessed on 15 June 2020).
25. Vick, S.G. *Planning, Design, and Analysis of Tailings Dams*; Wiley: New York, NY, USA, 1983; p. 369.
26. Bussière, B. Colloquium 2004: Hydrogeotechnical properties of hard rock tailings from metal mines and emerging geoenvironmental disposal approaches. *Can. Geotech. J.* **2007**, *44*, 1019–1052. [CrossRef]
27. Blight, G.; Bentel, G. The behaviour of mine tailings during hydraulic deposition. *J. S. Afr. Inst. Min. Metall.* **1983**, *83*, 73–86.
28. Long, K.R.; Singer, D.A. A Simplified Economic Filter for Open-Pit Mining and Heap-Leach Recovery of Copper in the United States. In *USGS Open-File Rep.*; 2001; 01-218. Available online: https://pubs.usgs.gov/of/2001/0218/pdf/of01-218.pdf (accessed on 15 September 2020).

29. b2bmetal (2020). Online Metal Marketplace with Requirements for Several Steel Grades Including Mechanical Properties, Chemical Composition and Grade Equivalents. Available online: http://www.b2bmetal.eu/en/pages/index/index/id/141/ (accessed on 20 November 2020).
30. Long, K.R. A Test and Re-Estimation of Taylor's Empirical Capacity-Reserve Relationship. *Nat. Resour. Res.* **2009**, *18*, 57–63. [CrossRef]
31. Dhawan, N.; Sadegh Safarzadeh, M.; Miller, J.D.; Moats, M.S.; Rajaman, R.K. Crushed ore agglomeration and its control for heap leach operations. *Miner. Eng.* **2013**, *41*, 53–70. [CrossRef]
32. ACH Minerals 2016. Appendix 2: Capital and Operating Cost Estimate—GR Engineering Services. Available online: https://consultation.epa.wa.gov.au/seven-day-comment-on-referrals/ravensthorpe-gold-copper-project/supporting_documents/CMS16331%20%20Referral%20%20Appendix%202%20Captial%20and%20Operating%20Cost%20Estimate.pdf (accessed on 29 July 2020).
33. Conić, V.; Stanković, S.; Marković, B.; Božić, D.; Stojanović, J.; Sokić, M. Investigation of optimal technology for copper leaching from old flotation tailings of the copper mine Bor (Serbia). *Metall. Mater. Eng.* **2020**, *26*, 209–222. [CrossRef]
34. Kuhn, K.; Meima, J.A. Characterization and Economic Potential of Historic Tailings from Gravity Separation: Implications from a Mine Waste Dump (Pb-Ag) in the Harz Mountains Mining District, Germany. *Minerals* **2019**, *9*, 303. [CrossRef]
35. Nevskaya, M.A.; Seleznev, S.G.; Masloboev, V.A.; Klyuchnikova, E.M.; Makarov, D.V. Environmental and Business Challenges Presented by Mining and Mineral Processing Waste in the Russian Federation. *Minerals* **2019**, *9*, 445. [CrossRef]
36. Figueiredo, J.; Vila, M.C.; Góis, J.; Pavani Biju, B.; Futuro, A.; Martins, D.; Dinis, M.L.; Fiúza, A. Bi-level depth assessment of an abandoned tailings dam aiming its reprocessing for recovery of valuable metals. *Miner. Eng.* **2019**, *133*, 1–9. [CrossRef]

NdFeB Magnets Recycling Process: An Alternative Method to Produce Mixed Rare Earth Oxide from Scrap NdFeB Magnets

Elif Emil Kaya [1,2,3], Ozan Kaya [4], Srecko Stopic [1,*], Sebahattin Gürmen [2] and Bernd Friedrich [1]

1. IME Process Metallurgy and Metal Recycling, RWTH Aachen University, 52056 Aachen, Germany; emil@tau.edu.tr (E.E.K.); bfriedrich@ime-aachen.de (B.F.)
2. Department of Metallurgical & Materials Engineering, Istanbul Technical University, 34469 Istanbul, Turkey; gurmen@itu.edu.tr
3. Department of Materials Science and Technology, Turkish-German University, 34820 Istanbul, Turkey
4. Department of Mechatronics Engineering, Istanbul Technical University, 34469 Istanbul, Turkey; kayaozan@itu.edu.tr
* Correspondence: sstopic@ime-aachen.de; Tel.: +49-176-7826-1674

Abstract: Neodymium iron boron magnets (NdFeB) play a critical role in various technological applications due to their outstanding magnetic properties, such as high maximum energy product, high remanence and high coercivity. Production of NdFeB is expected to rise significantly in the coming years, for this reason, demand for the rare earth elements (REE) will not only remain high but it also will increase even more. The recovery of rare earth elements has become essential to satisfy this demand in recent years. In the present study rare earth elements recovery from NdFeB magnets as new promising process flowsheet is proposed as follows; (1) acid baking process is performed to decompose the NdFeB magnet to increase in the extraction efficiency for Nd, Pr, and Dy. (2) Iron was removed from the leach liquor during hydrolysis. (3) The production of REE-oxide from leach liquor using ultrasonic spray pyrolysis method. Recovery of mixed REE-oxide from NdFeB magnets via ultrasonic spray pyrolysis method between 700 °C and 1000 °C is a new innovative step in comparison to traditional combination of precipitation with sodium carbonate and thermal decomposition of rare earth carbonate at 850 °C. The synthesized mixed REE- oxide powders were characterized by X-ray diffraction analysis (XRD). Morphological properties and phase content of mixed REE- oxide were revealed by scanning electron microscopy (SEM) and Energy-dispersive X-ray (EDX) analysis. To obtain the size and particle size distribution of REE-oxide, a search algorithm based on an image-processing technique was executed in MATLAB. The obtained particles are spherical with sizes between 362 and 540 nm. The experimental values of the particle sizes of REE-oxide were compared with theoretically predicted ones.

Keywords: rare earth elements; recycling; NdFeB; magnet; ultrasonic spray pyrolysis

1. Introduction

Rare earth elements (REEs) have a wide range of uses in technological products and applications. Due to the increased demand and supply risk, most REEs have been added to the list of critical metals. The production of REEs from primary resource causes environmental problems [1]. The recovery of REEs from waste materials is the most suitable strategy to find the solution of environmental problems and ensure the sustainability for production of REE raw materials in the future, according to an increased demand in industrial application. Most developed countries are importing REEs from China; 95% of REEs are supplied from China and in addition to this situation, export quotas of REEs applied by China have increased the export prices of REEs [2].

In order to produce rare earth elements oxides (REE-oxides), most researchers have studied different hydrometallurgical and pyrometallurgical strategies such as dry digestion [3], acid baking processes [4] and carbothermal reduction of ores and concentrates with subsequent leaching using strong acids [5] aiming at higher REE extractions. Demol et al. [4] found that sulfation reaction of monazite with acid was virtually complete after baking at 250 °C for 2 h, resulting in >90% solubilization of REEs, thorium and phosphate. To prevent silica gel formation and to increase the extraction efficiency of REEs, before leaching, the dry digestion process was performed with concentrated HCl [6]. In contrast to application of an acid baking, Ma et al. reported [7] that rare earth recovery from eudialyte concentrate is achieved by avoiding silica-gel formation using a dry digestion process at room temperature. Generally, a direct leaching process was also applied for the treatment of red mud to obtain a high REE extraction efficiency [8,9]. Because of the many disadvantages of direct leaching processes such as high consumption of leaching agents and non-selectivity [10], Borra et al. [11] reported that alkali roasting–smelting–leaching processes allow the recovery of aluminum, iron, titanium, and REEs from bauxite residue. Generally, recovery of REEs from secondary materials is a new possibility for production of these critical metals.

Therefore, recycling has considerable advantages over processing natural ores and concentrates on account of energy effectiveness and selectivity [12]. Neodymium iron boron magnets (NdFeB) are the most valuable REE secondary resource because they contain a high content (approximately 20%) of REEs, neodymium (Nd), dysprosium (Dy) and some REEs in minor quantities, such as praseodymium (Pr). Between 20 and 25% of REEs produced worldwide are used in the production of NdFeB. Increasing future production of hard disks, automotive applications, motors, speakers, air conditioners, electronic devices, electric bicycles and wind turbines provides a strong driving force for finding a new process for recycling spent NdFeB magnets [13–15]. Furthermore, an alternative product that can replace NdFeB magnets in today's technologies in terms of performance and cost has not been developed yet. Therefore, the recycling of spent NdFeB magnets is the most promising effective alternative for the solution of the supply problem of Nd, Dy and Pr.

Önal et al. [16] studied recycling of NdFeB magnets using sulfation, selective roasting and water leaching, enabling the production of a liquid with at least 98% rare earth purity. Furthermore, 98% extraction efficiency of REEs from NdFeB magnets was obtained by the acid-baking process with nitric acid [17]. After the acid baking process and subsequent water leaching of the treated concentrate, the produced suspension was filtrated in order to separate a pregnant leaching solution. To produce the REE oxides from leach liquor, all the proposed methods in the literature are completely based on precipitation methods by using various precipitation agents such as sodium carbonate and oxalic acid [18,19].

It is known that REE-carbonate or REE-oxalate can be produced from impurities present in sulfuric liquors using oxalic acid and sodium carbonate by a precipitation method [20,21]. It was reported that high purity REE-oxide (99.2%) was achieved using oxalic acid as a precipitation agent. Relatively lower purity RE-oxide was produced using sodium carbonate during precipitation [18]. The precipitation behavior of REEs with precipitation agents including oxalate, sulfate, fluoride, phosphate, and carbonate was examined using thermodynamic principles and calculations [22]. It was found that the pH of the system, types of the precipitation agent and present anions in the leach liquor have a noteworthy impact on the purity of the REE precipitants.

In contrast to the precipitation method, the production of nanosized REEs using an ultrasonic spray pyrolysis method is missing in the literature. Ultrasonic spray pyrolysis (USP) combines the ultrasound used for dispersing the precursor solution into droplets and chemical decomposition of the dissolved material inside the droplets at elevated temperatures, resulting in the formation of fine metallic, oxidic and composite powder [23–25]. This technique has been successfully used in the production of REE-oxide, the results of which are Y_2O_3, La_2O_3 Gd_2O_3, and CeO_2 [26–29]. The USP method enables synthesized spherical and fine REE-oxide in one-step. Moreover, the technique is capable of metal oxide with controllable chemical composition, particle size and morphology of particles by manipulating process parameters, which for the precursor type and concentration, reaction atmosphere, carrier gas flow rate, and reaction temperature [30–32]. In the present study, a new sustainable method was proposed for the production of mixed REE-oxide from REE-rich leach liquor. This proposed work summarizes the following operations: 1. Grinding and sieving; 2. Acid baking; 3. Calcination; 4. Leaching with water and 5. Ultrasonic spray pyrolysis, as shown at Figure 1.

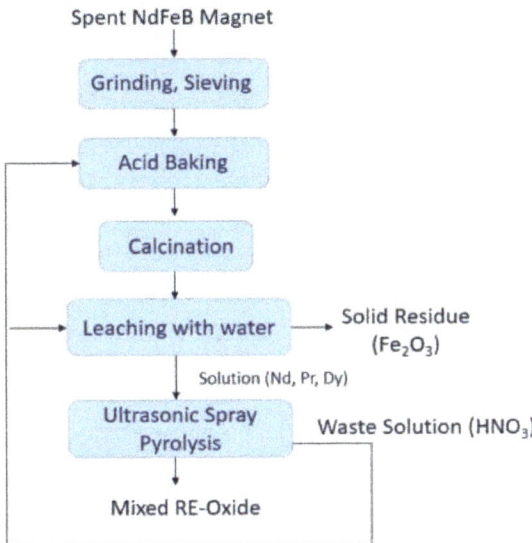

Figure 1. The proposed strategy for preparation of REE- oxides from spent NdFeB magnets.

In the final step of ultrasonic spray pyrolysis, the produced nitric acid shall be recycled and sent to the acid baking process. This study aims at investigating the conditions required to produce mixed REE-oxides in the combined hydrometallurgical process (acid baking with water dissolution and ultrasonic spray pyrolysis process). A literature review reveals that this information is currently not reported for the production of mixed RE-oxide using leach liquor. The proposed route promotes the enhancement of the circular economy of critical raw materials/REEs and could provide a high potential to increase resource efficiency for spent NdFeB magnets.

2. Experimental

2.1. Materials, Acid Baking, and Water Leaching

Waste NdFeB magnets used during the experiments were supplied in bulk form. Demagnetization was not necessary. Bulk and brittle NdFeB magnet pieces were crushed by jaw crusher Retsch BB 50, (Retsch GmbH, Haan, Germany) using dry ice to prevent magnet powders from catching fire. The crushing process was repeated three times to obtain the magnet powders to suitable powder's size. Nitric acid (65%) was used for acid baking without dilution and was purchased from VWR International GmbH, Darmstadt, Germany in analytical grade. All reagents were used without further purification. All solutions were prepared using deionized water. 16.6-gram magnet powders were dissolved in 500 mL of 2 molar HNO_3 acid solution to determine the chemical composition of the magnets. The chemical analysis of obtained solution was performed using ICP-OES analysis (SPECTRO ARCOS, SPECTRO Analytical Instruments GmbH, Kleve, Germany). Elemental composition of the NdFeB was determined by X-ray fluorescence (XRF) spectroscopy (Panalytical WDXRF spectrometer (Malvern Panalytical B.V., Eindhoven, The Netherlands)).

The extraction of REEs from NdFeB magnets was performed by nitric acid baking and a subsequent water leaching. The acid baking process was employed using highly concentrated HNO$_3$ (65%) with a 1:5 solid/liquid (S/L) ratio. Water was first added to the magnet powders to promote the ionization of the nitric acid before the acid baking process. After waiting 1 h, the mixture was calcined at 200 °C for 2 h. Water leaching experiments were performed in a 500 mL four-neck glass reactor equipped with a heating mantle and temperature controller (IKA Werke GmbH, Staufen im Breisgau, Germany) The leaching solution was kept under 550 revolutions/min agitation by a mechanical stirrer. Water leaching experiments were conducted with a 1:15 solid/liquid (S/L) ratio for 90 min. The leaching mixture was filtered using the filtering set up to separate the leaching solution from leach residue. Chemical content in the leach liquor was analyzed by ICP-OES to determine the purity of the leaching solution containing REE. The theoretical background of acid baking with nitric acid and water leaching process was reported elsewhere [17].

2.2. Ultrasonic Spray Pyrolysis Method for Production of RE-Oxide and Their Characterization

Very fine aerosol droplets were obtained from a leach solution using an ultrasonic atomizer (PRIZNano, Kragujevac, Serbia), with a frequency 1.75 MHz in an ultrasonic field obtained by 3 ultrasonic transducers. The aerosol was carried with nitrogen flow rate 1.0 L/min into in quartz tube (1.0 m length and 0.021 m diameter) between 700 °C and 1000 °C, placed in a Ströhlein Furnace, Selm, Germany. The flow rate of nitrogen was measured using special flowmeter gas unit (YOKOGAWA Deutschland GmbH, Ratingen). One step ultrasonic spray pyrolysis lab-scale horizontal equipment was shown in Figure 2. Experimental parameters were given in Table 1.

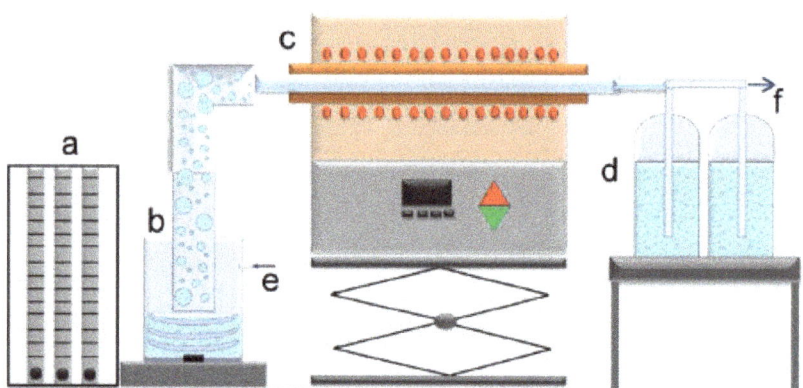

Figure 2. One step ultrasonic spray pyrolysis lab-scale horizontal equipment: (**a**)—gas flow regulation; (**b**)—ultrasonic aerosol generator; (**c**)—furnace with the wall heated reactor; (**d**)—collection bottles; e—gas inlet, f—gas outlet.

Table 1. Experimental parameters of ultrasonic spray pyrolysis method.

Samples Codes	Concentration of Nd(NO$_3$)$_3$ (g/L)	Concentration of Pr(NO$_3$)$_3$ (g/L)	Concentration of Dy(NO$_3$)$_3$ (g/L)	Reaction Temp (°C)	N$_2$ Flow Rate (L/min)	Ultrasonic Frequency (MHz)
S1	0.458	0.130	0.010	700	1.0	1.75
S2	0.458	0.130	0.010	800	1.0	1.75
S3	0.458	0.130	0.010	900	1.0	1.75
S4	0.458	0.130	0.010	1000	1.0	1.75

The SEM analysis of particles obtained by ultrasonic spray pyrolysis was performed at JSM 7000F by JEOL, (Construction year 2006, Japan) and EDX-analysis using Octane Plus-A by Ametek-EDAX, (construction year, 2015, USA) with Software Genesis V 6.53 by Ametek. XRD Analysis of RE-oxides powders was performed using Bruker D8 Advance with LynxEye detector (Bruker AXS, Karlsruhe, Germany). X-ray powder diffraction patterns were collected on a Bruker-AXS D4 Endeavor diffractometer in Bragg–Brentano geometry, equipped with a copper tube and a primary nickel filter providing Cu K$\alpha_{1,2}$ radiation (λ = 1.54187 Å).

3. Results and Discussion

Mixed RE-Oxide powders were synthesized by a one-step USP method from leach liquor. Thermodynamic investigations of a possible reaction were conducted by HSC software package 6.12 (Outotec, Espoo, Finland). Various reaction temperatures from 700 °C to 1000 °C were tested to investigate their role on the phase formation of RE-Oxide. The mixed REE-Oxide powders were characterized by X-ray diffraction analysis, scanning electron microscopy. To reveal size and size distribution of mixed RE-Oxide, SEM micrographs were examined via image-processing techniques in MATLAB (MathWorks, Natick, MA, USA).

3.1. Characterization of Scrap NdFeB Magnet

The magnet composition was determined by X-ray fluorescence (XRF). The major elements of the NdFeB magnet powder are Fe, Nd, Pr as major elements and the trace amounts of Mn, Co, Pd, Al and Si, as shown in Table 2.

Table 2. Chemical composition of NdFeB magnet powders determined by XRF.

Composition	Na_2O	Al_2O_3	SiO_2	MnO	Fe_2O_3	Co_3O_4	CuO
Concentration (%)	0.34	0.42	0.24	1.97	68.1	0.70	0.14
Composition	Ga_2O_3	As_2O_3	Nb_2O_5	PdO	Pr_2O_3	Nd_2O_3	Tb_4O_7
Concentration (%)	0.20	0.21	0.12	0.24	5.72	20.4	0.70

The contents of the NdFeB magnets were measured using inductively coupled plasma optical emission spectroscopy (ICP-OES). The ICP-OES analysis results of NdFeB magnet is given in Table 3.

Table 3. Chemical composition of NdFeB magnet powders sample.

Composition	B	Co	Cr	Cu	Dy
Concentration (mg/L)	278	245	<1	32.6	210
Composition	Fe	Mo	Nd	Ni	Pr
Concentration (mg/L)	210,000	<1	7580	<1	2340

ICP-OES analysis showed the presence of Fe, Nd, and Pr as the major elements and Cu and Co as minor elements.

X-ray diffraction (XRD) analyses were conducted to identify the phases in the NdFeB magnet powders. XRD analysis results of NdFeB magnet powders were given in Figure 3.

According to XRD patterns, the powder sample was well crystallized in the $Nd_2Fe_{14}B$ phase. The X-ray diffraction peaks could be indexed to the tetragonal structure with space group P42/mnm (JCPDS card 40-1028).

Scanning Electron Microscopy (SEM) analyses were performed to observe the morphology of the NdFeB magnet powders, as shown at Figure 4.

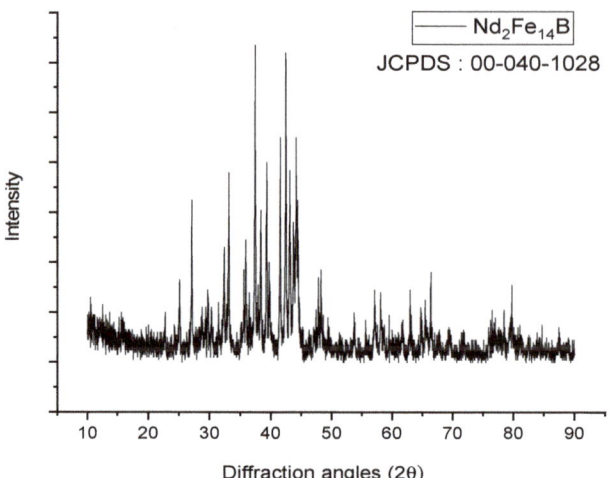

Figure 3. XRD pattern of NdFeB magnet powders.

Figure 4. SEM analysis of NdFeB magnet powders.

As shown at Figure 5, Energy Dispersive Spectroscopy (EDS) results demonstrate that NdFeB magnet powders primarily consist of Fe and Nd. EDS results are in good agreement with the ICP analysis results but due to the small amount of the other elements, they cannot be detected by EDS.

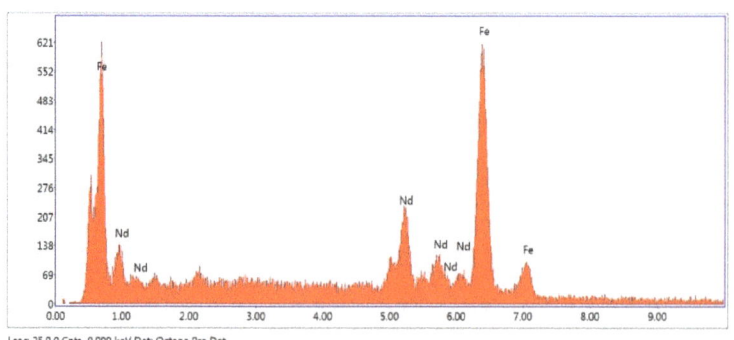

Figure 5. EDS-analysis of NdFeB magnet powders.

3.2. Production of REE-Oxide and Their Characterization

The production of mixed REE-oxides powder from scrap NdFeB magnet by nitric acid baking and water leaching followed by ultrasonic spray pyrolysis method was investigated. The concentration of metals ions in the leached solution were determined using ICP-OES analysis. The chemical composition of leach liquor obtained after the water leaching process is illustrated in Table 4. Leach liquor of the same chemical composition was used in all USP experiments.

Table 4. Chemical composition of the leach liquor.

Composition	B	Co	Cr	Cu	Dy
Concentration (mg/L)	80	30	<1	<1	100
Composition	Fe	Mo	Nd	Ni	Pr
Concentration (mg/L)	<1	<1	4580	<1	1300

Gibbs free energy change depending on reaction temperature was computed by HSC software (Outotec, Espoo, Finland), as shown at Figure 6. The formation of RE-oxides after evaporation of water in the furnace can be described as in the following equations:

$$2Nd(NO_3)_3 = Nd_2O_3 + 6NO_2 + 1.5O_2 \qquad (1)$$

$$2Dy(NO_3)_3 = Dy_2O_3 + 6NO_2 + 1.5O_2 \qquad (2)$$

$$2Pr(NO_3)_3 = Pr_2O_3 + 6NO_2 + 1.5O_2 \qquad (3)$$

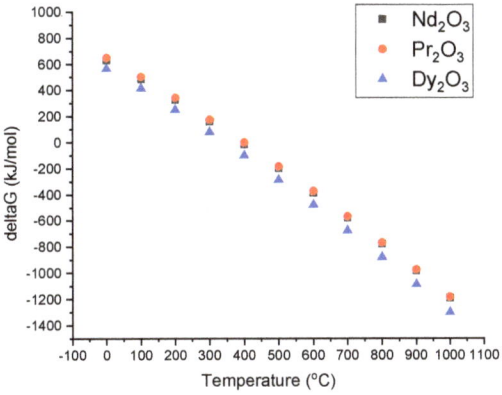

Figure 6. Gibbs free energy change depending on reaction temperature.

The Gibbs free energy for the temperature range of 0–1000 °C is exhibited in Figure 6. As can be seen, the Gibbs free energy is negative after 500 °C. This allows for RE-oxide to be formed by the thermal decomposition of leach liquor, which is energetically favored after 500 °C.

Figure 7 shows the XRD results for the samples synthesized at 700 °C, 800 °C, 900 °C and 1000 °C by ultrasonic spray pyrolysis method.

XRD analysis of powders obtained between 700 °C and 1000 °C confirmed the formation of a mixture of RE-oxides. The cubic structure of Nd_2O_3 with 20% of Pr_2O_3 was found between 700 °C and 800 °C as shown at Figure 7. Checks of the XRD Pattern for crystal structure leads to $Nd_{1.6}Pr_{0.4}O_3$ solid solution. An increase in temperature from 800 °C to 900 °C and 1000 °C leads to a mixture of cubic and trigonal structure of Nd_2O_3 with 20% of Pr_2O_3, as shown at Figure 7. An increase in temperature from 700 °C to 1000 °C increases the crystallinity of the obtained structure. Additionally, typical EDX-Analysis of powders was shown at Figure 8, confirming the presence of rare earth elements.

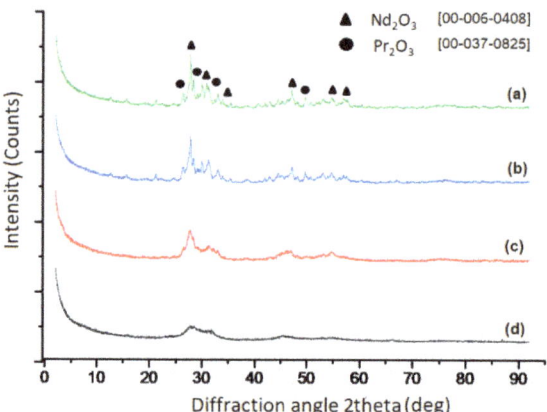

Figure 7. XRD analysis of RE-oxide powders synthesized with varying reaction temperatures (**a**) 1000 °C, (**b**) 900 °C, (**c**) 800 °C and (**d**) 700 °C.

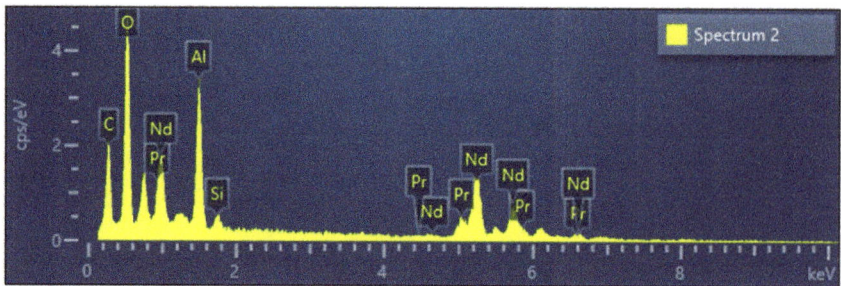

Figure 8. Typical EDS analyses of mixed REE-Oxide powders synthesized with varying reaction temperatures.

The morphological investigation of mixed RE-oxide produced by USP processes at different temperatures was conducted by SEM analysis. SEM analyses of the RE-oxide are illustrated in Figure 9.

As indicated in Figure 9, spherical RE-oxide was obtained at various reaction temperatures by the USP Method. The image-processing technique is one of the computational approaches widely getting implemented in various fields of material science. It is especially useful for interpreting the images as the results of SEM. The morphology and size of the RE-oxide nanoparticles were analyzed by SEM. Using SEM results, the morphological features of the RE-oxide nanoparticles, such as their diameter, were picked up by image processing and a particle search algorithm. The use of an image processing method algorithm is detailed in [32].

Applied image processing methods generate the black and white images from the original SEM images for determining the location and size of the RE-oxide nanoparticles. Since the particles are known to have spherical shapes [33], the Hough transform method was utilized for approximately defining the nanoparticles. The Hough transform draws new circles at the three boundary points. Then, the center of the circle is computed, with the junction point of new circles and diameter limits defined by the user. After the detecting RE-oxide nanoparticles, these particles were labelled with blue rough circles and their cumulative distribution results related to process conditions were achieved. The results are given in Figure 10.

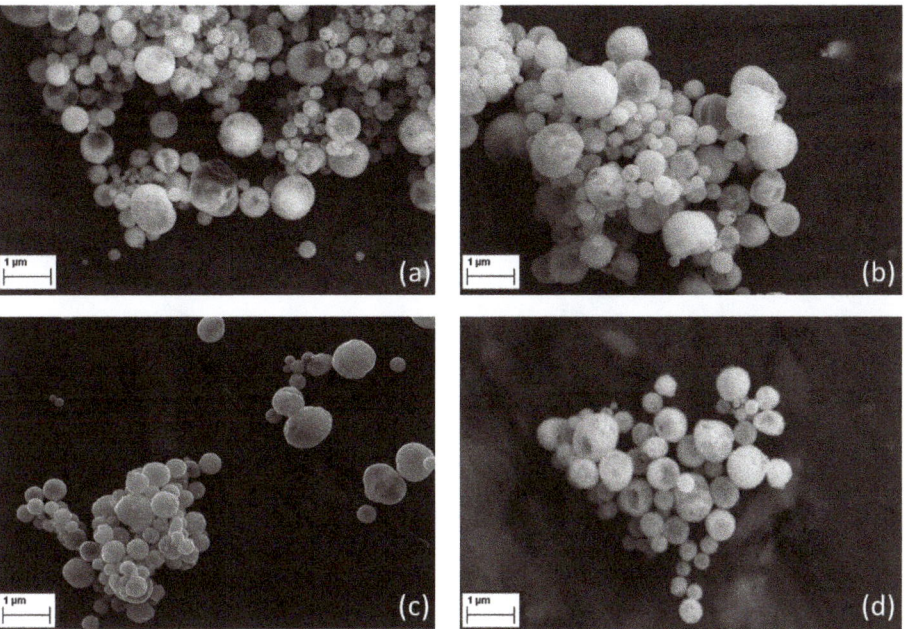

Figure 9. SEM analyses of mixed RE-Oxide powders synthesized with varying reaction temperatures (**a**) 700 °C; (**b**) 800 °C; (**c**) 900 °C; (**d**) 1000 °C.

Graph of the labelled nanoparticles were drawn, with cumulative distribution represented by the y-axis, and nanoparticle size represented by the x-axis, as seen in Figure 10a–d. The cumulative curve of RE-oxide nanoparticles whose sizes were calculated by an image-processing technique is represented by the blue dashed line. The mean values of RE-oxide nanoparticle size were calculated from SEM by the image-processing technique. These SEM results reveal that the particles of RE-oxide synthesized from a 0.6 g/L solution concentration at various reaction temperatures lay in the range of 200–700 nm. The mean particle size of RE-oxide synthesized at 700 °C, 800 °C, 900 °C and 1000 °C was found to be 362 nm, 417 nm, 468 nm and 540 nm, respectively.

The theoretical particle size of RE-oxides was calculated according to related equations. The formation of RE-oxides will be firstly defined via the diameter of aerosol droplet (d_d) as shown with Equation (4) proposed by Peskin and Raco [24]:

$$d_d = 0.34 \left(\frac{8\,\pi\sigma}{\rho_L f^2} \right)^{\frac{1}{3}} \quad (4)$$

where: f—ultrasound frequency; ρ_L—density of water solution; σ—surface tension.

Using the following values: for water solution: f- 1.75 MHz; ρ_L- 1.02 g/cm^3; σ- 0.07 J/m^2, the calculated aerosol droplet amounts 2.86 μm.

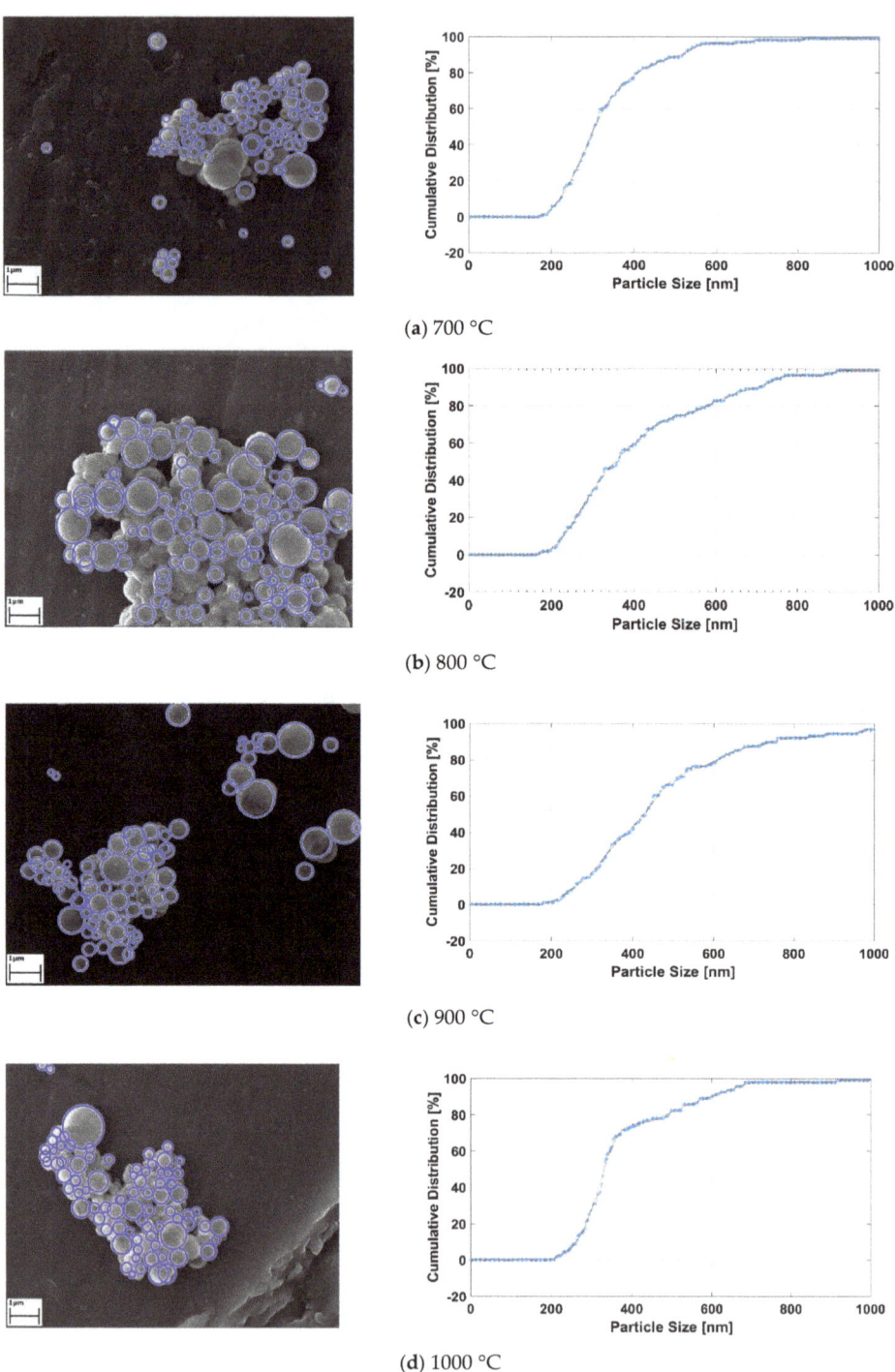

Figure 10. The detecting RE-oxide nanoparticles and theirs' cumulative curves related to USP-experiment conditions at (**a**) 700 °C, (**b**) 800°C, (**c**) 900°C and (**d**) 1000 °C.

The particle size *(dp)* depends on the droplet size and the concentration of the solution (C). This correlation between the concentration and other precursor characteristics and the final particle size, under the assumption that no precursor is lost in the process, can be described with the following Equation (5) derived via the Equation by Messing et al. [25]:

$$d_p = d_d \left(\frac{M_{REE-Nitrate}}{M_{REE-Oxide}} * \frac{C}{\rho} \right)^{0.33} \quad (5)$$

where the d_p is the diameter of the particle, the d_d is the diameter of the aerosol droplet calculated with Equation (4), Mp-molar mass of REE-nitrate (g/mol), the ρ is the density of REE-Oxide (Nd-, Pr-, and Dy-oxide), C is the concentration of the precursor solution.

Using the following values for molar mass of rare earth elements – nitrate (REE-nitrate) and rare earth elements- oxides (REE-oxides), densities of REE-oxides and concentrations of metals in solution (as shown in Table 4), the calculated values for particles sizes of REE-oxides using Equation (5) are presented in Table 5.

Table 5. Calculated theoretical particle size of RE-oxides using Equation (5).

REE-Nitrate	Nd(NO$_3$)$_3$	Pr(NO$_3$)$_3$	Dy(NO$_3$)$_3$
Molar mass of REE-nitrate (g/mol)	282.2	326.0	348.5
REE-oxides	Nd$_2$O$_3$	Pr$_2$O$_3$	Dy$_2$O$_3$
Density (g/cm^3)	7.2	6.9	7.8
Molar mass (g/mol)	336.5	329.8	373.0
Concentration of metal in solution (g/L)	0.458	0.130	0.010
Theoretical minimal particle size (nm)	108	76	31

The calculated minimal particle size (nm) amounts: 108, 76, 31 for Nd$_2$O$_3$, Pr$_2$O$_3$ and Dy$_2$O$_3$, total 215. The obtained values of particle sizes are compared with experimentally obtained values obtained by image process techniques. The differences between calculated and experimentally obtained values may be partially due to the approximate values used for surface tension and density of water solution, and mostly due to coalescence/agglomeration of aerosol droplets during transport to the furnace from an aerosol generator. Moreover, Equation (5) was based on the assumption of one particle per one droplet, and the influence of temperature on the mean particle size between 700 °C and 1000 °C was not taken into consideration.

4. Conclusions

Spherical particles of REE-oxides were produced from spent NdFeB magnets using a combined process and consists of: nitric acid baking process at 200 °C, water leaching, and ultrasonic spray pyrolysis between 700 °C and 1000 °C. Iron was removed from water solution using a hydrolysis process. XRD analysis of the obtained particles found a cubic and trigonal structure Nd$_2$O$_3$ with 20% Pr$_2$O$_3$, which is according to detected stoichiometry in solution after dissolution of spent NdFeB magnets. An increase in temperature from 700 °C to 1000 °C increases not only the crystallinity of the structure, but also the particle size between 362 and 540 nm. The minimal theoretical total particle size of prepared REE-oxides amounts to 215 nm. The differences between calculated and experimentally obtained values may be partially due to coalescence/agglomeration of aerosol droplets during transport to the furnace from an aerosol generator. Generally, we developed one combined environmentally friendly process for recovery of nanosized powder mixture of Nd$_2$O$_3$ and Pr$_2$O$_3$ from spent magnets and re-use of nitric acid. The final winning of the mixture of metallic Nd and Pr will be ensured using molten salt electrolysis [34].

Author Contributions: Conceptualization, E.E.K. and S.S.; funding acquisition, B.F. and S.G.; investigation, E.E.K.; methodology, E.E.K. and S.S.; supervision, S.G. and B.F.; writing—original draft, E.E.K., O.K. and S.S. All authors have read and agreed to the published version of the manuscript.

Funding: The research leading to these results has received funding from the AIF- German Federation of Industrial Research Associations, Germany and TÜBITAK- The Scientific of Technological Research Council of Turkey (Call identifier CORNET 29th Call) under grant agreement EN03193/20). The authors would like to greatly acknowledge TUBITAK/Turkey (Project No: 120N331) for financial support. Elif Emil Kaya would like to thank DAAD "Research stays of doctoral research assistants of the TDU at German partner universities" for financial support.

Institutional Review Board Statement: Not applicable.

Informed Consent Statement: Not applicable.

Data Availability Statement: Not applicable.

Conflicts of Interest: The authors declare no conflict of interest.

References

1. Kuang-Taek, R. Effects of rare earth elements on the environment and human health: A literature review. *Toxicol. Environ. Health Sci.* **2016**, *8*, 189–200.
2. Mancheri, N.A.; Sprecher, B.; Bailey, G.; Ge, J.; Tukker, A. Effect of Chinese policies on rare earth supply chain resilience. *Resour. Conserv. Recycl.* **2019**, *142*, 101–112. [CrossRef]
3. Ma, Y.; Stopic, S.; Gronen, L.; Obradovic, S.; Milivojevic, M.; Friedrich, B. Neural network modeling for the extraction of rare earth elements from eudialyte concentrate by dry digestion and leaching. *Metals* **2018**, *8*, 267. [CrossRef]
4. Demol, J.; Ho, E.; Senanayake, G. Sulfuric acid baking and leaching of rare earth elements, thorium and phosphate from a monazite concentrate: Effect of bake temperature from 200 to 800 °C. *Hydrometallurgy* **2018**, *179*, 254–267. [CrossRef]
5. Stopic, S.; Friedrich, B. Leaching of rare earth elements from bastnasite ore (third part). *Mil. Tech. Cour.* **2019**, *67*, 561–572. [CrossRef]
6. Stopic, S.; Friedrich, B. Deposition of silica in hydrometallurgical processes. *Mil. Tech. Cour.* **2020**, *68*, 65–78.
7. Ma, Y.; Stopic, S.; Friedrich, B. Hydrometallurgical treatment of a eudialyte concentrate for preparation of rare earth carbonate. *Johns. Matthey Technol. Rev.* **2019**, *63*, 2–13. [CrossRef]
8. Alkan, G.; Yagmurlu, B.; Cakmakoglu, S.; Hertel, T.; Kaya, S.; Gronen, L.; Stopic, S.; Friedrich, B. Novel approach for enhanced scandium and titanium leaching efficiency from bauxite residue with suppressed silica gel formation. *Nat. Sci. Rep.* **2018**, *8*, 5676. [CrossRef]
9. Alkan, G.; Yagmurlu, B.; Friedrich, B.; Ditrich, C.; Gronen, L.; Stopic, S.; Ma, Y. Selective silica gel scandium extraction from, iron –depleted red mud slags by dry digestion. *Hydrometallurgy* **2019**, *185*, 266–272. [CrossRef]
10. Davris, P.; Stopic, S.; Balomenos, E.; Panias, D.; Paspaliaris, I.; Friedrich, B. Leaching of rare earth elements from Eudialyte concentrate by supressing silicon dissolution. *Miner. Eng.* **2017**, *108*, 115–122. [CrossRef]
11. Borra, C.R.; Blanpain, B.; Pontikes, Y.; Binnemans, K.; van Gerven, T. Recovery of rare earths and other valuable metals from bauxite residue (red mud): A review. *J. Sustain. Metall.* **2016**, *2*, 365–386. [CrossRef]
12. Ayres, R.U.; Peiró, L.T. Material efficiency: Rare and critical metals. *Phylosophical Trans. R. Soc. A* **2013**, *371*, 20110563. [CrossRef]
13. Tao, X.; Huiqing, P. Formation cause, composition analysis and comprehensive utilization of rare earth solid wastes. *J. Rare Earths* **2009**, *27*, 1096–1102.
14. Sprecher, B.; Xiao, Y.; Walton, A.; Speight, J.; Harris, R.; Kleijn, R.; Kramer, G.J. Life cycle inventory of the production of rare earths and the subsequent production of NdFeB rare earth permanent magnets. *Environ. Sci. Technol.* **2014**, *48*, 3951–3958. [CrossRef]
15. Binnemans, K.; Jones, P.T.; Blanpain, B.; Van Gerven, T.; Yang, Y.; Walton, A.; Buchert, M. Recycling of rare earths: A critical review. *J. Clean. Prod.* **2013**, *51*, 1–22. [CrossRef]
16. Önal, M.; Borra, C.; Guo, M.; Blanpain, B.; van Gerven, T. Recycling of NdFeB magnets using sulfation, selective roasting and water leaching. *J. Sustain. Metall.* **2015**, *1*, 199–215. [CrossRef]
17. Önal MA, R.; Aktan, E.; Borra, C.R.; Blanpain, B.; Van Gerven, T.; Guo, M. Recycling of NdFeB magnets using nitration, calcination and water leaching for REE recovery. *Hydrometallurgy* **2017**, *167*, 115–123. [CrossRef]
18. Liu, Z.; Li, M.; Hu, Y.; Wang, M.; Shi, Z. Preparation of large particle rare earth oxides by precipitation with oxalic acid. *J. Rare Earths* **2008**, *26*, 158–162. [CrossRef]
19. Silva, R.G.; Morais, C.A.; Teixeira, L.V.; Oliveira, É.D. Selective precipitation of high-quality rare earth oxalates or carbonates from a purified sulfuric liquor containing soluble impurities. *Min. Metall. Explor.* **2019**, *36*, 967–977. [CrossRef]
20. Yun, Y.; Stopic, S.; Friedrich, B. Valorization of Rare Earth Elements from a steenstrupine concentrate via a combined hydrometallurgical and pyrometallurgical method. *Metals* **2020**, *10*, 248. [CrossRef]
21. Ma, Y.; Stopic, S.; Wang, X.; Forsberg, K.; Friedrich, B. Basic sulfate precipitation of zirconium from sulfuric acid leach solution. *Metals* **2020**, *10*, 1099. [CrossRef]

22. Han, K.N. Characteristics of precipitation of rare earth elements with various precipitants. *Minerals* **2020**, *10*, 178. [CrossRef]
23. Košević, M.; Stopic, S.; Cvetković, V.; Schroeder, M.; Stevanović, J.; Panic, V.; Friedrich, B. Mixed RuO_2/TiO_2 uniform microspheres synthesized by low-temperature ultrasonic spray pyrolysis and their advanced electrochemical performances. *Appl. Surf. Sci.* **2019**, *464*, 1–9. [CrossRef]
24. Peskin, R.L.; Raco, R.J. Ultrasonic atomization of liquids. *J. Acoust. Soc. Am.* **1963**, *34*, 1378–1381. [CrossRef]
25. Messing, G.; Zhang, S.; Jayanthi, G. Ceramic powder synthesis by spray pyrolysis. *J. Am. Ceram. Soc.* **1993**, *76*, 2707–2726. [CrossRef]
26. Emil, E.; Gürmen, S. Estimation of yttrium oxide microstructural parameters using the Williamson–Hall analysis. *Mater. Sci. Technol.* **2018**, *34*, 1549–1557. [CrossRef]
27. Yadav, A.A.; Lokhande, V.C.; Bulakhe, R.N.; Lokhande, C.D. Amperometric CO_2 gas sensor based on interconnected web-like nanoparticles of La_2O_3 synthesized by ultrasonic spray pyrolysis. *Microchim. Acta* **2017**, *184*, 3713–3720. [CrossRef]
28. Jung, D.S.; Hong, S.K.; Lee, H.J.; Kang, Y.C. Gd_2O_3: Eu phosphor particles prepared from spray solution containing boric acid flux and polymeric precursor by spray pyrolysis. *Opt. Mater.* **2006**, *28*, 530–535. [CrossRef]
29. Goulart, C.; Djurado, E. Synthesis and sintering of Gd-doped CeO_2 nanopowders prepared by ultrasonic spray pyrolysis. *J. Eur. Ceram. Soc.* **2013**, *33*, 769–778. [CrossRef]
30. Emil, E.; Alkan, G.; Gurmen, S.; Rudolf, R.; Jenko, D.; Friedrich, B. Tuning the morphology of ZnO nanostructures with the ultrasonic spray pyrolysis process. *Metals* **2018**, *8*, 569. [CrossRef]
31. Ardekani, S.R.; Aghdam AS, R.; Nazari, M.; Bayat, A.; Yazdani, E.; Saievar-Iranizad, E. A comprehensive review on ultrasonic spray pyrolysis technique: Mechanism, main parameters and applications in condensed matter. *J. Anal. Appl. Pyrolysis* **2019**, *141*, 104631. [CrossRef]
32. Kaya, E.E.; Kaya, O.; Alkan, G.; Gürmen, S.; Stopic, S.; Friedrich, B. New proposal for size and size-distribution evaluation of nanoparticles synthesized via ultrasonic spray pyrolysis using search algorithm based on image-processing technique. *Materials* **2020**, *13*, 38. [CrossRef] [PubMed]
33. Xu, L.; Oja, E.; Kultanen, P. A new curve detection method: Randomized Hough transform (RHT). *Pattern Recognit. Lett.* **1990**, *11*, 331–338. [CrossRef]
34. Cvetkovic, V.; Feldhaus, D.; Vukicevic, N.; Barudzija, T.; Friedrich, B.; Jovicevic, J. Investigation on the electrochemical behaviour and deposition mechanism of neodymium in NdF_3–LiF–Nd_2O_3 melt on Mo electrode. *Metals* **2020**, *10*, 576. [CrossRef]

Article

Electrochemical Investigation of Lateritic Ore Leaching Solutions for Ni and Co Ions Extraction

Jonas Mitterecker [1], Milica Košević [2], Srecko Stopic [1], Bernd Friedrich [1], Vladimir Panić [2,3,4], Jasmina Stevanović [2,3] and Marija Mihailović [2,*]

[1] Process Metallurgy and Metal Recycling, RWTH Aachen University, Intzestraβe 3, D-52072 Aachen, Germany; jonas.mitterecker@rwth-aachen.de (J.M.); SStopic@metallurgie.rwth-aachen.de (S.S.); BFriedrich@metallurgie.rwth-aachen.de (B.F.)

[2] Department of Electrochemistry, Institute of Chemistry, Technology and Metallurgy, National Institute of the Republic of Serbia, University of Belgrade, Njegoševa 12, 11000 Belgrade, Serbia; milica.kosevic@ihtm.bg.ac.rs (M.K.); panic@ihtm.bg.ac.rs (V.P.); jaca@tmf.bg.ac.rs (J.S.)

[3] Center of Excellence in Environmental Chemistry and Engineering-ICTM, University of Belgrade, Njegoševa 12, 11000 Belgrade, Serbia

[4] Department of Natural and Mathematical Science, State University of Novi Pazar, 36300 Novi Pazar, Serbia

* Correspondence: marija.mihailovic@ihtm.bg.ac.rs

Abstract: Determination of the extractive behavior of ionic species from lateritic ore leachates is complex, since the leachates are pregnant with tens of different ions in, as a rule, multiple oxide states. To examine the possible pathways of intrinsic electrochemical extraction of the crucial elements Ni and Co, it was necessary to make model solutions of these elements and to subject them to electrochemical examination techniques in order to obtain a benchmark. Beside Ni and Co, the model system for Fe had to be evaluated. Iron, as a dominant ore component by far, is the main interfering factor in the extraction processes of Ni and Co in rather low amounts from leaching solution. The leachate examination results were compared to separate model solutions, as well as to their combinations in concentrations and to pH values comparable to those of the leachate. The separation of the leachate components was initially performed by continuous increase in pH upon leaching with NaOH solution, and afterwards the pH-adjusted solutions were subjected to electrochemical investigation. With the purpose of connecting and quantifying the visual changes in leachate upon increase in pH, conductometric measurements were performed. Reactions of oxidation/precipitations were indicated, which led to the essential Fe removal by precipitation. Resulting solutions were found suitable for Ni and Co electrochemical extraction.

Keywords: leachate; metal ions extraction; selectivity; Fe removal; electrodeposition; conductometry

1. Introduction

Cobalt and nickel are considered both strategic and critical raw materials [1]. Nickel has already been used in batteries, such as nickel–cadmium or nickel–metal-hydride, but its use in lithium-ion batteries caused rising interest in this metal, since it is able to deliver a higher energy density and more storage capacity to these batteries. Nickel has become the most important for the lithium-ion battery cathodes, thus enabling the reduction the use of cobalt, which is scarce and more expensive [2]. Nevertheless, cobalt is still essential for the manufacturing of Li-ion batteries and its compounds for supercapacitor electrodes due to the high specific capacitance and high energy density as well as better cyclic stability of Co oxide [3]. With an outbreak of electric vehicle production, global demand for cobalt and nickel has been growing. Global demand for these metals in electric vehicle batteries from 2018 to 2025 is forecasted to increase by 10 times [4–6].

Along with increasing needs, there has been the growing interest to recover Ni and Co by hydrometallurgical processes, mainly focused on the improvement of leaching processes [7–10]. Investigation of lateritic ore leachates for Ni and Co extraction, including advanced electrochemical techniques, should make the processing of such ores economically and technologically feasible. The pregnant leach solution (PLS) of a lateritic ore contains plenty of elements gained from their native oxide state. Ni and Co may be extracted by electrochemical deposition, which is especially suitable due to the low content of cobalt in the ores [11–13]. There are, however, among all the others, high amounts of iron, which is undesired for the extraction processes of cobalt and nickel [14].

The electrochemical behavior of a leachate, i.e., PLS, was investigated by the usage of constant-potential electrodeposition of PLS metals (employing chronoamperometry (CA) method) and linear sweep voltammetry (LSV), as well as by conductometry (CM). In such a complex solution, containing dozens of different ions, many of them with multiple oxide states, it is hard to determine the behavior of ionic species separately. Moreover, the mutual interactions of those ions, as well as the changes in ion interference with the solution pH, make determination of exact leaching process parameters very difficult. To examine the extraction possibilities of these crucial elements, Ni and Co, it was necessary to make model solutions of these elements and to subject them to cyclic voltammetry (CV) and (CA/LSV) examination techniques in order to obtain benchmark. Beside Ni and Co, the model system for Fe had to be evaluated, as Fe is the main interfering factor of the Ni and Co extraction process from leaching solution. The leachate examination results were compared to separate model solutions containing desired metals, as well as their combination in concentration and pH values equal to those of the leachate. Oxidation, reduction and electrodeposition were investigated at −900 mV and at different pH. The characteristic pH values were selected according to visual changes in acidic leachate and the results of conductometric investigation upon stepped increase in pH. Conductometry as employed to quantify metal ions' transitions.

2. Materials and Methods

2.1. Leaching

Leaching of the lateritic ore was performed with 1.0 M sulfuric acid. The case-study ore (PT HUADI Nickel-Alloy Indonesia, Papanloe, Pa'jukukang, Bantaeng Regency, South Sulawesi 92461, Indonesia) composition is presented in Table 1. The composition of the ore was determined by X-ray fluorescence (AXIOS, PANalytical, Eindhoven, The Netherlands).

Table 1. Chemical analysis of the ore, mass %.

SiO_2	MgO	Fe_2O_3	NiO	Al_2O_3	CaO	MnO	Cr_2O_3	CuO	Co_3O_4	K_2O	TiO_2
11.70	5.81	70.70	3.80	4.27	0.13	0.73	1.78	<DL [1]	0.05	0.02	0.07

[1] DL–detection limit.

The mineral composition of the ore was determined by X-ray diffraction (XRD) analysis, which was performed on a RigakuMiniFlex 600 instrument with D/teXUltra 250 high-speed detector and an X-ray tube with a copper anode (Rigaku, Tokyo, Japan). Shooting conditions were: angle range 3–90°, step 0.02° and recording speed 10°/min. The voltage of the X-ray tube was 40 kV, and the current was 15 mA. The identification of minerals was performed in the Match! Software, and the obtained diffractograms were compared with the data from the ICDD database. The limit of detection of XRD analysis is about 1%. The minerals rhomboclase ($HFe(SO_4)_2 \cdot 4(H_2O)$) and getite (FeOOH) were identified.

The leaching was carried out at elevated temperature in a round-bottom flask under reflux. The mechanical stirring unit was introduced through an upper opening. The flask was placed inside a temperature controlled heating unit (SAF Wärmetechnik GmbH, Weinheimer Str. 2A, 69509 Mörlenbach, Germany). The automated heating unit kept the temperature at a constant level over the time of the experiment. To automate the heating, a thermocouple was constantly measuring the temperature during the trial. The ore concentrate leaching was carried out with a solid to liquid ratio of 1 g:5 mL, at the temperature of 70 °C and a stirring speed of 300 rpm for 120 min. The obtained leaching solution was filtered through quantitative filter paper (Grade 391-blue spot, LLG, Meckenheim, Germany). The leachate was subjected to a pH change by addition of appropriate portions of 10 M NaOH solution (a high concentration of NaOH is applied in order to consider the changes in ion concentrations as negligible upon PLS neutralization). Specific changes in the system, e.g., occurrence of precipitates, color change, etc., at different pH values were evaluated.

The starting leachate was of pH 0.5, and characteristic pH values according to specific visual changes are as follows: 1.5, 2.3, 3.7 and 5.3. Chemical analysis of the leachate at different pH was conducted using ICP-OES (Spectro Arcos, SPECTRO Analytical Instruments GmbH, Boschstr. 10, 47533 Kleve, Germany).

2.2. Preparation of Model Solutions

The following model solutions: $CoSO_4$, $NiSO_4$, $FeSO_4$, $CoSO_4$–$NiSO_4$ and $CoSO_4$–$NiSO_4$–$FeSO_4$ were prepared according to the metal concentrations in PLS measured by ICP-OES. Model solutions were prepared by dissolving $CoSO_4$, $NiSO_4$ and $FeSO_4$ in 1 M sulfuric acid using p.a. chemicals and double distilled water. Concentrations of Ni, Co and Fe ions, as found in NaOH-treated PLS and listed in Table 2, were persistent in all of the model solutions. A base solution of pH 3.7 was prepared without metal ions using H_2SO_4.

Table 2. Overview of the pH and concentrations (mmol/L) of the leachate and model solutions.

pH	Co	Ni	Fe
3.7	0.3424	13.5	4.8
5.3	0.2661	12.3	0.0464

The selected characteristic pH values were 3.7 and 5.3 according to visual changes during the leaching process and conductometric investigation.

2.3. Electrochemical Methods

In order to connect and quantify the visual changes in leachate upon increase in pH, and to separate the precipitations of metal ions from neutralization of unreacted acid, conductometric measurements (CM) were performed. The leachate resistance was measured continuously upon addition of NaOH solution portions at room temperature (22 ± 1 °C) in a home-made conductometric cell consisting of the two Pt parallel plates at fixed distance. pH was measured independently by standard digital pH-meter. The alternating voltage of 10 mV r.m.s. amplitude around open circuit reading (few mV differences between the plates) at 10 kHz was applied to induce directed ion motion. The applied frequency corresponds to the minimum of imaginary cell impedance, which was checked by recording the impedance spectra. The resistance reading was taken from the resistance/time dependences as a stable reading for at least 1 min, while single resistance measurement at 10 kHz took less than 1 s. The input voltage signal and the cell response were generated and evaluated by Biologic SP-200 potentiostat (BioLogic SAS, Grenoble, France). In addition, 1:1 Diluted (with 1.0 M H_2SO_4) native pregnant leach solution (PLS) and PLS with initial addition of 5 mL of H_2O_2 to 50 mL of PLS (PLSox) were subjected to CM.

Other electrochemical measurements (CA, LSV) were performed in a standard three-electrode system using potentiostat/galvanostat, Biologic model SP-200 (Bio-Logic SAS, Grenoble, France). The three-electrode cell consisted of a SCE reference electrode (all potentials are given in SCE scale), platinum wire as a counter electrode and working electrode.

The working electrodes were as follows:

1. Glassy carbon electrode (GC) was the working electrode for cyclic voltammetry (CV) measurements in PLSs, $CoSO_4$ and $NiSO_4$ model solutions, as well as in base solution. The working electrode was mechanically polished consecutively on emery papers of

the following grades: 280, 360, 800 and 1000, and then on polishing cloths (Buehler Ltd., Lake Bluff, IL, USA) impregnated with alumina of 1, 0.3 and 0.05 µm grades;
2. For the examination of the electrochemical potentiostatic extracts from PLSs at different pH, working electrodes were the alloys deposited onto polished GC;
3. To model the electrochemical dissolution by LSV, working electrodes were Ni–, Co–, Fe–, Ni–Co and Ni–Co–Fe alloys deposited on GC from the model solutions. Before each electrodeposition, the GC surface was mechanically polished with a polishing cloth (Buehler Ltd., Lake Bluff, IL, USA) impregnated with a water suspension of alumina powder (0.05 µm grade). Characterization of alloy composition was performed by alloy electrochemical dynamic dissolution using a linear sweep voltammetry (LSV) technique.

CV measurements were performed in the native PLS (pH 0.5) as well as in NaOH-treated PLSs at pH 3.7 and 5.3. Both $NiSO_4$ and $CoSO_4$ model solutions were also subjected to the CV measurements. Anodic CV branches were recorded upon hold at the cathodic potential limit for 2 min. In order to check the hydrogen evolution reaction's (HER) contribution to the CV currents in leachates, the base solution was also subjected to CV measurements. All CV measurements were conducted at 25.0 °C at a sweep rate of 10 mV s^{-1}. The metal deposition was carried out potentiostatically at −0.9 V for 60 min, at 1000 rpm under N_2 bubbling. Obtained deposits were electrochemically checked using linear sweep voltammetry (LSV). LSV measurements were conducted in a phosphate buffer solution (pH 7.2) with an anodic sweep rate of 10 mV s^{-1} at 25 °C. A schematic overview of the experimental procedure is shown in Figure 1.

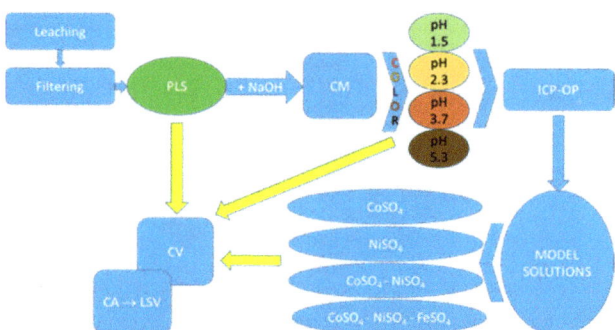

Figure 1. Schematic overview of the experimental procedure.

All electrochemical experiments were conducted in N_2-saturated solutions.

3. Results and Discussion

Visual color changes were detected upon addition of the portions of 10 M NaOH. Natively green PLS at pH 0.5 changed its color to orange-yellow at pH 1.3 and finally reddish when pH 2.1 is reached. With an increasing volume of NaOH added, brownish flakes occurred; stable solid brown precipitate was registered at pH 2.8. NaOH and H_2O_2 addition led to color changes at different pH values in comparison to the case without H_2O_2 addition: orange-yellow color appeared at pH 1.5, red color at pH 2.7 and brown color with solid precipitates at pH 4.9.

3.1. Conductometric Measurements

The intrinsic ionic transitions registered by conductometric measurements during neutralization by NaOH are illustrated by Figures 1 and 2, which show the changes in resistance and incremental resistance, respectively, with a volume of 10 M NaOH added. Incremental resistance is gained as the first derivative of the curves from Figure 2.

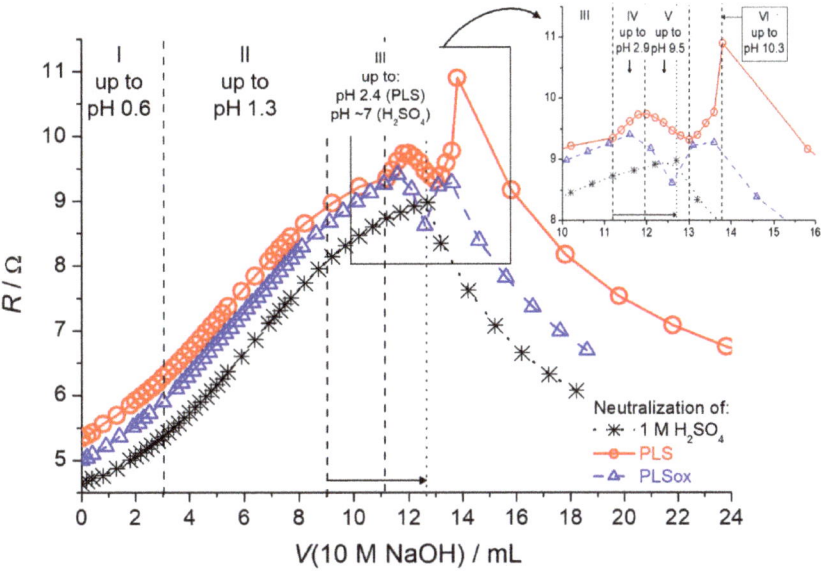

Figure 2. The changes in resistance in 1.0 M H$_2$SO$_4$, PLS and PLSox upon neutralization by NaOH gained by impedance conductometric measurements. Input voltage: 10 mV r.m.s., 10 kHz; 22 ± 1 °C.

Figure 2 shows generally higher resistances of PLS and PLSox with respect to 1 M H$_2$SO$_4$ as a leaching medium, since some portion of the acid was spent for dissolution of lateritic ore. On the other hand, PLSox is of a bit lower resistance down the curve, which indicates that addition of peroxide causes the oxidation of the ions to attain higher, and, consequently, more conductive states. Up to pH 2.4, the resistance increases, whereas the shape of the curves appears quite similar, with obvious changes in resistance increments per volume of NaOH added. This enables the recognition of the three neutralization phases (I–III), with mid phase II of the most pronounced incremental resistance change (Figure 3). As presented in Figure 3, the three investigated solutions are of similar incremental resistance changes in phases I–III. These phases are then to be recognized as neutralization phases of the excess of acid as a leaching agent. The neutralization of the sulfuric acid ends at around pH 7, indicated as a sudden change from increase to decrease in resistance. In the cases of PLS and PLSox, this transition finally takes place above pH 10, since NaOH is used for precipitation of ions in pH range 2.4–10 through phases IV–VI, as indicated in the inset of Figure 2 in details.

Neutralization of the ions (phases IV–VI) appears much faster than that of the excess of leaching agent because the incremental resistance in phases I–III is lower. In a narrow pH range of 2.4–2.9, there is a unique and fast increase in resistance for PLS, followed by resistance decrease (phase V), which is common for PLS and PLSox. The incremental resistance decrease is much higher for PLSox. It could be that phase IV corresponds to neutralization of lower oxidation states since it is hardly distinguishable in the case of PLSox. Consequently, phase V could be related to the post-neutralization state, after neutralizations of ions in higher oxidation states (the process is much faster for PLSox with respect to PLS).

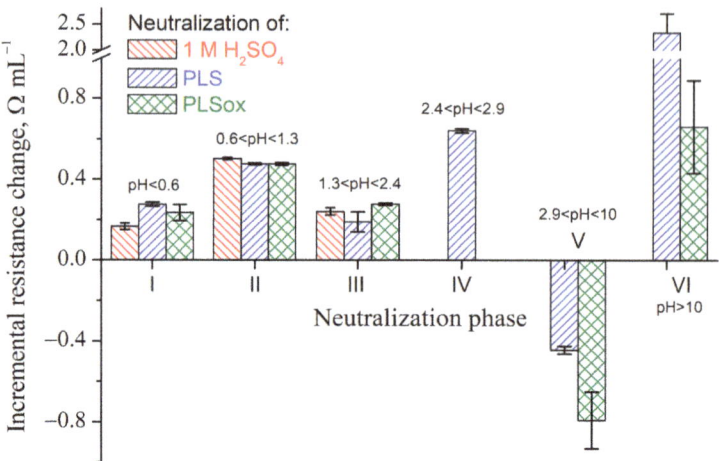

Figure 3. The changes in resistance per volume of 10M NaOH added calculated as first derivative of the curves from Figure 2; specific phases of neutralization, I–VI, as well as corresponding pH ranges, are indicated according to intersections of tangent lines. The standard deviations of linearization are presented as error caps.

According to conductometric considerations, it follows that precipitation of the main ions in PLS is occurring when pH reaches ca. 2.4, with precipitations of Fe species as the most abundant component. This is in accordance with the visual appearance of the massive brownish precipitate and corresponding Pourbaix diagram for Fe, which reports the transitions of metal ions to oxohydroxide ions (the change in PLS color from green to orange-yellow was observed at pH below 2) and finally hematite, roughly in the pH range of 1–5 in a rather acidic environment. According to Figure 2 and Pourbaix diagram, precipitation of Fe species should end in the pH range 2.9–5, whereas the precipitation of other ionic components should take place between pH 5 and 10 upon addition of NaOH solution in volumes between 12.5 and 13 mL (Figure 2, inset—transition from decrease to increase in resistance, phase VI). According to Pourbaix diagrams for Ni and Co, precipitations of corresponding hydroxides/oxides should not start at pH below 6 and 9, respectively. In addition, Ni and Co ions in the solution are stable at the electrode potentials negative to 0.5 V_{SVE}. These considerations introduce a possibility to separate Fe from Ni and Co by Fe precipitation in the pH range of 3–6 and, afterwards, electrochemical extraction of Ni/Co. The extracts can be analyzed by electrochemical dissolution up to 0.5 V_{SVE}, and even beyond, but with generation of the Ni/Co oxides/hydroxides in the electrode–electrolyte interphase. Bearing in mind these findings, the pH values of PLS were set for electrochemical investigations as reported in Table 2.

3.2. Basic Electrochemical Behavior of PLS

Cyclic voltammograms in native PLS (pH 0.5) and PLSs at pH 3.7 and 5.3 are shown in Figure 4. Reversible Fe^{3+}/Fe^{2+} redox transition dominates the CV response of PLS at pH 0.5, 3.7 and 5.3 (Figure 4a,b), indicating the excess of 3+ state by higher currents of the reduction peak at around −0.15 V with respect to the anodic counterpart at around 0.9 V. On the other hand, the CVs of the NaOH treated solutions (Figure 4b) are of considerably lower currents due to the removal of iron from the solution. The onset of Fe deposition and a side reaction of hydrogen evolution (HER) in the solution of pH 3.7 take place at around −0.8 V. The presence of the two weak anodic shoulders can be seen at around 0.0 and 0.2 V. These shoulders are transferred to the peak at 0.0 V upon increase in pH to 3.7. A small peak at the potential of 0.7 V indicates that there are still some traces of iron in the solution

(Table 2). At pH 5.3, this peak shifted catholically to 0.56 V, while new reversible peaks at around 0.9 V arose due to transitions in the Ni/Co oxides.

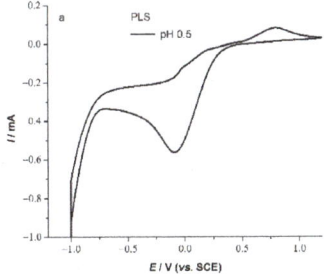 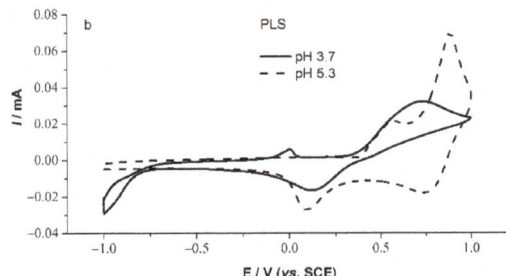

Figure 4. Cyclic voltammograms of (**a**) native PLS at pH 0.5 and (**b**) NaOH-treated PLSs (pH 3.7 and 5.3), v =10 mV s^{-1}. The anodic branches are recorded upon hold at the cathodic limit for 2 min.

Figure 5 shows cyclic voltammograms of NiSO$_4$ and CoSO$_4$ model solutions of different pH and concentrations (Table 2), compared to CV of base solution. Base solution of pH 3.7 was prepared by setting the pH of 1.0 M H$_2$SO$_4$ with NaOH.

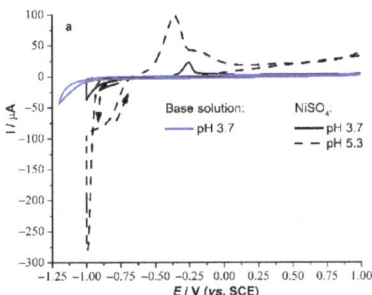 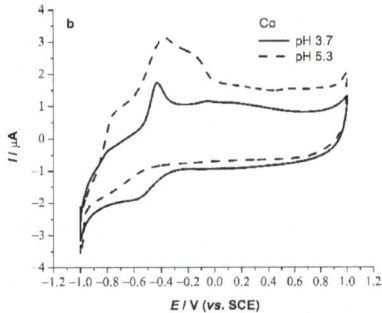

Figure 5. CV of GC in N$_2$-saturated (**a**) NiSO$_4$, compared to base and (**b**) CoSO$_4$ solutions, v = 10 mV/s. The anodic branches are recorded upon hold at the cathodic limit for 2 min.

Although there was no pronounced hydrogen evolution even at pH 3.7 down to −1.0 V (Figure 5, blue line), the nickel electrodeposition process depended considerably on the pH value of the solution. It should be taken into account that slightly different Ni^{2+} and Co^{2+} concentrations at pH 3.7 and 5.3 solutions (Table 2) could also have an impact on the electrodeposition processes. However, despite lower concentration at pH 5.3, both NiSO$_4$ and CoSO$_4$ at pH 5.3 exhibited more intense anodic peaks compared to those in pH 3.7 solutions. At lower pH, HER, as a side reaction, takes place in model solutions on metal deposits in comparison to the base solution where HER takes place on GC. HER thus spends considerable current, and consequently a lower amount of metal is being deposited. This can be observed as the less intensive anodic oxidation/dissolution peaks. In addition, due to more intense HER, i.e., profound H$^+$ consumption, there is local pH increase near the electrode surface. If the pH value is high enough, hydrolysis of the metal ion as an additional reaction may occur that can lead to the precipitation of the metal hydroxides. Consequently, different phases of a metal in a deposit may occur [15]. Unlike Ni, Co electrodeposition process is less sensitive to pH due to rather low Co concentrations.

LSV curves of the deposits obtained potentiostatically at −0.9 V from PLSs and model solutions are shown in Figures 5–8. Multiple peaks indicate multiple phases of the deposits and/or subsequent processes in which dissolved metal ions participate.

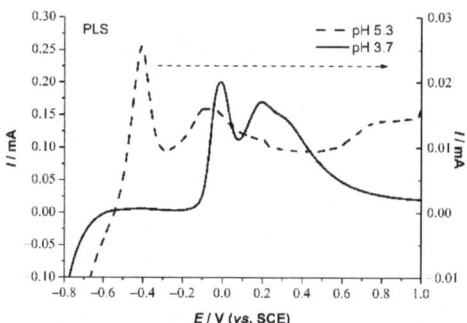

Figure 6. LSV curves in phosphate buffer solution (pH 7.2, 25°C) of deposits obtained after CAs at −900 mV in PLSs.

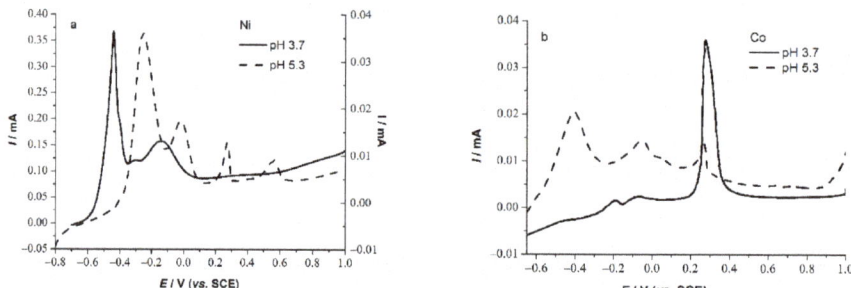

Figure 7. LSV curves in phosphate buffer solution (pH 7.2, 25 °C) of deposits obtained after CAs at −900 mV in (**a**) NiSO$_4$ and (**b**) CoSO$_4$.

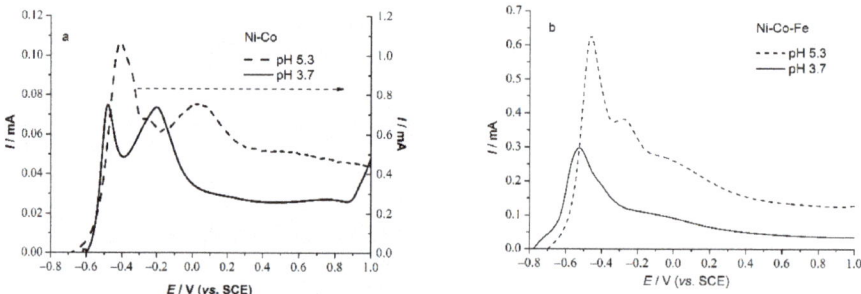

Figure 8. LSV curves in phosphate buffer solution (pH 7.2, 25 °C) of deposits obtained after CAs at −900 mV in (**a**) NiSO$_4$–CoSO$_4$ and (**b**) NiSO$_4$–CoSO$_4$–FeSO$_4$.

Ni- curves at different pH (Figure 7a) showed typical LSV behavior in phosphate buffer with two main anodic peaks below 0.0 V assignable to Ni$\Longrightarrow$Ni^{2+} transition. The peaks at more positive potentials appear suppressed at pH 5.3 due to formation of NiO/Ni(OH)$_2$ film, [16,17]. The presence of Ni$\Longrightarrow$Ni^{2+} transition-related peaks in Ni-pH3.7 and Ni-pH5.3 LSV curves suggest that Ni was deposited in a metallic form at both pH values.

The shape of the Co-pH5.3 LSV curves indicates that cobalt was deposited in a mainly metallic form at pH 5.3 (Figure 7b), with suppression of the peaks at more negative potentials at lower pH. Namely, an anodic peak around −0.410 V, related to Co$\Longrightarrow$Co(OH)$_2$ transition, occurred in Co-pH5.3 LSV response. Additional anodic peaks at −0.08 and 0.260 V correspond to Co(OH)$_2$$\Longrightarrow$Co$_3O_4$ and Co$_3$O$_4$$\Longrightarrow$CoOOH transitions, respectively.

Onset of CoOOH$\Longrightarrow$CoO$_2$ transition can be seen at 0.9 V. On the other hand, surface of a deposit obtained at pH 3.7 consisted mostly of Co$_3$O$_4$ due to a dominant peak around 0.3 V, representing the Co$_3$O$_4$$\Longrightarrow$CoOOH transition [18].

These differences in active/passive behavior of the surfaces of Co and Ni deposits can be explained by different surface morphology due to different model solutions concentrations and pH. Owing to the different electrodeposition process (Figure 5), different Ni and Co phases can be potentiostatically formed [19–21]. At lower pH, side HER may have a larger impact on the morphology of metal deposition. An increase in pH value near the electrode due to HER, i.e., H$^+$ consumption, may lead to the precipitation of the hydroxide [15].

Comparing the starting leachate of pH 0.5 with 111.61 mmol/L of Fe and the solution of pH 5.3, where Fe content was negligible (Table 2.), the goal of Fe removal is reached, as commented further in Figure 4. LSV–Ni–Co consisted of the two main peaks associated to the Ni$\Longrightarrow$Ni^{2+} and Co$\Longrightarrow$Co^{2+} transitions. Like Ni–Co curves, Figure 8, the pH-5.3 PLS curve, Figure 6, consists of the two anodic peaks around −0.4 V. Furthermore, Ni–Co and leachate curves consisted of a broad peak around 0 V, while the peak present in the Ni-curve (Figure 7) was at −0.17 V. This anodic shifting can be assigned to the influence of Co on the Ni–Co LSV curve, which causes the appearance of the broad peak between −0.2 and 0.2 V present in sole Co-curve (Figure 7b). Compared to the Ni–Co curve (Figure 8a), LSV curve related to deposit obtained from Ni–Co–Fe model solution (Figure 8b) consisted of mainly of a peak at around −0.4 V. A peak at 0 V, seen in LSV of both Ni–Co and leachate at pH 3.7 and pH 5.3, is not evident. It seems that the presence of a small amount of Fe can contribute to the suppression of the abovementioned peak (Figure 8).

The shape of the leachate and Ni–Co curves (Figures 6 and 8) resembles the shape of both Ni and Co-pH 5.3 curves (Figures 7 and 8), suggesting that Ni–Co dominates leachate deposit composition. A peak at around −0.4 V, related to M$\Longrightarrow$M^{2+} transition, is present in all of these curves except the PLS-pH3.7 curve. Peaks present in PLS-pH3.7 curve at 0 and 0.2 V correspond to the peaks registered in Ni-pH3.7 and both Co and Ni–Co curves (Figure 8a).

Both Fe-pH5.3 and Fe-pH3.7 (Figure 9) curves contain the peaks related to transitions of different hydroxide forms, indicating that deposits are in the state of the hydroxide rather than metallic Fe [22]. LSV–Fe curves for both Fe-pH5.3 and Fe-pH3.7 (Figure 9) exhibit peaks at 0.0 and 0.2 V which are suppressed at LSV–Ni–Co–Fe curves.

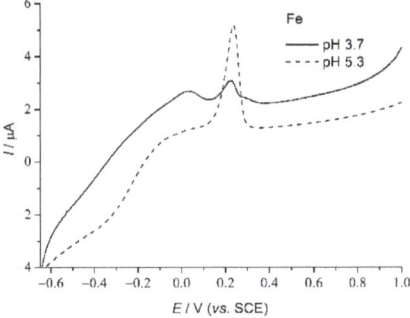

Figure 9. LSV curves in phosphate buffer solution (pH 7.2, 25°C) of deposits obtained after CAs at −900 mV in FeSO$_4$.

LSV results are in accordance with Table 2, which shows that Fe was removed from the leachate at the pH 5.3. However, although Fe was almost completely removed, its presence slightly influenced the peaks at the Ni–Co–Fe LSV curve (Figure 7). The LSV curve related to the deposit obtained from leachate at this pH value is in good agreement with the curve of the corresponding Ni–Co model deposit. Fe ion removal is a precondition in the Ni and Co extraction process from leaching solution, since the presence of Fe is detrimental in those ions' extraction process from PLS. Extracted Ni and Co, i.e., electrodeposited in metallic form, can be valorized from lateritic ore leaching solution.

4. Conclusions

There is a possibility to separate Fe from Ni and Co in leachates of lateric ores by Fe precipitation in the pH range 3–6, and afterwards to electrochemically extract Ni/Co. The extracts can be subjected to electrochemical dissolution up to 0.5 V_{SVE}, and even further with generation of the Ni/Co oxides/hydroxides in the electrode–electrolyte interphase.

Reversible Fe^{3+}/Fe^{2+} redox transition dominates the CV response of native PLS at pH 0.5, indicating the excess of the 3+ state by higher currents of a reduction CV peak. On the other hand, the CVs of the NaOH-treated leaching solutions are of considerably lower currents, indicating a removal of Fe from the solution. The onset of Fe deposition and a side HER in solution with pH 3.7 takes place at around −0.8 V.

Fe was almost successfully removed from the leachate at the pH 5.3. The LSV curve related to the deposit obtained from leachate at this pH value is in good agreement with the curve of the corresponding Ni–Co model deposit.

The presence of Ni$\Longrightarrow$Ni^{2+} transition related peaks in Ni-pH3.7 and Ni-pH5.3 LSV curves suggests that Ni was deposited in a metallic form at both pH values.

The shape of the Co-pH5.3 LSV curves indicates that cobalt was deposited in a mainly metallic form only at pH 5.3 (Figure 7b). Namely, the anodic peak around -0.410 V, related to Co$\Longrightarrow$Co(OH)$_2$ transition, occurred at Co-pH5.3 LSV response.

Author Contributions: Conceptualization, M.M., J.S., S.S., V.P. and B.F.; methodology, S.S. and V.P.; formal analysis, J.M. and M.K.; investigation, J.M., M.K. and M.M.; resources, J.S.; data curation, J.M. and M.K.; writing—original draft preparation, J.M. and M.K.; writing—review and editing, M.M. and V.P.; supervision, J.S. and B.F.; project administration, M.M. All authors have read and agreed to the published version of the manuscript.

Funding: This research was supported by the Science Fund of the Republic of Serbia, Program DIASPORA, #GRANT No.6463002, CAPTAIN.

Data Availability Statement: Not applicable.

Acknowledgments: The authors would like to thank Katarina Božić, University of Belgrade, Institute of Chemistry, Technology and Metallurgy, National Institute of the Republic of Serbia, for conductometric measurements assistance.

Conflicts of Interest: The authors declare no conflict of interest. The funders had no role in the design of the study; in the collection, analyses or interpretation of data; in the writing of the manuscript, or in the decision to publish the results.

References

1. Slack, J.F.; Kimball, B.E.; Shedd, K.B. Cobalt. In *Critical Mineral Resources of the United States—Economic and Environ Mental Geology and Prospects for Future Supply*; Schulz, K.J., DeYoung, J.H., Seal, R.R., Bradley, D.C., Eds.; USGS Pubs: Reston, VA, USA, 2017; Volume 1082, pp. F1–F40.
2. Bloomberg. Available online: https://www.bloomberg.com/news/articles/2021-02-25/musk-says-nickel-is-biggest-concern-for-electric-car-batteries (accessed on 12 December 2021).
3. Wood, M.; Li, J.; Ruthera, R.E.; Du, Z.; Self, E.C.; Meyer, H.M., III; Daniel, C.; Belharouak, I.; Wood, D.L., III. Chemical stability and long-term cell performance of low-cobalt, Ni-Rich cathodes prepared by aqueous processing for high-energy Li-Ion batteries. *Energy Stor. Mater.* **2020**, *24*, 188–197. [CrossRef]
4. Deetman, S.; Pauliuk, S.; van Vuuren, D.P.; van der Voet, E.; Tukker, A. Scenarios for demand growth of metals in electricity generation technologies, cars, and electronic appliances. *Environ. Sci. Technol.* **2018**, *52*, 4950–4959. [CrossRef] [PubMed]
5. Davis, J.R. (Ed.) *Nickel, Cobalt, and Their Alloys, ASM Specialty Handbook*; ASM International: Novelty, OH, USA, 2000.
6. Statista. Available online: https://www.statista.com/statistics/967700/global-demand-for-nickel-in-ev-batteries/ (accessed on 12 December 2021).
7. Banza, A.N.; Gock, E.; Kongolo, K. Base metals recovery from copper smelter slag by oxidising leaching and solvent extraction. *Hydrometallurgy* **2002**, *67*, 63–69. [CrossRef]

8. Altundogan, H.S.; Boyrazli, M.; Tumen, F. A study on the sulphuric acid leaching of copper converter slag in the presence of dichromate. *Miner. Eng.* **2004**, *17*, 465–467. [CrossRef]
9. Song, S.; Sun, W.; Wang, L.; Liu, R.; Han, H.; Hu, Y.; Yang, Y. Recovery of cobalt and zinc from the leaching solution of zinc smelting slag. *J. Environ. Chem. Eng.* **2019**, *7*, 102777.
10. Khalid, M.K.; Hamuyuni, J.; Agarwal, V.; Pihlasalo, J.; Haapalainen, M.; Lundström, M. Sulfuric acid leaching for capturing value from copper rich converter slag. *J. Clean. Prod.* **2019**, *215*, 1005–1013. [CrossRef]
11. Meshram, P.; Prakash, U.; Bhagat, L.; Abhilash; Zhao, H.; van Hullebucsh, E.D. Processing of Waste Copper Converter Slag Using Organic Acids for Extraction of Copper, Nickel, and Cobalt. *Minerals* **2020**, *10*, 290. [CrossRef]
12. Potysz, A.; Lens, P.N.; van de Vossenberg, J.; Rene, E.R.; Grybos, M.; Guibaud, G.; van Hullebusch, E.D. Comparison of Cu, Zn and Fe bioleaching from Cu-metallurgical slags in the presence of *Pseudomonas fluorescens* and *Acidithiobacillus thiooxidans*. *Appl. Geochem.* **2016**, *68*, 39–52. [CrossRef]
13. Yang, Z.; Rui-lin, M.; Wang-dong, N.; Hui, W. Selective leaching of base metals from copper smelter slag. *Hydrometallurgy* **2010**, *103*, 25–29. [CrossRef]
14. Elias, M. Nickel laterite deposit—Geological overview, resources and exploitation. In *Giant Ore Deposits: Characteristics, Genesis and Exploration*; Cooke, D., Pongratz, J., Eds.; Special Publication 4; University of Tasmania, Centre for Ore Deposit Research: Hobart, Australia, 2002; Volume 6, pp. 205–220.
15. Koza, J.M.; Uhlemann, M.; Gebert, A.; Schultz, L. The effect of a magnetic field on the pH value in front of the electrode surface during the electrodeposition of Co, Fe and CoFe alloys. *J. Electroanal. Chem.* **2008**, *617*, 194–202. [CrossRef]
16. Macdonald, D.D.; Liang, R.Y.; Pound, B.G. An Electrochemical Impedance Study of the Passive Film on Single Crystal Ni(111) in Phosphate Solutions. *J. Electrochem. Soc.* **1987**, *134*, 2981. [CrossRef]
17. Martini, E.M.A.; Amaral, S.T.; Muller, I.L. Electrochemical behaviour of Invar in phosphate solutions at pH 6.0. *Corros. Sci.* **2004**, *46*, 2097–2115. [CrossRef]
18. Chivot, J.; Mendoza, L.; Mansour, C.; Pauporte´, T.; Cassir, M. New insight in the behaviour of Co–H_2O system at 25–150 °C, based on revised Pourbaix diagrams. *Corros. Sci.* **2008**, *50*, 62–69. [CrossRef]
19. Yu, Y.; Sun, L.; Ge, H.; Wei, G.; Jiang, L. Study on Electrochemistry and Nucleation Process of Nickel Electrodeposition. *Int. J. Electrochem. Sci.* **2017**, *12*, 485–495. [CrossRef]
20. Tian, L.; Xu, J.; Qiang, C. The electrodeposition behaviors and magnetic properties of Ni–Co films. *Appl. Surf. Sci.* **2011**, *257*, 4689–4694. [CrossRef]
21. Yu, Y.; Song, Z.; Ge, H.; Wei, G.; Jian, L. Electrochemical mechanism of cobalt film electrodeposition process. *Mater. Res. Innov.* **2016**, *20*, 280–284. [CrossRef]
22. Richardson, J.A.; Abdullahi, A.A. *Corrosion in Alkalis, Reference Module in Materials Science and Materials Engineering*; Elsevier: Amsterdam, The Netherlands, 2018. [CrossRef]

Article

Selenate Adsorption from Water Using the Hydrous Iron Oxide-Impregnated Hybrid Polymer

Vesna Marjanovic [1,*], Aleksandra Peric-Grujic [2], Mirjana Ristic [2], Aleksandar Marinkovic [2], Radmila Markovic [1], Antonije Onjia [2] and Marija Sljivic-Ivanovic [3]

1 Mining and Metallurgy Institute Bor, Zeleni Bulevar 35, 19210 Bor, Serbia; radmila.markovic@irmbor.co.rs
2 Faculty of Technology and Metallurgy, University of Belgrade, Karnegijeva 4, 11120 Belgrade, Serbia; alexp@tmf.bg.ac.rs (A.P.-G.); risticm@tmf.bg.ac.rs (M.R.); marinko@tmf.bg.ac.rs (A.M.); onjia@anahem.org (A.O.)
3 Vinča Institute of Nuclear Sciences, University of Belgrade, 12-14 Mike Petrovića Street, 11351 Belgrade, Serbia; marijasljivic@vin.bg.ac.rs
* Correspondence: vesna.marjanovic@irmbor.co.rs; Tel.: +381-69-631736

Received: 10 November 2020; Accepted: 2 December 2020; Published: 4 December 2020

Abstract: Hybrid adsorbent, based on the cross-linked copolymer impregnated with hydrous iron oxide, was applied for the first time for Se(VI) adsorption from water. The influence of the initial solution pH, selenate concentration and contact time to adsorption capacity was investigated. Adsorbent regeneration was explored using a full factorial experimental design in order to optimize the volume, initial pH value and concentration of the applied NaCl solution as a reagent. Equilibrium state was described using the Langmuir model, while kinetics fitted the pseudo-first order. The maximum adsorption capacity was found to be 28.8 mg/g. Desorption efficiency increased up to 70%, and became statistically significant with the reagent concentration and pH increase, while the applied solution volume was found to be insignificant in the investigated range. Based on the results obtained, pH influence to the adsorption capacity, desorption efficiency, Fourier transform infrared (FTIR) and X-ray diffraction (XRD) analysis of loaded adsorbent, it was concluded that the outer- and inner-sphere complexation are mechanisms responsible for Se(VI) separation from water. In addition to the experiments with synthetic solutions, the adsorbent performances in drinking water samples were explored, showing the purification efficiency up to 25%, depending on the initial Se(VI) concentration and water pH. Determined sorption capacity of the cross-linked copolymer impregnated with hydrous iron oxide and its ability for regeneration, candidate this material for further research, as a promising anionic species sorbent.

Keywords: macroporous polymer; goethite; factorial design; desorption

1. Introduction

Selenium in the environment is naturally present in rocks and soils in a few oxidation states: as selenite, selenate, selenide and elemental Se. It is a trace element in natural deposits of the ore containing the other minerals, such as sulfides of heavy metals [1]. Besides the naturally present Se, its concentration increase in the environment is caused by human activities, especially mining, coal combustion, pesticides production, agriculture, etc. [2]. Nowadays, a large amount of diverse types of wastewater containing harmful chemicals are generated by the industry, and the water crisis is caused by untreated wastewater disposal [3]. Selenium is present in the effluents of the final phases of ore processing, mainly as selenite (at low pH value), or selenate (at high pH value).

Several selenium chemical derivatives such as selenomethionine, selenocysteine, selenate and selenite are the major sources of dietary selenium, out of which selenomethionine is the most widely consumed [4]. This element is necessary as a micronutrient in the form of selenoproteins, and extremely important for many biological functions, like the formation of thyroid hormones, DNA synthesis, antioxidant defense, fertility and reproduction. Many different classes of naturally and synthetic organoselenium compounds have been explored as the antiproliferative agents, and the field is constantly emerging, with several compounds demonstrating a pronounced cytotoxic activity against cancer cells compared to the non-transformed ones [5]. However, selenium intake in excessive amounts can be extremely toxic for living organisms, depending on the concentration, but also on the chemical form and other dietary components involved. In general, contradictory opinions about its toxicity and necessity caused differences in permitted values of this element in drinking water legislation: due to the World Health Organization [6] and European Commission (EC) [7] the drinking water limit for selenium is 10 µg L^{-1}, while the US Environmental Protection Agency (EPA) standard limit is 50 µg L^{-1} [8].

Many processes for selenium removal from water, including the chemical reduction, adsorption, bioremediation, phytoremediation, and electrochemical methods, were extensively studied. It is known that sorption plays an important role in transport and control of the target metal contaminants in the ecosystem [9]. Bearing in mind that selenate in soils and sediments preferentially reacts with ferric Fe(III) oxides and hydroxides [10], the synthetic and natural iron-based adsorbents were applied for selenate separation.

Recently, the ER/DETA/FO/FD adsorbent was synthesized in a two step processes: by amination of cross-linked macroporous polymer (ER) with diethylenetriamine (DETA) in the first step and impregnation with hydrous iron oxide (FO), in the second step. The obtained material, after freeze drying (FD) process, has shown a good adsorption potential towards As oxyanions [11]. In this study, the performances of ER/DETA/FO/FD for Se(VI) removal from water were investigated.

The main aim of this work was to investigate the effect of mostly studied parameters: pH, contact time and concentration of selenate to sorption capacity, as well as to examine the possibility of loaded adsorbent regeneration and process optimization by varying desorption reagent pH, volumes and concentrations. Furthermore, the possibility of proposed adsorbent utilization for Se(VI) removal from drinking water was tested as well.

2. Materials and Methods

The adsorption experiments were undertaken using the cross-linked macroporous polymer impregnated with hydrous iron oxide, _-FeOOH (ER/DETA/FO/FD). It was previously applied as an efficient adsorbent for arsenate, and its synthesis and characterization were reported previously [11]. The efficiency of the chosen adsorbent for selenium removal has been investigated in the synthetic and real water solution. Synthetic solutions of Se(VI) were prepared by dissolving the appropriate amount of Na_2SeO_4 (p.a. purity grade, Sigma Aldrich), in mili-Q water. The real solutions were prepared in the same way as the synthetic solutions, using drinking water instead of mili-Q, in order to simulate drinking water samples with elevated Se(VI) concentration. Drinking water was collected from the water supply network in Belgrade, Serbia.

The adsorption experiments were performed using 4 mg of adsorbent added to 25 mL of synthetic solution in polyethylene flasks, at room temperature (22 °C). The flasks were shaken on the orbital Heldoph shaker (Heidolph North America, Wood Dale, IL, USA) (at a constant speed of 170 rpm. The sorption process was investigated as a function of contact time (15–500 min), initial pH value (2–11) and initial selenium concentration (0.1–5 mg L^{-1}), while the temperature and solid/liquid ratio were kept constant.

After defined contact time, in adsorption and desorption experiments, the suspensions were filtered through 0.45 µm filter and Se concentration in solution was measured using an inductively coupled plasma mass spectrometry ICP-MS (Agilent 7700 Series, Agilent Technologies, Inc. Tokyo, Japan). In drinking water samples, beside Se, the concentrations of Ca, Mg, Na, K, Pb, Fe, Cu, Zn, Ni, Mg were measured; pH values were monitored using the pH-meter of WTW Ino Lab.

The amount of Se adsorbed (mg g^{-1}) at time t (q_t) and in the equilibrium (q_e) were calculated using Equations (1) and (2), where c_0, c_t and c_e are the initial selenium concentration, selenium concentration in solution after appropriate adsorption time (t) and in equilibrium (c_e), respectively; V is a solution volume and m is an adsorbent mass.

$$q_t = (c_0 - c_t)V/m \quad (1)$$

$$q_e = (c_0 - c_e)V/m \quad (2)$$

The desorbed amount (Q_{des}) is calculated as the amount of selenium desorbed from one gram of spent adsorbent (Equation (3)):

$$Q_{des} = c_{des} \cdot V_{des}/M \quad (3)$$

where c_{des} is Se concentration in desorption solution, V_{des} is volume of desorption solution and M is the weight of spent adsorbent. Finally, desorption efficiency (%) is calculated as a ratio of desorbed amount (Q_{des}) and initially sorbed amount q_e, multiplied by 100.

X-ray diffraction (XRD) analysis and Fourier transform infrared (FTIR) analysis were used to determine the mineralogical and surface composition of Se loaded adsorbent. XRD analysis was undertaken using a small-angle x-ray scattering (SAXS) diffractometer (Rigaku Smartlab, Austin, TX, USA). within 2θ range 10–90 with 0.05 step size. The FTIR analysis has been performed using a Nicolet IS 50 FTIR Spectrometer (Thermo Fisher Scientific, Waltham, MA, USA) operating in the attenuated total reflection (ATR) mode in the region 400–4000 cm^{-1} and resolution of 4 cm^{-1} with 32 scans.

In order to investigate the possibility of adsorbent regeneration, the series of desorption experiments were performed. NaCl, as a non-aggressive reagent, was chosen in the experimental part. The desorption efficiency was investigated as a function of selected process variables: leaching solution concentration, pH value and solution volume (three levels), using a full factorial experimental design (Table 1). A high level of volume was 25 mL as in the adsorption experiment while the lower volumes were tested in order to investigate the desorption process that generates a lower amount of waste liquid.

Table 1. Process variables and their levels.

Variable Code	Variable	Levels		
		Low	Medium	High
A	Initial pH	7	-	11
B	Concentration (mol L^{-1})	0	-	0.5
C	Volume (mL)	5	15	25

The experimental design matrix was generated using the Minitab software (Released 13.0, State College, PA, USA). The values of process variables were specified in each experiment. The aqueous solutions, used in these experiments, were obtained using mili-Q water and NaCl and NaOH (p.a. purity). The other experimental conditions were constant: temperature (22 °C), agitation rate (170 rpm) and contact time (300 min). The experiments were undertaken in triplicate and the average values of response function were considered in the statistical analysis. The experiments were performed in a random order to assure that the uncontrolled factors do not affect the results and to evaluate the experimental errors properly. The results interpretation and analysis were also undertaken using the Minitab software.

3. Results

3.1. Effect of pH

Investigation of the effect of initial pH value (pH$_i$) onto Se sorption efficiency has shown that it increases the efficiency (Figure 1) at pH between 2 and 4. Further increase in pH resulted in a significant decrease of sorption efficiency.

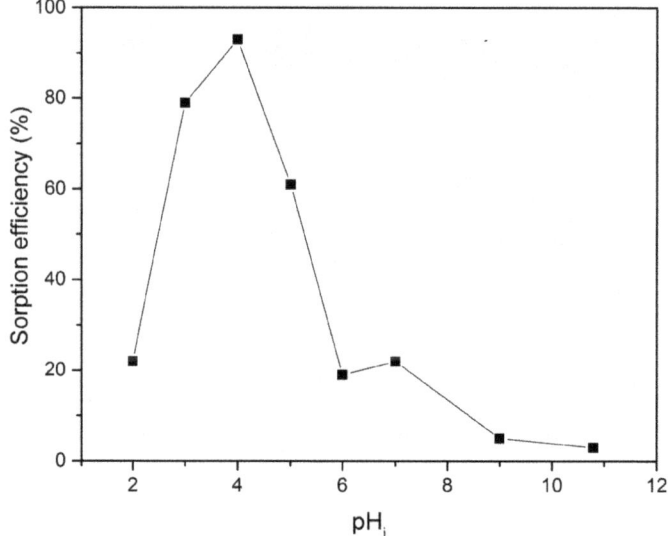

Figure 1. The effect of initial pH value (pH_i) onto sorption efficiency (%); experimental conditions: $c_0 = 1$ mg L^{-1}, temperature 22 °C, stirring rate 170 rpm, contact time 300 min.

3.2. X-ray Diffraction (XRD) Analysis

The results of XRD analysis of the ER/DETA/FO/FD before adsorption and Se loaded ER/DETA/FO/FD are shown in Figure 2. Peaks, characteristic for goethite, are observed at the 2θ value of 21.2, 33.2, 36.6 and 53.2, referring to the ICDD PDF2 No. 81-0464. Similar peaks were observed after adsorption, indicating that the crystallinity of material did not significantly change due to the adsorption of selenate.

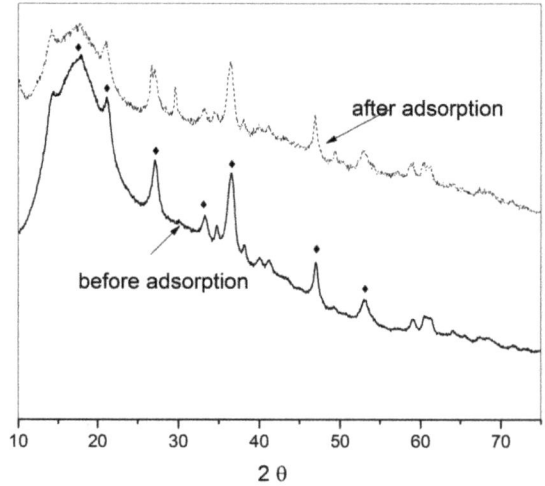

Figure 2. X-ray diffraction (XRD) records of the ER/DETA/FO/FD adsorbent before and after adsorption (experimental conditions for preparation of Se loaded sample: Se initial concentration 5 mg L^{-1}, ambient temperature, solid/liquid ratio 0.16 g L^{-1}, contact time 300 min). Symbol (♦) denotes characteristic peaks of goethite.

3.3. Process Kinetics

The dependence of sorption capacity on a contact time is shown in Figure 3. The removal was faster at the beginning of the process what can be attributed to the higher number of free active sites, as well as to the more intensive driving force for the mass transfer. The sorption capacity obtained in equilibrium was around 5.8 mg g^{-1} which is more than 90% of the total Se(VI) amount. The equilibration was attained after 300 min.

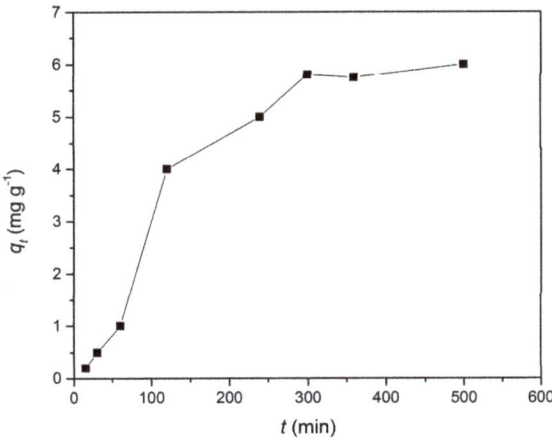

Figure 3. The effect of contact time onto Se(VI) adsorption using the ER/DETA/FO/FD; experimental conditions: c_0 = 1 mg L^{-1}, temperature 22 °C, stirring speed 170 rpm, pH = 4, sorbent dose 0.16 g L^{-1}.

3.4. Effect of the Initial Se(VI) Concentration—Adsorption Isotherms

The effect of the initial adsorbate concentration on the adsorbed Se(VI) amount was studied; the initial Se concentrations varied in the range of 0.1 to 5 mg L^{-1}. With the increase of initial Se(VI) concentration in solution, the sorbed amount in the solid phase increased, as well as the equilibrium concentration (Figure 4). According to the Gils classification [12], the isotherm obtained is of the L-type.

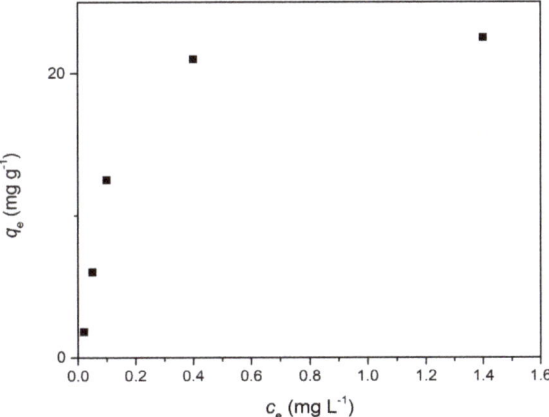

Figure 4. Adsorption isotherm for selenate by the ER/DETA/FO/FD adsorbent at pH = 4; contact time 300 min, initial concentration range from 0.1 to 5 mg/L, sorbent dose 0.16 g L^{-1}.

3.5. Fourier Transform Infrared (FTIR) Spectra of Adsorbent

In order to analyze the surface of material after sorption, the FTIR spectra of Se unloaded and loaded adsorbent were recorded (Figure 5). The broad band at ~3370 cm^{-1} belongs to the hydroxyl stretching region. It is more intense on loaded adsorbent, due to the adsorption of hydroxyl ions. In the region 1100–2000 cm^{-1}, the position of bands after adsorption has not been significantly changed; nevertheless, some bands became more intense. Compared to the spectrum a), two new bands at 950 and 852 cm^{-1} appeared, and they can be assigned to the bands of adsorbed SeO$_4$ species. The sharp peak at 852 cm^{-1} is similar to the spectra of selenate in solution [13], indicating an outer-sphere complex present at the adsorbent–solution interface; an oxyanion retains its hydration shell and does not form a direct chemical bond with the surface, but a complex involving electrostatic forces.

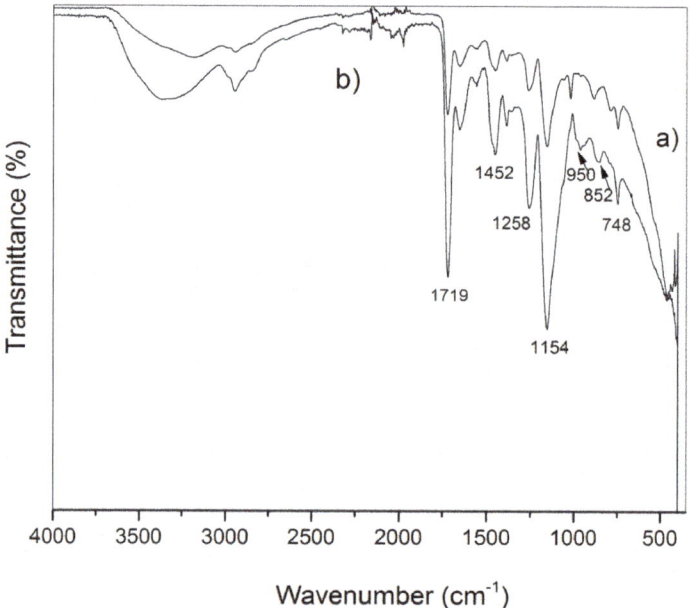

Figure 5. Fourier transform infrared (FTIR) spectrum of adsorbent (**a**) before and (**b**) after Se adsorption.

3.6. Regeneration Studies

Since a high sorption capacity of the investigated sorbent was achieved, the regeneration studies were performed in order to determine the possibility of its reuse The full factorial experimental design was used in this study, since this method has proven to be suitable for the adsorption process investigations [14,15]. The effect of desorption solution pH, concentration and volume onto desorption efficiency was studied. The experimental design matrix was generated on the basis of two levels of factor (pH and concentration) and three levels of solution volume and comprised 12 runs (Table 2).

Desorption efficiency (Figure 6) was insignificant when the mili-Q water, pH = 7 was used (runs 7, 8, 9). The process was more efficient (50–70%) when solutions with a higher pH or NaCl concentration were used.

Table 2. Experimental design matrix.

Run	A	B	C	Initial pH	C (NaCl) mol L^{-1}	V (mL)	Final pH
1	2	2	3	11	0.5	25	10.9
2	1	2	1	7	0.5	5	7.0
3	2	2	1	11	0.5	5	10.8
4	2	1	3	11	0	25	11.1
5	2	1	1	11	0	5	10.5
6	2	2	2	11	0.5	15	10.7
7	1	1	1	7	0	5	6.2
8	1	1	3	7	0	25	6.6
9	1	1	2	7	0	15	6.6
10	1	2	2	7	0.5	15	6.8
11	2	1	2	11	0	15	10.9
12	1	2	3	7	0.5	25	6.9

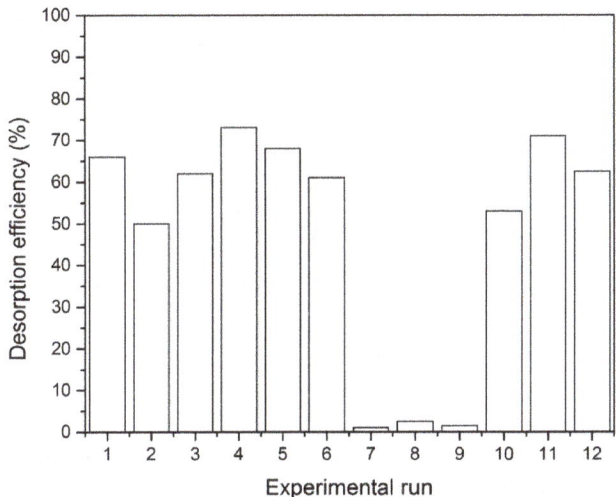

Figure 6. Desorption efficiency obtained for each experimental run (loaded adsorbent from adsorption studies after equilibration with Se, c_o 5 mg L^{-1}). The experimental conditions of each run are given in Table 2.

3.7. Separation of Selenate from Spiked Drinking Water Samples

In this study, series of spiked drinking water samples were prepared with different Se concentrations (1 mg L^{-1}, 2 mg L^{-1} and 5 mg L^{-1}) and initially adjusted pH values at 4 and 7. Concentrations of selected cations are presented in Table 3.

Table 3. Concentrations of selected metal cations in drinking water samples (mg L^{-1}).

Ca	Mg	K	Fe	Pb	Cu	Zn	Ni	Mn
49.1 ± 4.5	8.8 ± 0.8	4.8 ± 0.5	<0.01	<0.02	0.15 ± 0.03	<0.05	<0.01	<0.01

The concentration of analyzed cations did not change during adsorption due to a repulsion of these ions by a positively charged sorbent surface. The adsorption experiments with drinking water at pH 4 have shown that the removal efficiency of applied adsorbent was significantly lower compared to the efficiency obtained using Se(VI) in synthetic (distilled) water (Table 4).

Table 4. The removal efficiency of selenate by the ER/DETA/FO/FD adsorbent from drinking water.

Selenate Solution in	pH	Se initial Concentration (mg L^{-1})	Removal Efficiency (%)
Synthetic water	4	1	91
		2	77.5
		5	71
Drinking water	4	1	25
		2	17.5
		5	13
Drinking water	7	1	14
		2	8
		5	6.5

4. Discussion

The pH value of solution is an important factor in adsorption processes since it affects the metal species distribution, as well as the protonation of the adsorbent surface groups. Selenium(VI) distribution as a function of pH, obtained using the MINTEQ software, showed a coexistence of $HSeO_4^-$ and SeO_4^{2-} at pH < 4, while at pH ≥ 4 SeO_4^{2-} is dominantly present in solution. Even the surface of adsorbent, studied in this research, is positively charged until pH_{pzc} 8.8 [11], and a highest sorption capacity was obtained at pH 4 (Figure 1). In order to explain this phenomenon, the reactions taking place at the adsorbent surface (cross-linked copolymer modified with hydrous ferric oxide, mainly in the form of goethite (Figure 2), are presented with the appropriate constants (K_{a1} and K_{a2}) [16]. It is obvious that both of them are pH dependent.

= Fe-OH is neutral, while = $FeOH_2^+$ and = Fe-O$^-$ are in their protonated and deprotonated forms, respectively.

$$= FeOH_2^+ \rightarrow H^+ + = Fe-OH \cdots K_{a1} = \frac{[H^+][= FeOH]}{[= FeOH_2^+]} = 10^{-6.5}$$

$$= FeOH_2^+ \rightarrow H^+ + = Fe-O \cdots K_{a2} = \frac{[H^+][= FeO^-]}{[= FeOH]} = 10^{-9}$$

At pH = 4, $FeOH_2^+$ is dominant at solid surface [16] and SeO_4^{2-} in the liquid phase; with the pH increase, in the pH range 5–8, $FeOH_2^+$ concentration decreases, as well as the adsorbent sorption capacity for selenate ions. In addition, the adsorption of Se(VI) on to a positively charged adsorbent is diminished in a highly acidic media, at pH < 2 (Figure 1) due to the dominantly occurring $HSeO_4^-$ ion. Based on the results obtained, it can be concluded that SeO_4^{2-} ions are bonded to $FeOH_2^+$ present on the surface.

The relation between q_t and time (t) was analyzed using the well-known kinetic models in the linear forms:

Pseudo-first order [17]:

$$\ln\frac{(q_e - q_t)}{q_e} = -k_1 \cdot t \quad (4)$$

Pseudo-second order [18]:

$$\frac{t}{q_t} = \frac{1}{k_2 \cdot q_e^2} + \frac{t}{q_e} \quad (5)$$

Intraparticle diffusion model [19]:

$$q_t = k_d \cdot t^{0.5} + c \quad (6)$$

where k_1, k_2 and k_d denote constants of the pseudo-first, pseudo-second and intraparticle diffusion model, respectively. The parameters of the model were calculated from the slope-intercept form. Furthermore, the determination coefficient (R^2), F-value and p-value were calculated in order to evaluate the accuracy of the applied models (Table 5).

Table 5. Kinetic model parameters.

Model	Parameters			R^2	F	p
Pseudo-second order model	q_e 13.99 mg g^{-1}	h 3.54 × 10^{-2} mg (g min)$^{-1}$	k_2 2.53 × 10^{-3} g (mg min)$^{-1}$	0.651	7.52	0.052
Pseudo-first order model	q_e 7.24 mg g^{-1}	k_1 9.16×10^{-3} min^{-1}		0.986	363	7.3 × 10^{-6}
Intraparticle diffusion model	k_d 0.374 mg g^{-1} min$^{-1/2}$			0.952	100	1.7 × 10^{-4}

Higher R^2 and F-value and lower p-value ($p < 0.05$) could point out the models which are suitable for data description. Also, the experimentally determined q_e value and that obtained from the plot of t/q_t vs. t differ significantly, indicating that the process mechanism does not follow the pseudo-second order model, but the pseudo-first. The plot of q_t versus $t^{0.5}$ (Equation (6)) is linear, but does not pass through the origin, indicating that an intra-particle diffusion is not the mechanism of sorption [19].

The maximum sorption capacity, defined from the plateau part, was 22.5 mg g^{-1}. Sorption isotherms were fitted using the most utilized models, Langmuir and Freundlich. One of the mostly used linear forms of the Langmuir model [20] is:

$$\frac{c_e}{q_e} = \frac{c_e}{q_m} + \frac{1}{q_m K_L} \quad (7)$$

while, the linearization of the Freundlich [21] model gives:

$$ln q_e = ln K + 1/n \cdot ln c_e \quad (8)$$

where q_e (mg g^{-1}) and c_e (mg L^{-1}) denote the equilibrium concentration of ions in the solid and liquid phase, respectively, q_m (mg g^{-1}) is the maximum sorption capacity, K_L (L g^{-1}) is the Langmuir constant related to the energy of adsorption, n and K_F are the Freundlich isotherm parameters.

Linear fitting of functions c_e/q_e vs. c_e, and $ln q_e$ vs. $ln c_e$ gave the parameters of the Langmuir and Freundlich models, respectively (Table 6).

Table 6. Parameters of the Langmuir and Freundlich isotherm model, obtained for selenate adsorption by the ER/DETA/FO/FD.

Model	Model Parameter	Value
Langmuir	K_L	2.981 L mg^{-1}
	q_m	28.8 mg g^{-1}
	F	6.69
	p	0.081
	R^2	0.78
Freundlich	n	1.246
	K_F	30.377 mg^{1-n}L^{3n}g^{-1}
	F	5.62
	p	0.14
	R^2	0.736

Additionally, the higher F- and R^2 values, as well as a lower p-value have shown that the Langmuir model is more suitable for the experimental data description than the Freundlich model, indicating that a homogeneous surface of adsorbent is covered with a single layer of adsorbed molecules [22].

The effect of process variables (A—initial pH, B—NaCl concentration, C—desorption solution volume) onto system responses (percentage of the desorbed amounts and final pH values) were evaluated using the statistical software. Also, the interactions between variables were considered, since they might have a significant effect on system response. The coefficients in the equation were calculated using the second order regression model (Equation (9)), giving the information about the effect of process variables.

$$Y = \beta_1 A + \beta_2 B + \beta_3 C + \beta_{12} AB + \beta_{13} AC + \beta_{23} BC + \varepsilon, \tag{9}$$

where: Y—system response; A, B, C—independent variables, meaning the initial pH, salt concentration and volume of desorption solution; AB, AC, BC—interactions terms; β_1, β_2, β_3, β_{12}, β_{13}, β_{23} —regression coefficients; ε—residual.

In addition, the analysis of variance (ANOVA) was used in order to define the statistically significant factors and/or their interactions (Table 7). The results have shown that a desorption efficiency was significantly influenced by the factors A and B, and their interaction (AB) at $p < 0.05$. On the other hand, the volume of the leaching reagent did not play any important role in the investigated desorption process.

Table 7. Analysis of variance (ANOVA) regression analysis for Se(VI) desorption data.

Variable	System Response			
	Desorption (%)		Final pH	
	F	p	F	p
A	1574	0.001	20195	<0.001
B	634.4	0.002	48.48	0.02
C	16.89	0.056	21.20	0.045
AB	944.0	<0.001	63.84	0.015
AC	0.19	0.839	2.84	0.260
BC	6.68	0.13	43.00	0.030
-	s = 1.691 R^2 = 99.94% R^2 (adj) = 99.66%		s = 0.0505 R^2 = 99.99% R^2 (adj) = 99.95%	

Considering the final pH values as a system response, all of the investigated variables and their interactions were statistically significant except the AC interaction. The obtained R^2 values were >99.9%, indicating that the experimental data for both system responses could be explained with a high accuracy.

Visualization of the results obtained was undertaken using the *Main effect plots* and *3D surface plots* (Figure 7). The increase of each process variable provoked the increase of system response. The effect of initial pH of leaching solution had the highest effect on desorption efficiency and final pH.

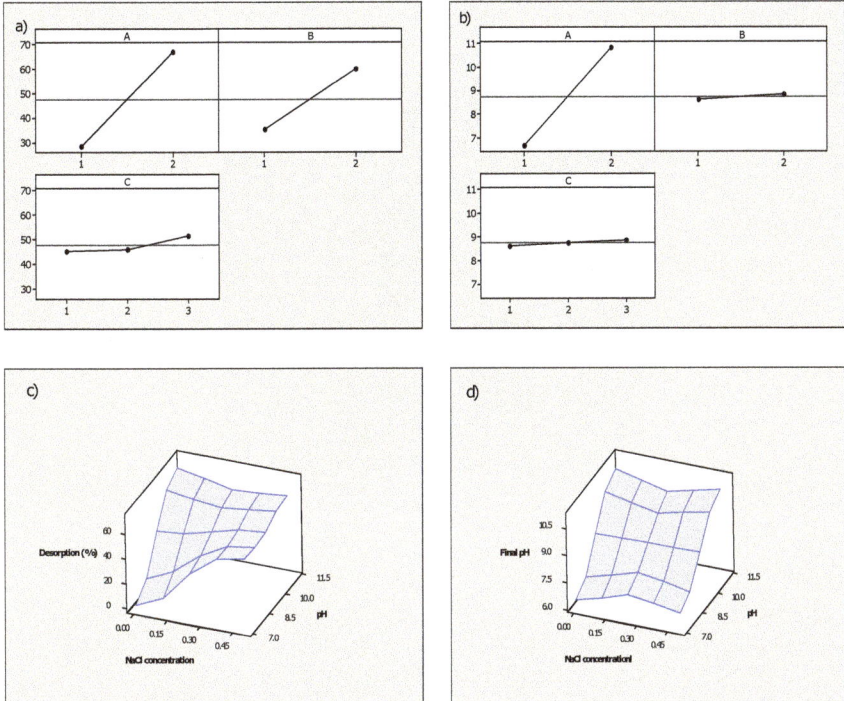

Figure 7. The main effect plots for (**a**) desorption efficiency and (**b**) final pH. Three-dimensional surface plots for (**c**) desorption efficiency and (**d**) final pH.

The removal efficiency of applied adsorbent was lower in spiked drinking water solutions, compared to the adsorption from distilled water (pH 4), Table 4. This is in line with the reported results where a significant adsorption suppression was observed when phosphate and sulphate anions coexisted [15,23]. In part, a non-specific sorption through electrostatic attraction between the negatively charged anions and positive sites on the sorbent obviously takes place. Efficiencies at pH = 7 were about 50% lower in comparison with those obtained at pH = 4.

On the basis of the summarized experimental results, the mechanism of selenate adsorption by the cross-linked macroporous polymer impregnated with hydrous iron oxide (ER/DETA/FO/FD) is proposed. The highest removal efficiency and maximum adsorption capacity were obtained at pH = 4, due to the formation of a complex between $FeOH_2^+$ at the solid surface and SeO_4^{2-}, dominantly present in the liquid phase at this pH. The maximum observed desorption efficiency was 72.5%, at pH = 11, indicating that about 30% of selenate is irreversibly adsorbed, probably due to the formation of the inner-sphere complex (Equation (10)), known as incompletely reversible. On the other hand, the outer-sphere complexation (Equations (11) and (12)), largely electrostatic, is a reversible process responsible for desorption. In the outer-sphere complex, the ion retains its hydration sphere and attaches to the surface via electrostatic forces, whereas the inner-sphere complex is partially dehydrated and directly bound to the surface [24].

$$>FeOH + H^+ + SeO_4^{2-} \rightarrow >FeSeO_4^- + H_2O \tag{10}$$

$$>FeOH + H^+ + SeO_4^{2-} \rightarrow >FeOH_2^+ - SeO_4^{2-} \tag{11}$$

$$2>FeOH + 2H^+ + SeO_4^{2-} \rightarrow (>FeOH_2^+)_2 - SeO_4^{2-} \tag{12}$$

A possible schematic presentation of the adsorbed selenate species on the FEG–SEM (field-emission gun—scanning electron microscopy) image of ER/DETA/FO/FD is given in Figure 8.

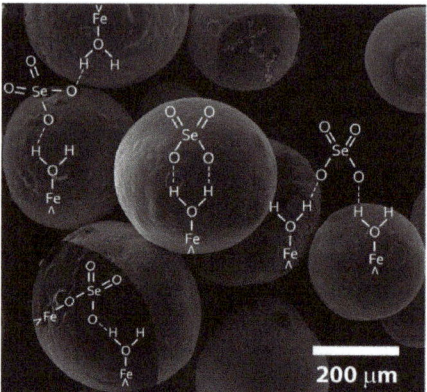

Figure 8. Structures of selenate surface complexes on macroporous polymer impregnated with hydrous iron oxide (ER DETA/FO/FD).

5. Comparative Evaluation of Different Adsorbents for Se(VI) Removal

The adsorbent investigated in this study is compared with the iron oxide-modified adsorbents applied for the removal of selenate from water (Table 8). It is obvious that the adsorption capacity of material investigated has a higher value than those obtained using the other adsorbents.

Table 8. Experimental conditions and uptake capacities of various iron containing sorbents for the selenate removal from water.

Adsorbent	pH	Se(VI) Concentration Range	Adsorbent Dosage	Volume (ml)	Contact Time (min)	Temperature	Maximum Sorption Capacity (mg g^{-1})	Reference
Natural hematite from Cerro del Hierro (Spain)	4	3×10^{-6} and 5×10^{-4} mol dm^{-3}	100 mg	20	-	room	0.18	[14]
Natural goethite from Cerro del Hierro (Spain)	4	3×10^{-6} and 5×10^{-4} mol dm^{-3}	100 mg	20	-	room	0.24	[14]
Synthetic Jacobsite (MnFe$_2$O$_4$) NM	4	0.25–10 mg L^{-1}	10 mg	4	15	room	0.76	[25]
Fe$_3$O$_4$ nanomaterials produced by non microwave-assisted synthetic techniques	4	0.25–10 mg L^{-1}	10 mg	4	15	room	1.43	[15]
Fe$_3$O$_4$ nanomaterials produced by non microwave-assisted synthetic techniques	4	0.25–10 mg L^{-1}	10 mg	4	15	room	2.37	[15]
Iron (Fe^{3+}) oxide/hydroxide nanoparticles sol (NanoFe)	4	12 ppm	15–635 mg L^{-1}	-	1	room	15.1	[26]
Low-Cost Goethite Nanorods	7.2	~0.500 mg L^{-1}	0.05–1 g L^{-1}	100	360	room	4.75	[27]
Iron oxide impregnated hybrid polymer	4	0.1–5 mg L^{-1}	0.16 g L^{-1}	25	300	room	28.8	This study

6. Conclusions

In this study, the adsorption properties of cross-linked macroporous polymer impregnated with hydrous iron oxide for the removal of Se(VI) ions were tested. Based on the results concerning the adsorption processes, the following conclusions were drawn:

- the investigated process was pH dependent, with the best performances at pH 4;
- a pseudo-first model was the most appropriate for the kinetic data description;
- experimentally determined maximum adsorption capacity of the investigated adsorbent towards Se(VI) was found to be 22.5 mg/g, while the value calculated using the Langmuir model was 28.8 mg/g, depicting its prominent adsorption potential.

Furthermore, the desorption process of adsorbed Se(VI) ions was investigated using the full factorial design as a function of leaching solution pH, NaCl concentration and applied volume. It was observed that the most important factor for desorption efficiency was the interaction between solution pH and concentration, followed by these factors solely. The effect of used leaching solution volume was insignificant in the investigated range. The increase of solution concentration and/or pH, increased the desorption efficiency up to 70% showing that the adsorbent is partly regenerative. The mechanism of selenate sorption by sorbent examined in this research probably includes the outer- and inner-sphere surface complex formation and the process rate determines the inner-sphere formation as a slower one.

Author Contributions: Individual contributions of authors are as following: conceptualization, M.S.-I. and A.M.; methodology, M.R., A.P.-G.; software, A.O.; validation, V.M. and R.M.; formal analysis, V.M., R.M. and A.O.; investigation, V.M. and M.R.; resources, A.M., V.M. and R.M.; data curation, V.M.; writing—original draft preparation, V.M. and M.S.-I; writing—review and editing, M.R. and A.P.-G.; visualization, M.R. and A.P.-G.; supervision, M.S.-I.; project administration, M.R. All authors have read and agreed to the published version of the manuscript.

Funding: This research received no external funding.

Acknowledgments: Financial support for this study was partly provided by the Ministry of Education, Science and Technological Development of the Republic of Serbia (Contract No. 451-03-68/2020-14/200135).

Conflicts of Interest: The authors declare no conflict of interest.

References

1. Khamkhash, A.; Srivastava, V.; Ghosh, T.; Akdogan, G.; Ganguli, R.; Aggarwal, S. Mining-Related Selenium Contamination in Alaska, and the State of Current Knowledge. *Minerals* **2017**, *7*, 46. [CrossRef]
2. He, Y.; Xiang, Y.; Zhou, Y.; Yang, Y.; Zhang, J.; Huang, H.; Shang, C.; Luo, L.; Gao, J.; Tang, L. Selenium Contamination, Consequences and Remediation Techniques in Water and Soils: A Review. *Environ. Res.* **2018**, *164*, 288–301. [CrossRef] [PubMed]
3. Noreen, S.; Mustafa, G.; Ibrahim, S.M.; Naz, S.; Iqbal, M.; Yaseen, M.; Javed, T.; Nisar, J. Iron Oxide (Fe_2O_3) Prepared via Green Route and Adsorption Efficiency Evaluation for an Anionic Dye: Kinetics, Isotherms and Thermodynamics Studies. *J. Mater. Res. Technol.* **2020**. [CrossRef]
4. Hariharan, S.; Dharmaraj, S. Selenium and Selenoproteins: It's Role in Regulation of Inflammation. *Inflammopharmacology* **2020**, *28*, 667–695. [CrossRef] [PubMed]
5. Gandin, V.; Khalkar, P.; Braude, J.; Fernandes, A.P. Organic Selenium Compounds as Potential Chemotherapeutic Agents for Improved Cancer Treatment. *Free Radic. Biol. Med.* **2018**, *127*, 80–97. [CrossRef]
6. WHO. *Guidelines for Drinking-Water Quality*, 4th ed.; Incorporating the 1st Addendum; WHO: Geneva, Switzerland, 2012; ISBN 9789241549950.
7. EU. Commission Directive (EU) 2015/1787 of 6 October 2015 amending Annexes II and III to Council Directive 98/83/EC on the quality of water intended for human consumption. *Off. J. Eur. Union* **2015**, *L260*, 6–18.
8. EPA. National Primary Drinking Water Regulations Contaminant MCL or Potential Health Effects from Common Sources of Contaminant Public Health TT 1 (mg/L) 2 Long-Term 3 Exposure above the MCL in Drinking Water Goal (mg/L) 2. 2009. Available online: https://www.epa.gov/sites/production/files/2016-06/documents/npwdr_complete_table.pdf (accessed on 22 September 2020).

9. Veneu, D.M.; Yokoyama, L.; Cunha, O.G.C.; Schneider, C.L.; Monte, M.B.D.M. Nickel Sorption Using Bioclastic Granules as a Sorbent Material: Equilibrium, Kinetic and Characterization Studies. *J. Mater. Res. Technol.* **2019**, *8*, 840–852. [CrossRef]
10. Peak, D.; Sparks, D.L. Mechanisms of Selenate Adsorption on Iron Oxides and Hydroxides. *Environ. Sci. Technol.* **2002**, *36*, 1460–1466. [CrossRef]
11. Taleb, K.; Markovski, J.; Milosavljević, M.; Marinović-Cincović, M.; Rusmirović, J.; Ristić, M.; Marinković, A. Efficient Arsenic Removal by Cross-Linked Macroporous Polymer Impregnated with Hydrous Iron Oxide: Material Performance. *Chem. Eng. J.* **2015**, *279*, 66–78. [CrossRef]
12. Giles, C.H.; Smith, D.; Huitson, A. A General Treatment and Classification of the Solute Adsorption Isotherm. I. Theoretical. *J. Colloid Interface Sci.* **1974**, *47*, 755–765. [CrossRef]
13. Wijnja, H.; Schulthess, C.P. Vibrational Spectroscopy Study of Selenate and Sulfate Adsorption Mechanisms on Fe and Al (Hydr)oxide surfaces. *J. Colloid Interface Sci.* **2000**, *229*, 286–297. [CrossRef] [PubMed]
14. Rovira, M.; Giménez, J.; Martínez, M.; Martínez-Lladó, X.; de Pablo, J.; Martí, V.; Duro, L. Sorption of Selenium(IV) and Selenium(VI) onto Natural Iron Oxides: Goethite and Hematite. *J. Hazard. Mater.* **2008**, *150*, 279–284. [CrossRef] [PubMed]
15. Gonzalez, C.M.; Hernandez, J.; Peralta-Videa, J.R.; Botez, C.E.; Parsons, J.G.; Gardea-Torresdey, J.L. Sorption Kinetic Study of Selenite and Selenate onto a High and Low Pressure Aged Iron Oxide Nanomaterial. *J. Hazard. Mater.* **2012**, *211–212*, 138–145. [CrossRef] [PubMed]
16. Cumbal, L.; Sengupta, A.K. Arsenic Removal Using Polymer-Supported Hydrated Iron(III) Oxide Nanoparticles: Role of Donnan Membrane Effect. *Environ. Sci. Technol.* **2005**, *39*, 6508–6515. [CrossRef]
17. Lagergren, S. Zur Theorieder Sogennanten Adsorption Geloester Stoffe. *Kungliga Svenska Vetenskapsakademiens Handlingar* **1898**, *24*, 1–39.
18. Ho, Y.S.; Mckay, G. Pseudo-second Order Model for Sorption Processes. *Process Biochem.* **1999**, *34*, 451–465. [CrossRef]
19. Weber, W.; Morris, C. Kinetics of Adsorption on Carbon from Solution. *J. Sanit. Eng. Div.* **1963**, *89*, 31–60.
20. Langmuir, I. The Adsorption of Gases on Plane Surfaces of Glass, Mica and Platinum. *J. Am. Chem. Soc.* **1918**, *40*, 1361–1403. [CrossRef]
21. Freundlich, H. *Capillary and Colloid Chemistry*; Methuen and Co. Ltd.: London, UK, 1926.
22. Alguacil, F.J.; García-Díaz, I.; Baquero, E.E.; Largo, O.R.; López, F.A. Oxidized and Non-Oxidized Multiwalled Carbon Nanotubes as Materials for Adsorption of Lanthanum (III) Aqueous Solutions. *Metals* **2020**, *10*, 765. [CrossRef]
23. Chan, Y.T.; Liu, Y.T.; Tzou, Y.M.; Kuan, W.H.; Chang, R.R.; Wang, M.K. Kinetics and Equilibrium Adsorption Study of Selenium Oxyanions onto Al/Si and Fe/Si Coprecipitates. *Chemosphere* **2018**, *198*, 59–67. [CrossRef]
24. Payne, T.E.; Brendler, V.; Ochs, M.; Baeyens, B.; Brown, P.L.; Davis, J.A.; Ekberg, C.; Kulik, D.A.; Lutzenkirchen, J.; Missana, T.; et al. Guidelines for Thermodynamic Sorption Modelling in the Context of Radioactive Waste Disposal. *Environ. Model. Softw.* **2013**, *42*, 143–156. [CrossRef]
25. Gonzalez, C.M.; Hernandez, J.; Parsons, J.G.; Gardea-Torresdey, J.L. A Study of the Removal of Selenite and Selenate from Aqueous Solutions Using a Magnetic Iron/Manganese Oxide Nanomaterial and ICP-MS. *Microchem. J.* **2010**, *96*, 324–329. [CrossRef]
26. Zelmanov, G.; Semiat, R. Selenium Removal from Water and its Recovery Using Iron (Fe3+) Oxide/Hydroxide-Based Nanoparticles Sol (NanoFe) as an Adsorbent. *Sep. Purif. Technol.* **2013**, *103*, 167–172. [CrossRef]
27. Amrani, M.A.; Ghaleb, A.M.; Ragab, A.E.; Ramadan, M.Z.; Khalaf, T.M. Low-cost Goethite Nanorods for as (III) and Se(VI) Removal from Water. *Appl. Sci.* **2020**, *10*, 7237. [CrossRef]

Publisher's Note: MDPI stays neutral with regard to jurisdictional claims in published maps and institutional affiliations.

© 2020 by the authors. Licensee MDPI, Basel, Switzerland. This article is an open access article distributed under the terms and conditions of the Creative Commons Attribution (CC BY) license (http://creativecommons.org/licenses/by/4.0/).

Article

Synergism Red Mud-Acid Mine Drainage as a Sustainable Solution for Neutralizing and Immobilizing Hazardous Elements

Hugo Lucas [1], Srecko Stopic [1,*], Buhle Xakalashe [2], Sehliselo Ndlovu [3] and Bernd Friedrich [1]

[1] IME Process Metallurgy and Metal Recycling, RWTH Aachen University, 52056 Aachen, Germany; HLucas@metallurgie.rwth-aachen.de (H.L.); bfriedrich@ime-aachen.de (B.F.)
[2] Pyrometallurgy Division, MINTEK, Private Bag X3015, Randburg 2125, South Africa; buhlex@mintek.co.za
[3] School of Chemical and Metallurgical Engineering, University of the Witwatersrand, Johannesburg 2000, South Africa; Sehliselo.Ndlovu@wits.ac.za
* Correspondence: sstopic@ime-aachen.de; Tel.: +49-17678261674

Abstract: Acid mine drainage (AMD) and red mud (RM) are frequently available in the metallurgical and mining industry. Treating AMD solutions require the generation of enough alkalinity to neutralize the acidity excess. RM, recognized as a waste generating high alkalinity solution when it is in contact with water, was chosen to treat AMD from South Africa at room temperature. A German and a Greek RM have been evaluated as a potential low-cost material to neutralize and immobilize harmful chemical ions from AMD. Results showed that heavy metals and other hazardous elements such as As, Se, Cd, and Zn had been immobilized in the mineral phase. According to European environmental standards, S and Cr, mainly present in RM, were the only two elements not immobilized below the concentration established for inert waste.

Keywords: acid mine drainage; red mud; neutralization; immobilization; precipitation

1. Introduction

Acid Mine Drainage (AMD) is a term for wastewaters from mining processes. AMD is a sulfate-based solution concentrating several metallic ions from ores [1,2]. During mining operations, rocks containing sulfide minerals are fragmented, increasing the surface area exposed to water and air, favoring the generation of AMD at high rates [3]. Tabelin et al. [4] mentioned that contaminated debris contains several hazardous elements such as arsenic (As), selenium (Se), and boron (B), and heavy metals like cadmium (Cd), copper (Cu), lead (Pb), and zinc (Zn). Pollution generated by these naturally contaminated rocks could pose severe problems to living organisms, including surrounding populations living close to mine and disposal sites.

Park et al. [5] mentioned different remediation strategies like neutralization, adsorption, ion exchange, membrane technology, biological mediation, and electrochemical for reducing AMD negative environmental impacts on human health and ecosystems. Nonetheless, complex techniques require high labor, energy, and maintenance costs besides a continuous supply of chemicals and long-term monitoring of AMD on ecosystems and underground waters.

Igarashi et al. [6] treated AMD using a lab-scale continuous ferrite process flow setup removing Cu and As in the first sludge, which is stable in standard leaching tests. Magnetic magnesium-ferrites and magnetite were generated when dissolved Si was low. Nevertheless, the treatment could not neutralize all toxic metallic ions.

AMD from South Africa originates from sulfide conglomerates stored on deposits where the rain rinses the acid and metals such as uranium out of the dumps [3]. Additionally, South Africa has severe limitations on freshwater, with an average rainfall of under 450 mm per year [7]. AMD from South African mines containing around 3500 mg/l [3]

of sulfates with pH values between 2 and 3 favor the dissolution of several metals that can infiltrate groundwater deposits and rivers pathways in areas close to the deposits. Iron sulfides resulting from pyrite (FeS_2) oxidation in contact with air enter in the solution under Fe^{3+} and naturally precipitate in river pathways creating a bright orange trail [8].

AMD waters can be classified by the content of acid and dissolved metals with the Ficklin diagram [9]. At the surface, rain or surface water dilute AMD increase pH values driving to predominantly precipitate aluminates favoring the adsorption of other metallic ions [10]. On the other hand, heavy metals are generally removed by Fe precipitation. Al precipitation only becomes significant when the iron content is low, but this is rarely the case.

Mwewa et al. [11] have studied the synthesis of poly-alumino-ferric sulfate coagulant from acid mine drainage by precipitation reaching the recovery of Fe and Al at pH 5.0, about 99.9 and 94.7% for Fe and Al, respectively. An increase of pH-value up to 7 leads to the overall Al recovery of 99.1%. Although Al precipitation was 99.1% at pH 7, the precipitate formed at pH 5 was chosen for coagulant production due to the reduced risks to co-precipitate other impurities in substantially higher concentrations. Dissolution of precipitate in 5.0% (w/w) sulfuric acid produced a coagulant containing 89.5% Fe and 10.0% Al, which is comparable to the PFS commercial coagulant. This process can be easily integrated into existing AMD treatment plants, providing revenue and minimizing treatment costs, as well as reducing the sludge volume by 95.0%.

Keller et al. [12] have studied the effectiveness of fly ash and RM as strategies for sustainable AMD management. Because of the high alkalinity, German RM is the most promising precipitation agent achieving the highest pH-values. Coal fly ash is less efficient than RM for neutralization and precipitation. Temperature increases adsorption kinetics. In this study, the authors found a maximum pH-value of 6.0 was reached by adding 100 g German RM at 20 °C to AMD-water with an initial pH value of 1.9. RM removed 99% of Al present as hydroxide for a pH 5.0. Some rare-earth elements as Y and Ce precipitate in contact with Greek RM with an efficiency of 50 and 80%, respectively, at 60 °C in 5 min.

Our main aim is to perform a neutralization of AMD using RM from Greece and Germany and to discuss the possibility of recovering valuable metals such as Al, Zn, Mn, and rare earth elements (REE). Using RM leads to the preparation of wastewater for returning into processing or releasing to the environment. Another aim of neutralization with alkaline material is to formulate solid waste materials for further metal winning processes in order to immobilize the hazardous metals.

2. Materials, Characterization and Experimental Procedure

2.1. Materials

The constituent concentration (µg/L) from the eluate obtained from these two bauxite residues: Red Mud from Greece (RM.Gr) and Red mud from Germany (RM.De) will be compared with the concentration of anions and cations in AMD, as shown in Table 1.

The pH of the AMD wastewater ranges around 2.0 in contrast to both red mud with ones above 10. Table 2 compares pollutants' concentration in mg/kg (see Equation (1)) and EU standardized limits to define hazardous, non-hazardous, and inert waste.

For assessing the synergistic effect of AMD combined with RM, materials were mixed using different ratios, keeping the lines of the compliance test EN 124557 but using AMD instead of distilled water.

Kaußen and Friedrich [13] described the mineralogy of the RM.De. No sufficient reference data (known crystallographic measurements) exist to characterize the missing 1.5% accounts unknown or amorphous from bauxite residue mineralogical composition. In the current X-ray diffraction (XRD) full profile fitting mineral phase quantification, it was not possible to quantify amorphous content. Amorphous content can be determined in phase quantification when a known quantity of an internal standard such as corundum is added to the sample. Result showed in Table 3.

Table 1. Concentration of pollutants in µg/L of AMD, RM.Gr (Red Mud from Greece) and RM.De (Red mud from Germany) (RM eluates obtained from the elution test based on EN 12457-2).

Element	AMD	RM.Gr	RM.De
F [mg/L]	8.5	7.5	2.6
Cl [mg/L]	2.8	4.8	0.8
NO_3 [mg/L]	<0.1	0.3	0.3
SO_4 [g/L]	16.6	0.1	<0.1
As [µg/L]	15.4	389.5	6.7
Ba [µg/L]	1.9	<0.1	<0.1
Cd [µg/L]	67.8	0.1	<0.1
Cr [µg/L]	34.6	18.4	555.2
Cu [µg/L]	216.2	<5	<5
Mo [µg/L]	<0.1	19.1	<0.1
Ni [µg/L]	982.2	0.2	<0.1
Pb [µg/L]	8.4	<0.1	<0.1
Sb [µg/L]	<0.1	<0.1	<0.1
Se [µg/L]	1.3	11.8	4.0
V [µg/L]	0.6	2,623.4	202.9
Zn [µg/L]	7,250.3	<0.1	<0.1
pH	2.0	10.3	11.3

Table 2. Leaching pollutants from RM in contact with distillate water in mg/kg based on EN 12457-2 compared with the EU standardized limits.

Element [mg/kg]	Limits for Inert Waste	Limits for Non-hazardous Waste	Limits for Hazardous Waste	RM Greece	RM Germany
Cl	800	15,000	25,000	48.07	7.70
F	10	150	500	74.57	25.83
SO_4	1000	20,000	50,000	1100	37.27
Ni	0.4	10	40	<0.01	<0.01
Pb	0.5	10	50	<0.01	<0.01
Sb	0.06	0.7	5	<0.01	<0.01
Se	0.1	0.5	7	0.12	0.04
Zn	4	50	200	<0.10	<0.10
As	0.5	2	25	3.89	0.07
Ba	20	100	300	<0.01	<0.01
Cd	0.04	1	5	<0.01	<0.01
Cr total	0.5	10	70	0.18	5.55
Cu	2	50	100	<0.05	<0.05
Hg	0.01	0.2	2	<0.01	<0.01
Mo	0.5	10	30	0.19	<0.01

In addition to phases indicated by XRD, Alkan et al. [14] performed a **Q**uantitative **E**valuation of **M**inerals by **S**canning Electron Microscopy (QEMSCAN) analysis of Greek red mud. They revealed the presence of large amounts of Fe-, Ca-, Al-, and Si-mixed oxide in red mud, where a certain amount of TiO_2 is entrapped. Due to the heterogeneous nature of this complex oxide, chemical composition and stoichiometry vary through the volume. Therefore, a crystalline phase could not be fully assigned. Varying compositions imply that this complex oxide may be an aggregate or intergrowth of several oxides inherent from Bayer Process.

Table 3. RM mineralogy and quantification [12,13].

Mineral [wt%]	RM.Gr	RM.De
Cancrinite [$Na_6Ca_{1.5}Al_6Si_6O_{24}(CO_3)_{1.6}$]	15	–
Perovskite [$CaTiO_3$]	4.5	–
Hematite [Fe_2O_3]	30	44
Boehmite [$AlO(OH)$]	3	13
Goethite [$FeO(OH)$]	9	–
Anatase [TiO_2]	0.5	5
Calcium aluminium iron silicate hydroxide[$Ca_3AlFe(SiO_4)(OH)_8$]	17	–
Quartz [SiO_2]	2	–
Rutile [TiO_2]	0.5	3
Calcite [$CaCO_3$]	4	–
Chamosite [$(Fe^{2+},Mg)_5Al(AlSi_3O_{10})(OH)_8$]	4	–
Diaspore [$AlO(OH)$]	9	–
Gibbsite [$Al(OH)_3$]]	2	15
Sodalite [$Na_4(SiAl)_3O_{12}Cl$]	–	7
Nepheline [$Na_3KAl_4Si_4O_{16}$]	–	7
Albite [$NaAlSi_3O_3$]	–	4
Katoite [$Ca_3Al_2(SIO_4)_{1.5}(OH)_6$]	–	2
%OH-species	12.3	14.0

2.2. Characterization of the Studied Materials

The AMD sample was collected from Mpumalanga, South Africa. All sampling and laboratory analysis was performed in accordance with recognized global standards such as the International Standards Organization (ISO). After sampling and laboratory analysis in South Africa, all samples were sent to Germany. The AMD water was characterized using ICP-OES analysis (SPECTRO ARCOS, SPECTRO Analytical Instruments GmbH, Kleve, Germany) and solid samples by X-ray fluorescence (Axios FAST, Malvern Panalytical GmbH, Kassel, Germany). The AMD was first filtrated in order to remove the formed precipitate, but AMD was not acidified. The solid samples were ground up before the X-ray diffraction analysis (XRD) analysis. Bauxite residue, employed during AMD-treatment as the main raw material, was provided by Aluminum of Greece plant, Metallurgy Business Unit, Mytilineos S.A. (AoG). The sample was first homogenized by using laboratory sampling procedures (riffling method), and then a representative sample was dried in a static furnace at 105 °C for 24 h. Subsequently, the material was milled using a vibratory disc mill, and the sample was fully characterized.

Chemical analyses of major and minor elements were executed via the fusion method (1000 °C for 1 h with a mixture of $Li_2B_4O_7$/KNO_3 followed by direct dissolution in 10% HNO_3 solution) through a Perkin Elmer 2100 Atomic Absorption Spectrometer (AAS) (Perkin Elmer Inc., Waltham, MA, USA), a Spector Xepos Energy Dispersive X-ray Fluorescence Spectroscope (ED-XRF, SPECTRO, Kleve, Germany), a Thermo Fisher Scientific X-series 2 Inductively Coupled Plasma Mass Spectrometer (ICP-MS) (Perkin Elmer, Inc., Waltham, MA, USA), and a Perkin Elmer Optima 8000 Inductively Coupled Plasma Optical Emission Spectrometer (ICP-OES) (Perkin Elmer, Inc., Waltham, MA, USA). In contrast, the loss of ignition (LOI) of the sample was provided by differential thermal analysis (DTA), using a SETARAM TG Labys-DS-C (SETARAM Instrumentation, Caluire, France) system in the temperature range of 25–1000 °C with a 10 °C/min-heating rate, in air atmosphere.

Mineralogical phases were detected by XRD using a Bruker D8 Focus powder diffractometer with nickel-filtered CuKα radiation (λ = 1.5405 Å) coupled with XDB Powder Diffraction Phase Analytical System version 3.107, which evaluated the quantification of mineral phases via profile fitting specifically for bauxite ore and bauxite residues.

2.3. Methodology

As previously mentioned, two types of RM were tested in this study, one from aluminum plants in Greece and another from Germany. For the compliance test EN 124557-2, 90 g of dried RM was mixed with 900 ml of distilled water (L/S 10:1) using PTFE (polytetrafluorethylene) bottles. An end-over-end tumbler was utilized to agitate the samples for 24 h at 6–8 rpm. Leachate volumes are calculated according to Equation (1):

$$L = \left(10\left[\frac{l_{H_2O}}{kg}\right] - \frac{M_c}{100}\left[\frac{l_{H_2O}}{kg.\%}\right]\right) \times M_D \quad (1)$$

where:
- L is the volume of used leaching agent (in L);
- M_D is the dry mass of the test portion (in kg);
- M_c is the moisture content ratio (in %).

After 24 h, samples were filtered with a vacuum filtration device, and eluates obtained were post filtered using 0.45 µm membrane filters. Solutions obtained from the elution tests based on the standard EN 12457-2 were analyzed using a Metrohm ion chromatography IC 881 Compact IC Pro (METROHM, Riverview, FL, USA) and an inductively coupled plasma mass spectrometer Agilent 8800 ICP-MS Triple Quad and atomic fluorescence spectrometer for Hg (Analytic Jena GmbH, Jena, Germany). These chemical results in µg/L were used as a reference to compare the anions and cations in solution with AMD.

In order to check the compliance of RM with EU standardized limits, constituent leached from bauxite residues as a function of the original input material is calculated from the following Equation (2):

$$A = C \times \left[\frac{L}{M_D} + \frac{M_c}{100}\right] \quad (2)$$

where:
- A is the release of a constituent at L/S = 10 (in mg/kg of dry matter);
- C is the concentration of a particular constituent in the eluate (in mg/L);
- L is the volume of leachate used (in L);
- M_D is the dry mass of the test portion (in kg);
- M_c is the moisture content ratio (in %).

Mixing AMD/RM was carried out and analyzed using ratios of 90:10, 80:20, 70:30, and 60:40. Each experiment was repeated four times under the same conditions (24 h, 22 °C). Eluates obtained after filtering were compared with the concentration of anions and cations in AMD and eluates from RM under the standard EN 124557-2. Results were presented using the following notation:
- RM10%: RM 10 wt% + AMD 90 wt%
- RM20%: RM 20 wt% + AMD 80 wt%
- RM30%: RM 30 wt% + AMD 70 wt%
- RM40%: RM 40 wt% + AMD 60 wt%
- AMD: 100% solution of acid mining rain
- RM 1:10 H2O: EN 124557-2 standard elution test with distillate water with liquid/solid ratio 1:10.

After filtering, filter cakes were dried at 80 °C for 72 h and then comminuted below 90 µm employing a planetary ball mill (Fritsch PULVERISETTE 6) for chemical characterization. Chemical composition was analyzed using a PAN analytical WDXRF spectrometer (Malvern Panalytical GmbH, Kassel, Germany) with different internal programs. "Omnian" was used for the quantitative analysis of bulk compounds and "Pro-trace" was used for trace element analysis (lower ppm range). Solutions were also analyzed using IC and ICP-MS technics (Perkin Elmer, Inc., Waltham, MA, USA).

3. Results and Discussion

3.1. pH Influence

According to the elution test EN 12457-2 carried out on these two bauxite residues, these materials cannot be considered inert waste according to the EU landfill directive 1999/31/EC. Besides the leaching outside limits of As, Cr, F, and SO_4, RM is highly alkaline (see RM1:10H2O in Figure 1) and needs to be disposed of in special reservoirs.

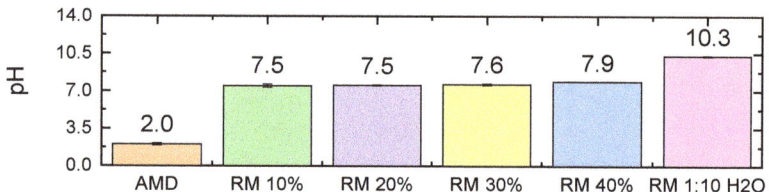

Figure 1. Influence of an addition of Greek RM at pH-value of acid mine drainage (AMD) in 24 h at room temperature.

From the mineralogical quantification carried out by Kaußen and Friedrich [13] on RM.De and Keller et al. [12] on RM.Gr, it can be highlighted that German RM contains 1.7% more hydroxide species than the Greek (see Table 1), explaining its higher pH-value, as shown at Figure 2.

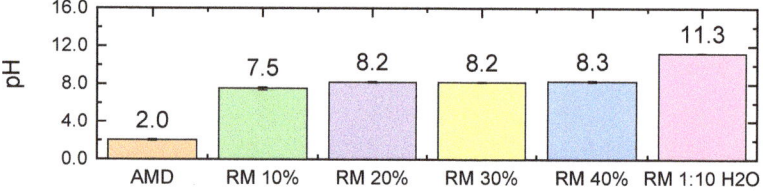

Figure 2. Influence of an addition of German RM at pH-Value of AMD in 24 h at room temperature.

A higher concentration of hydroxides increases RM's ability to react with H^+ cations allowing for improved precipitation of metal ions, many of which are considered hazardous to living organisms.

Previous experiments of Paradis et al. [15] showed that RM has a good neutralization capacity for a short time, but the long-term neutralization potential is uncertain. Thus, brine was added to RM to verify if it can improve long-term alkalinity retention of RM. Therefore, our experiments take time in 24 h but under continuous agitation, which for standard elution test such as EN 12457-2 is one of the worst-case scenarios for natural leaching. Neutralization of both RM and AMD is evidenced by a final pH above 8 and Fe and S precipitation in the mineral phase (see Table 4). The increase of RM over AMD is only preceded by a slight increase of final alkalinity but maintaining pH values between 7.5 and 8.

Table 4. Chemical analysis of compounds after elation tests.

wt%	SiO$_2$	Al$_2$O$_3$	CaO	Na$_2$O	TiO$_2$	Fe	S
RM.Gr	6.3	16.3	8.1	2.8	5.2	32.5	0.04
RM.Gr.10	6.2	13.9	10.6	2.6	5.0	34.1	3.39
RM.Gr.20	7.4	16.3	10.2	2.2	5.7	34.6	2.12
RM.Gr.30	7.6	16.8	10.0	2.7	5.6	35.5	1.34
RM.Gr.40	7.9	16.7	10.0	2.9	5.0	34.3	0.95
RM.De	12.3	16.1	6.4	9.1	10.8	20.7	0.02
RM.De.10	13.0	17.0	6.2	8.0	9.9	25.9	1.53
RM.De.20	15.3	19.4	6.4	7.7	10.4	21.5	0.70
RM.De.30	12.7	14.8	6.7	7.5	11.5	28.5	0.50
RM.De.40	11.7	12.7	6.9	7.0	12.3	31.4	0.42

Notation: RM.Gr.XX: mineral obtained after mixing XX wt% RM.Gr and (100-XX) wt% AMD.

3.2. Influence of RM-Addition on Anions Concentration

Figure 3 depicts different anion concentration from both Greek RM (figures on the left side) and German RM (figures on the right side). Addition of 10 wt% RM.Gr decreases the concentration of sulfate ions from 16.8 to 5.4 g/L, as shown in Figure 3a. RM.De shows the same behavior as the Greek, but with concentration falling around 10 g/L, so it was less effective to precipitate sulfate ions (Figure 3e). The leachate limit in terms of sulpfate anions becoming hazardous is 5.0 g/L (equivalent to 50,000 mg/kg according to Equation (2)) according to the EU landfill directive 1999/31/EC. An increase in the RM ratio from 10 to 40 % does not lead to a change of anion concentration.

Fluoride anions are present in both AMD and RMs (Figure 3d,h). Results showed a net positive synergistic effect on the AMD-RM mix, preventing these ions from being in the solution.

In contrast to sulfate and fluoride ions, chlorides anions increase in both RMs when the RM ratio increase from 0.1 to 0.4 (see Cl in Figure 3c,g. RM.Gr released two or three times more Cl anions than RM.De. According to XRD results, there is 0.07% Cl in the Greek RM against 0.2% in German RM. Even though there is less Cl available in the Greek RM than in the German, it releases more Cl anions suggesting that it might relate to its mineralogy. Cl in the German RM is concentrated on Sodalite (see Table 3), which is not in the Greek case.

Nitrate anions showed particular behaviors. For one side, its concentration increases with both RMs in contact with AMD compared to the content in AMD and its release from RM in contact with pure water. Perhaps the increase of these anions is related to air penetration in the eluates by the constant mechanical agitation and the reaction between the different compound present in RM with AMD. Gas dissolution in liquids is inversely proportional to an increase in temperature and ions in dissolution. Precipitating most of the ions when the pH is close to neutral values may explain the increase of NO$_3$ anions in both RMs. It was surprising that both RMs follow different trends, with an NO$_3$ concentration increase in RM.Gr raising the RM ratio from 0.1 to 0.4. The NO$_3$ concentration decreased using RM.De instead by using the same RM ratios. Maybe this comportment can be explained by the interaction of different microbial communities that present AMD with the different minerals present in both RMs [16–19].

Nevertheless, after combining different RM and AMD ratios, the concentration of fluoride, nitrate, and chloride ions in the obtained eluates was below the concentration limits for inert waste, 1 and 80 mg/L, respectively, according to Landfill Directive 1999/31/EC.

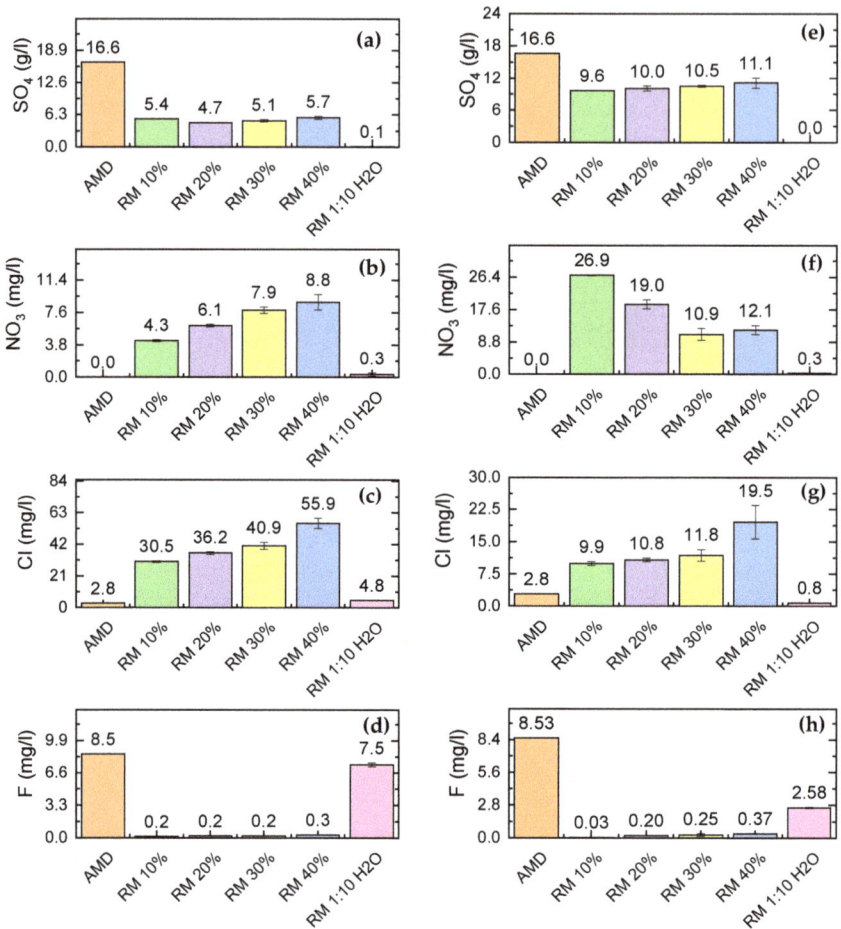

Figure 3. Anion concentration in AMD. RM.Gr (**left**) and RM.De (**right**), (**a,e**) SO_4, (**b,f**) NO_3, (**c,g**) Cl, and (**d,h**) F.

3.3. Influence of RM-Addition on Cation Concentration

Several metallic cations are considered hazardous due to their negative impacts on living organisms and ecosystems. Most of the national and international legislation undergo the content and leaching of these elements: As, Ba, Cd, Cr, Cu, Hg, Mo, Ni, Pb, Sb, Se, V, and Zn.

3.3.1. AMD:RM Negative Synergism

(a) Chromium

AMD has a Cr concentration of 34.6 µg/L; on the one hand, RM from Greece leaches out above 18.4 µg/L, and on the other hand, the German releases a concentration 30 times higher (555 µg/L).

The addition of Greek RM in a ratio of 0.1 to AMD favors the precipitation and immobilization of Cr in the mineral phase. For ratios higher than 0.3, this positive effect is only not neutralized but, on the contrary, enhanced to 271 µg/L. AMD in contact with RM.De depicts the same negative behavior, but Cr in the solution is ten times higher than RM.Gr, as shown at Figure 4.

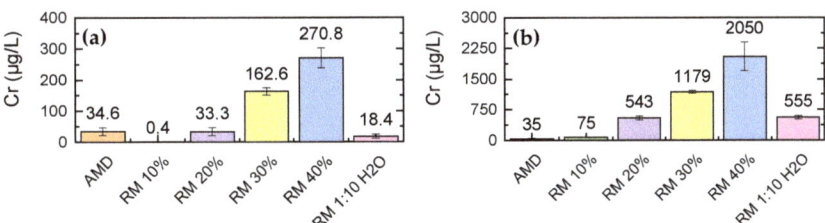

Figure 4. Effect on Cr ions in solution: (**a**) RM.Gr, and (**b**) RM.De.

(b) Barium

No leaching of Ba has been seen in both RMs under standard elution tests. Ba concentration in AMD stands at 1.9 µg/L, and despite low concentrations in both solutions (AMD and RM1:10H$_2$O), mingling both residues had a negative effect when the RM ratio debased 0.3 for Greek and 0.2 for German RM. For a ratio RM/AMD equal to 0.4, Ba in eluates reaches 51 and 77 times its AMD concentration regarding Greek and German RM, respectively (Figure 5).

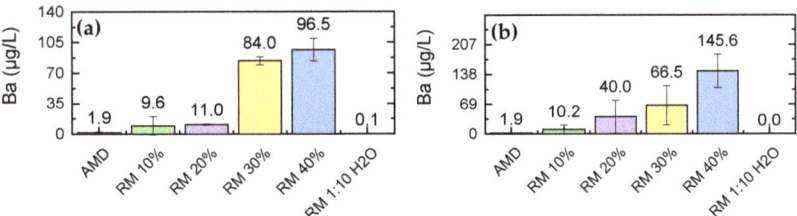

Figure 5. Effect on Ba ions in solution: (**a**) RM.Gr, and (**b**) RM.De.

(c) Antimony

As shown in Figure 6, Sb, mainly present in RM, has a concentration well below the limit established for inert waste (0.06 mg/kg or its equivalent in the eluate 600 µg/L). German RM (Figure 6b) tends to increase Sb, but concentrations are still low enough to be considered harmful.

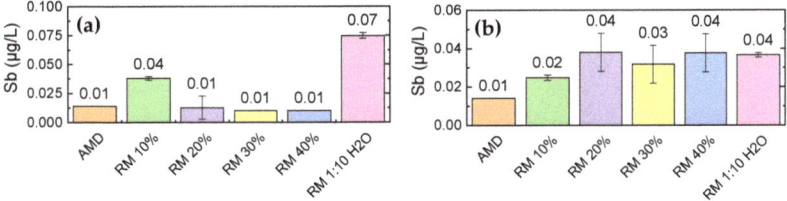

Figure 6. Effect on Sb ions in solution: (**a**) RM.Gr, and (**b**) RM.De.

(d) Selenium

Se in RM is leaching out 0.12 mg/kg (12 µg/L in solution) under standard conditions in RM.Gr (Figure 7a) and 0.04 mg/kg (4 µg/L in solution) in RM.De (Figure 7b). In both cases, concentrations are in agreement with the limit of inert waste (see Table 2). AMD enhance the release of Se in harmful concentrations, reaching a maximum of 1005 µg/L in RM.Gr and 82 µg/L in RM.De when the RM ratio is 40%.

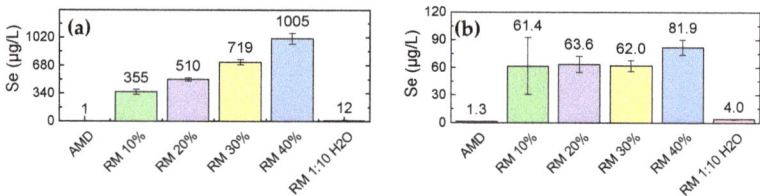

Figure 7. Effect on Se ions in solution: (**a**) RM.Gr, and (**b**) RM.De.

3.3.2. AMD:RM Positive Synergism: RM over AMD

Compared to standard leaching conditions, eluates obtained from the two analyzed RM showed Cd, Cu, Pb, and Zn cations within the inert waste limits. RM depicts positive effects on neutralizing these cations from AMD. Regardless, the utilized type of RM, a concentration of 10%, was enough to neutralize Cd, Pb, and Zn almost completely from AMD (see Figures 8–10). The only exception was Cd from RM.De, where a ratio of 0.2 was needed for complete precipitation (Figure 8a).

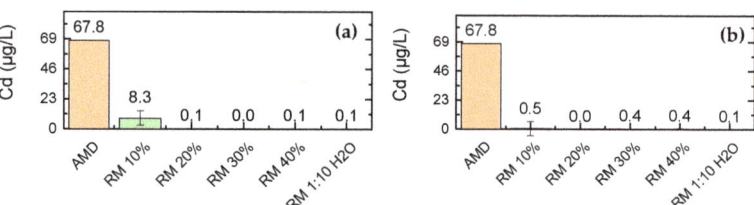

Figure 8. Effect on Cd ions in solution: (**a**) RM.Gr, and (**b**) RM.De.

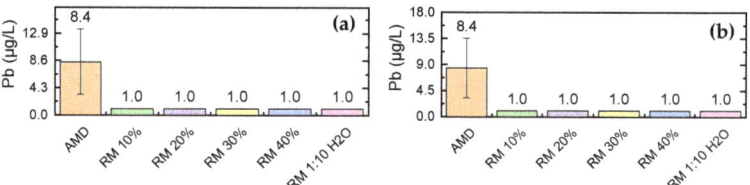

Figure 9. Effect on Pb ions in solution: (**a**) RM.Gr, and (**b**) RM.De.

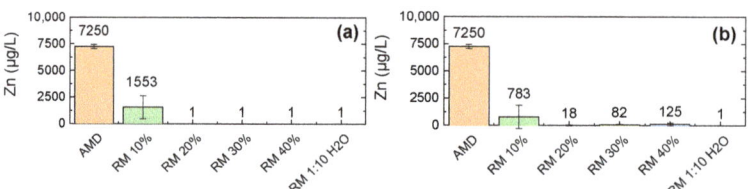

Figure 10. Effect on Zn ions in solution: (**a**) RM.Gr, and (**b**) RM.De.

Fe, Cr, Zn, Ni, and Cu are the cations with the higher concentration in AMD, and except Cr, RM is effective in precipitating them. However, perhaps due to its higher alkalinity, German RM is twice as effective as the Greek. This effect is shown in Figure 11 by Ni behavior, where the concentration decreases with increasing RM ratio but at a slower rate using Greek bauxite residues.

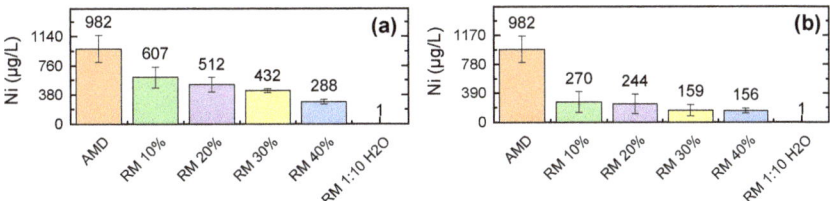

Figure 11. Effect on Ni ions in solution: (**a**) RM.Gr, and (**b**) RM.De.

An RM ratio of 0.1 showed to be adequate to precipitate Cu, but increasing the ratio beyond 0.1 decreases the RM efficiency to neutralize Cu cations from AMD (see Figure 12).

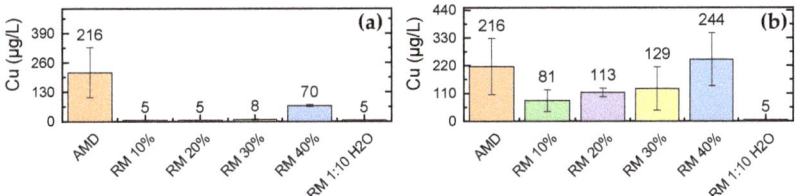

Figure 12. Effect on Cu ions in solution: (**a**) RM.Gr, and (**b**) RM.De.

For an RM ratio of 0.4, the eluate contains 244 µg/L in RM.De against 70 µg/L when Greek RM is utilized.

Besides its higher alkalinity, RM also contains elements outside the established limits to be considered an inert waste. Standard elution tests showed that As leaches out 3.9 mg/kg (389 µg/L) in RM.Gr and 0.07 mg/kg (6.7 µg/L) in RM.De (see Tables 1 and 2). According to these results, Greek RM releases As under the limits of non-hazardous waste and the German below the limits of inert waste (see limits in Table 2). AMD also contains As (13 µg/L) but in a concentration that cannot be considered harmful. In both cases, RM proved to be effective to immobilize As in the mineral phase (see Figure 13).

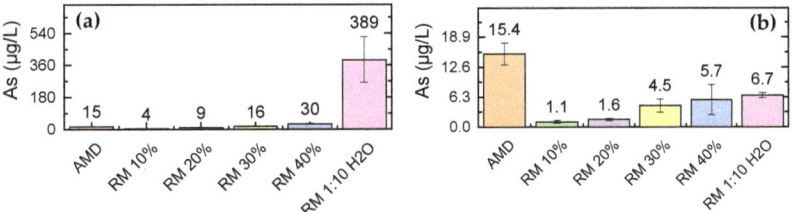

Figure 13. Effect on As ions in solution: (**a**) RM.Gr, and (**b**) RM.De.

Mo concentration in AMD is extremely low but is naturally leaching out from RM in contact with distilled water. Greek RM releases around 200 times more Mo than the German (Figure 14); nevertheless, both were underneath the limit of 0.5 mg/kg for inert waste (see Table 2).

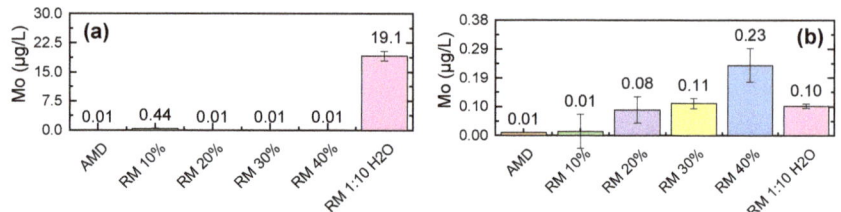

Figure 14. Effect on Mo ions in solution: (**a**) RM.Gr, and (**b**) RM.De.

For RM ratios higher than 0.1, AMD fully immobilizes Mo. In German RM, Mo's solubility increased with the RM ratio; nevertheless, the concentration was still below 50 µg/L to be considered outside the inert waste limits (see Table 2).

Vanadium (V) as a critical metal is not officially included in the list of chemical elements present in waste that have to be monitored to decide its classification and treatment. Nevertheless, studies have demonstrated associations between V and an increased risk of several pathologies like hypertension, dysrhythmia, and cancer, besides other health risks [20]. According to the standard EN 12457-2, V in RM leached out 2.6 and 0.2 mg/L in Greek and German RM, respectively. Di Carlo et al. [21] showed that the V release in RM is outside phytotoxic levels representing a risk to the environment. Like As behavior, AMD favors neutralization and immobilization of V in the RM mineral phase (as shown in Figure 15).

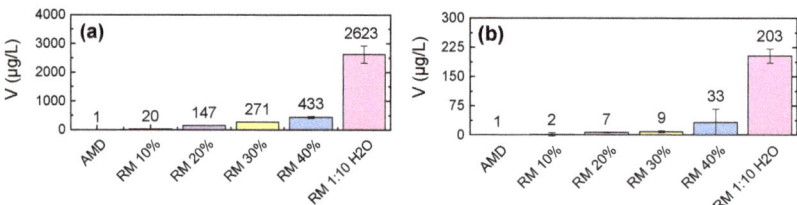

Figure 15. Effect on V ions in solution: (**a**) RM.Gr, and (**b**) RM.De.

3.4. Discussion of Concentration of Elements in the Mineral Phase

Fe and S were the main elements in AMD, and tend to precipitate in the mineral phase when pH increases by adding RM (see Table 4). Before mixing both residue types, Fe-oxide represented 46 wt% and 32 wt% in Greek and German RMs, respectively. Nevertheless, these concentrations rise to 50 wt% for the Greek and 31.4 wt% for the German when the RM ratio goes to 0.4. The starting S content in RM is 0.1% in Greek and 0.04% in German RM. These concentrations reach a maximum of 3.4% in RM.Gr and 1.5% in RM.De for an RM ratio of 0.1.

As shown in Figure 16, Ni is slightly enriched in RM when the higher ratio is utilized to neutralize AMD. These results are also consistent with Ni's decreasing profile in the eluates (as shown previously in Figure 12). Figure 16b shows a constant V concentration in the German RM and a slight decrease in the Greek for RM ratios 0.1 and 0.2. As shown in Figure 16, there is almost no V in AMD, and this element is immobilized in RM when AMD reacts with it. Therefore, the decrease of Ni is explained by the precipitation of other elements such as S and Fe.

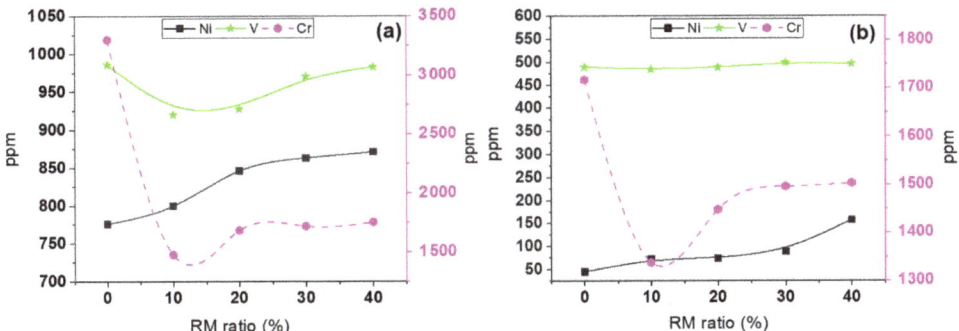

Figure 16. Concentration of Ni, V, Zn in the mineral phase: (**a**) Greek RM and (**b**) German RM.

Cr was one of the few exceptions in the synergistic approach of mixing AMD and RM. Cr content diminishes in both RMs (Figure 16) while increasing in the eluates (Figure 5). Besides the negative environmental impact of releasing Cr in the eluates, the selectivity with which this element is released while most of the other components precipitate allows us to imagine a method to recover Cr from RM.

Figure 17 shows the impact of mixing AMD with RM for several strategic and valuable metals such as Y, La, Ce, Nd, and Nb. For Y, La, and Nb, no significant change has been seen in their concentration when the RM ratio increases. Nd has a smooth decline compared with the original content in RM. On the other hand, Ce shows a marked increase in the mineral phase of both RMs, suggesting that this element was present in AMD.

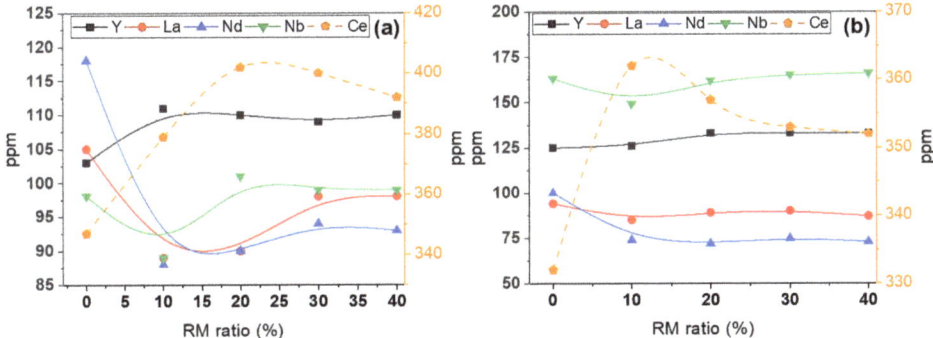

Figure 17. Concentration of Y, La, Nd, Nb, and Ce in the mineral phase: (**a**) Greek RM and (**b**) German RM.

4. Conclusions

The following conclusions are found in this work:

- AMD has benefited from RM's alkaline character, favoring a complete immobilization of fluoride ions and a substantial reduction of sulfates. Nevertheless, the decrease of sulfates was not enough to reach a concentration of 5 g/L, which is the limit for not being considered hazardous according to the EU Landfill Directive 1999/31/EC.
- NO_3 concentration by using Greek or German RM to neutralize AMD showed different trends, and these results should be deeply studied, especially the interaction of AMD microbial communities with RM.
- There is a net positive synergism on mixing an alkaline waste such as RM with AMD from an environmental perspective. RM over AMD was efficient to neutralize Cd, Pb,

- Ni, Cu, and Zn. On the other hand, AMD was effective to immobilize As, Mo, and V from RM.
- Regarding Cr, Ba, Se, and Sb, the mix offers selective separation despite a negative increment of these ions in the eluates. Nevertheless, Ba and Sb's increments were low compared to legally established limits to consider waste as harmful. The selective dissolution of Cr and the immobilization of most of the metallic ions in the mineral phase can be a strategy to explore in order to recover this element.
- Valuable elements present in RM such as La, Nd, and Nb tend to remain in the mineral phase. Elements such as Ce and Y present in AMD precipitates under the effect of RM enriching the mineral phase. Several authors had explored a pyrometallurgical treatment of RM to recover pig iron and enhance the content of critical raw material (CRM) in the final slag [22–24]. This approach can be beneficial to increase both pig iron and CRM from the filter cakes produce after coagulating AMD ions into an RM matrix.
- Except for an increase of Cr in solution, both RM showed a positive net effect in decreasing and immobilizing the primary metal ions considered hazardous to human life and ecosystems. M. Cvijovic et al. [25] proposed improved chemical treatment of surface water and sludge application as a compost. As shown in their work, the metals with content over the maximum limit (mg/kg), 169 Ni, 69 Cr, and 5.7 Pb, can be reduced by zeolite, which is maybe a solution for removal of Cr from our solution.

Author Contributions: Conceptualization, H.L. and S.S.; funding acquisition, S.N. and B.F.; investigation, H.L.; methodology, H.L., S.S. and B.X.; supervision, S.N. and B.F.; Writing—original draft, H.L. and S.S. All authors have read and agreed to the published version of the manuscript.

Funding: This research and Article Processing Charges (APC) were funded by the International Office of the BMBF in Germany, AddWater Project (Grant number. 01DG17024).

Data Availability Statement: Not applicable.

Acknowledgments: We would like to thank the International Office of the BMBF in Germany for the financial support in the AddWater Project: (No. 01DG17024) in cooperation with our colleagues from the University of Witwatersrand, Johannesburg, South Africa, funded through the NRF in South Africa (Ref: GERM160705176077).

Conflicts of Interest: The authors declare no conflict of interest.

References

1. WHG. *Federal Water Act*; German Federal Ministry of Justice and Consumer Protection: Berlin, Germany, 2009; Chapter 32.b.
2. Knight, J.; Rogerson, C.M. *The Geography of South Africa*, 1st ed.; Springer International Publishing: Cham, Germany, 2019; Chapter 4; pp. 27–31.
3. McCarthy, T.S. The impact of acid mine drainage in South Africa. *S. Afr. J. Sci.* **2011**, *107*, 5–6. [CrossRef]
4. Tabelin, C.B.; Igarashi, T.; Tabelin, M.V.; Park, I.; Opiso, E.M.; Ito, M.; Hiroyoshi, N. Arsenic, selenium, boron, lead, cadmium, copper and zinc in naturally contaminated rocks: A review of their sources, modes of enrichment, mechanisms of release and mitigation strategies. *Sci. Total Environ.* **2018**, *645*, 1522–1553. [CrossRef]
5. Park, I.; Tabelin, C.B.; Jeon, S.; Li, X.; Seno, K.; Ito, M.; Hiroyoshi, N. A review of recent strategies for acid mine drainage prevention and mine tailings recycling. *Chemosphere* **2019**, *219*, 588–606. [CrossRef]
6. Igarashi, T.; Herrera, P.S.; Uchiyama, H.; Miyamae, H.; Iyatomi, N.; Hashimoto, K.; Tabelin, C.B. The two-step neutralization ferrite-formation process for sustainable acid mine drainage treatment: Removal of copper, zinc and arsenic and the influence of coexisting ions on ferritization. *Sci. Total Environ.* **2020**, *715*, 136877. [CrossRef]
7. Bwapwa, J.K. A Review of Acid Mine Drainage in a Water-Scarce Country: Case of South Africa. *Environ. Manag. Sustain. Dev.* **2017**, *7*, 1. [CrossRef]
8. Singer, P.C.; Stumm, W. Acidic Mine Drainage: The Rate-Determining Step. *Science* **1970**, *167*, 1121–1123. [CrossRef]
9. Plumlee, G.S.; Smith, K.S.; Montour, M.R.; Ficklin, W.H.; Mosier, E.L. Geologic Controls on the Composition of Natural Waters and Mine Waters Draining Diverse Mineral-Deposit Types. In *The Environmental Geochemistry of Mineral Deposits*; Reviews in Economic Geology; Society of Economic Geologists: Littleton, CO, USA, 1999; Volume 6, Chapter 19; pp. 373–435.
10. Lim, J.; Yu, J.; Wang, L.; Jeong, Y.; Shin, J.H. Heavy Metal Contamination Index Using Spectral Variables for White Precipitates Induced by Acid Mine Drainage: A Case Study of Soro Creek, South Korea. *IEEE Trans. Geosci. Remote Sens.* **2019**, *57*, 4870–4888. [CrossRef]

11. Mwewa, B.; Stopic, S.; Ndlovu, S.; Simate, G.S.; Xakalashe, B.; Friedrich, B. Synthesis of Poly-Alumino-Ferric Sulphate Coagulant from Acid Mine Drainage by Precipitation. *Metals* **2019**, *9*, 1166. [CrossRef]
12. Keller, V.; Stopic, S.; Xakalashe, B.; Ma, Y.; Ndlovu, S.; Mwewa, B.; Simate, G.; Friedrich, B. Effectiveness of Fly Ash and Red Mud as Strategies for Sustainable Acid Mine Drainage Management. *Minerals* **2020**, *10*, 707. [CrossRef]
13. Kaussen, F.; Friedrich, B. Phase characterization and thermochemical simulation of (landfilled) bauxite residue ("red mud") in different alkaline processes optimized for aluminum recovery. *Hydrometallurgy* **2018**, *176*, 49–61. [CrossRef]
14. Alkan, G.; Diaz, F.; Gronen, L.; Stopic, S.; Friedrich, B. A mineralogical assessment on bauxite residue (red mud) after acidic leaching for titanium recovery. *Metals* **2017**, *7*, 458. [CrossRef]
15. Paradis, M.; Duchesne, J.; Lamontagne, A.; Isabel, D. Using red mud bauxite for the neutralization of acid mine tailings: A column leaching test. *Can. Geotech. J.* **2006**, *43*, 1167–1179. [CrossRef]
16. Dai, Z.; Guo, X.; Yin, H.; Liang, Y.; Cong, J.; Liu, X. Identification of Nitrogen-Fixing Genes and Gene Clusters from Metagenomic Library of Acid Mine Drainage. *PLoS ONE* **2014**, *9*, e87976. [CrossRef]
17. Kaksonen, A.H.; Puhakka, J.A. Sulfate Reduction Based Bioprocesses for the Treatment of Acid Mine Drainage and the Recovery of Metals. *Eng. Life Sci.* **2007**, *7*, 541–564. [CrossRef]
18. Sun, W.; Xiao, E.; Krumins, V.; Dong, Y.; Li, B.; Deng, J.; Wang, Q.; Xiao, T.; Liu, J. Comparative Analyses of the Microbial Communities Inhabiting Coal Mining Waste Dump and an Adjacent Acid Mine Drainage Creek. *Microb. Ecol.* **2019**, *78*, 651–664. [CrossRef]
19. Oshiki, M.; Ishii, S.; Yoshida, K.; Fujii, N.; Ishiguro, M.; Satoh, H.; Okabe, S. Nitrate-Dependent Ferrous Iron Oxidation by Anaerobic Ammonium Oxidation (Anammox) Bacteria. *Appl. Environ. Microbiol.* **2013**, *79*, 4087–4093. [CrossRef] [PubMed]
20. Gummow, B. Vanadium: Environmental Pollution and Health Effects. In *Encyclopedia of Environmental Health*; Elsevier: Amsterdam, The Netherlands, 2011; pp. 628–636. [CrossRef]
21. Di Carlo, E.; Boullemant, A.; Courtney, R. Ecotoxicological risk assessment of revegetated bauxite residue: Implications for future rehabilitation programmes. *Sci. Total Environ.* **2020**, *698*, 134344. [CrossRef]
22. Alkan, G.; Yagmurlu, B.; Gronen, L.; Dittrich, C.; Ma, Y.; Stopic, S.; Friedrich, B. Selective silica gel free scandium extraction from Iron-depleted red mud slags by dry digestion. *Hydrometallurgy* **2019**, *185*, 266–272. [CrossRef]
23. Borra, C.R.; Blanpain, B.; Pontikes, Y.; Binnemans, K.; Van Gerven, T. Recovery of Rare Earths and Other Valuable Metals From Bauxite Residue (Red Mud): A Review. *J. Sustain. Met.* **2016**, *2*, 365–386. [CrossRef]
24. Valeev, D.; Zinoveev, D.; Kondratiev, A.; Lubyanoi, D.; Pankratov, D. Reductive Smelting of Neutralized Red Mud for Iron Recovery and Produced Pig Iron for Heat-Resistant Castings. *Metals* **2019**, *10*, 32. [CrossRef]
25. Cvijović, M.; Murić, M.; Čudić, V. Improved chemical treatment of Sušica surface water, Zlatibor area, and sludge application. *Vojn. Glas.* **2020**, *68*, 293–320. [CrossRef]

Article

Basic Sulfate Precipitation of Zirconium from Sulfuric Acid Leach Solution

Yiqian Ma [1,2], Srecko Stopic [1,*], Xuewen Wang [3], Kerstin Forsberg [2] and Bernd Friedrich [1]

[1] Institute of Process Metallurgy and Metal Recycling (IME), RWTH Aachen University, Intzestraße 3, 52056 Aachen, Germany; yiqianm@kth.se (Y.M.); bfriedrich@ime-aachen.de (B.F.)
[2] Department of Chemical Engineering, KTH Royal Institute of Technology, Teknikringen 42, 11428 Stockholm, Sweden; kerstino@kth.se
[3] School of Metallurgy and Environment, Central South University, Lushan South Road 932, Changsha 410083, China; wxwcsu@163.com
* Correspondence: sstopic@ime-aachen.de; Tel.: +49-2419-5860

Received: 29 July 2020; Accepted: 12 August 2020; Published: 13 August 2020

Abstract: H_2SO_4 was ensured to be the best candidate for Zr leaching from the eudialyte. The resulting sulfuric leach solution consisted of Zr(IV), Nb(V), Hf(IV), Al(III), and Fe(III). It was found that ordinary metal hydroxide precipitation was not feasible for obtaining a relatively pure product due to the co-precipitation of Al(III) and Fe(III). In this reported study, a basic zirconium sulfate precipitation method was investigated to recover Zr from a sulfuric acid leach solution of a eudialyte residue after rare earth elements extraction. Nb precipitated preferentially by adjusting the pH of the solution to around 1.0. After partial removal of SO_4^{2-} by adding 120 g of $CaCl_2$ per 1L solution, a basic zirconium sulfate precipitate was obtained by adjusting the pH to ~1.6 and maintaining the solution at 75 °C for 60 min. Under the optimum conditions, the loss of Zr during the SO_4^{2-} removal step was only 0.11%, and the yield in the basic zirconium sulfate precipitation step was 96.18%. The precipitate contained 33.77% Zr and 0.59% Hf with low concentrations of Fe and Al. It was found that a high-quality product of ZrO_2 could be obtained from the basic sulfate precipitate.

Keywords: zirconium; eudialyte; hydrometallurgy; basic sulfate precipitation

---

1. Introduction

Zirconium metal is widely used in the areas of atomic energy, metallurgy, the military, petrochemicals, aerospace, new materials science as well as in medicine [1]. Zirconium is also an important alloying element in steel industry [2]. The major source for zirconium production is zircon ($ZrSiO_4$), which is highly stable, requiring high temperature sintering for decomposition [3]. Eudialyte is a complex Na-Ca-zirconosilicate mineral that can be a potential commercial source of zirconium [4,5]. The content of Zr (5–10%) in eudialyte is much lower than in zircon, but it can be easily decomposed using acid [6–8].

The typical empirical chemical formula for eudialyte is $Na_4(Ca, Ce, Fe)_2ZrSi_6O_{17}(OH, Cl)_2$, but it displays a wide range of chemical compositions [6]. Eudialyte is also rich in Fe, Al, Mn, Ti, K, Nb and contains significant quantities of rare earth elements (REE) [4,6,9]. As such, the comprehensive extraction of the valuable metals must be considered for practical eudialyte processing [6,10]. REE extraction from eudialyte sometimes takes priority, since these elements are more valuable [11].

To date, acid decomposition of eudialyte has been extensively studied [7,8,12–14]. Sulfuric acid (H_2SO_4) is considered to be the best candidate for Zr leaching from the eudialyte or eudialyte residue after REE extraction. However, the leaching process is not selective, so many impurities remain in the sulfuric acid leach solution; such as sodium, iron, and aluminum [6,8,15,16]. Data concerning the recovery of Zr from the sulfate media containing these impurities are scarce in the general literature.

A complete treatment process to produce a Zr product can only be found in a few feasibility studies [17]. Lebedev et al. studied the ordinary precipitation by adding Na_2CO_3, but the resulting Zr carbonate was found to be a mixture of Zr(IV), Hf(IV), Nb(V), Fe(III), and Al(III) carbonates [6]. Further separation and purification are indispensable for the production of Nb, Hf, or Zr products. Tasman Metal Ltd. has studied the exploitation and utilization of the eudialyte from Norra Kärr, Sweden. Solvent extraction followed with precipitation was used to recover Zr from the sulfate media [18]. However, the details of the solvent and process parameters were not reported. Ion exchange can also be employed to recover Zr in eudialyte processing, but the process has a lot of operations and is very complicated [16].

On the other hand, Hf is in the same period with Zr in the periodic table, so the chemical properties of these two elements are nearly identical due to the lanthanide contraction [19]. Hf is usually present in trace quantities in minerals that contain Zr [20]. The Hf concentration in eudialyte is about 0.2%, and it follows with Zr during the treatment of eudialyte [6,11,16]. Commercial-grade Zr product containing Hf can be used in chemical process industries, but for use as a cladding material, the Zr product must be Hf-free due to their varied neutron-absorbing properties [21].

This reported study focused on the extraction process following acid leaching. Based on the solution chemistry, a basic zirconium sulfate precipitation was investigated to recover Zr from the sulfuric acid leach solution of a eudialyte residue, anticipating the production of a precipitate with a low impurity level. Furthermore, preparation of ZrO_2 from the basic sulfate precipitate was also attempted.

2. Materials and Methods

2.1. Materials and Analysis

As a resource for REE in the EURARE project, which was funded by the European Commission for the development of a sustainable exploitation scheme for Europe's rare earth ore deposits, eudialyte ore was mined in South Greenland, and after beneficiation, the eudialyte concentrate was the initial material for REE extraction. After REE extraction, the eudialyte residue was known to contain attractive quantities of Zr. H_2SO_4 was used to leach Zr from the eudialyte residue, and a detailed description of leaching process has been reported [10] by our group. The composition of the resulting sulfuric acid leach solution is listed in Table 1. Among the metal ions, iron ions existed only in the form of Fe(III), because H_2O_2 was added in the leaching process. This study focused on the subsequent step of Zr recovery from the sulfate leachate. As can be seen from the data in Table 1, the concentration of SO_4^{2-} in the leach solution was high and the main impurities in the leach solution were found to be Al and Fe. The $CaCl_2$, Na_2CO_3, and HCl used in this work were of analytical grade, and all aqueous solutions were prepared using distilled water. The elemental content of the solution was determined by inductively coupled plasma emission spectroscopy (ICP) using a PS-6 PLASMA SPECTROVAC (Baird, Waltham, MA, USA). The composition of the precipitate was determined by X-ray fluorescence (AXIOS, PANalytical, Eindhoven, The Netherlands), and Zr in the basic sulfate precipitate was analyzed from solution after it was solubilized. The content of the basic sulfate precipitate was also characterized using X-ray diffraction (XRD, PANalytical X'PERT-PRO diffractometer, Malvern PANalytical, Eindhoven, The Netherlands) and Fourier-transform infrared spectroscopy (FT-IR, Spectrum 100, Perkin Elmer, Inc., Waltham, MA, USA). The prepared ZrO_2 was characterized by XRD and the composition of ZrO_2 was evaluated using ICP analysis. The data of chemical compositions and concentrations in this paper are the averages of 3 analysis results.

Table 1. Chemical composition of experimental solutions, g/L.

Element	Zr	Hf	Fe	Al	Nb	Si	Ca	SO_4^{2-}	pH
Acid leach solution	10.95	0.26	2.44	10.55	0.48	<0.001	0.7	117.50	0.57
Solution after precipitation of Nb	10.83	0.26	2.44	10.55	0.10	<0.001	0.6	116.12	0.94

2.2. Thermodynamic Analysis and Methods

Metal precipitation can be defined as a process where metal ions are reacted with other compounds to form a low solubility product. Metal hydroxide precipitation is the most common example of this [22]. The log [Me^{n+}]-pH diagram based on the solubility product constants of metals hydroxide at 298.15 K is shown in Figure 1, where Nb^{5+} in the solution is readily hydrolyzed at low pH, and the pH ranges for the hydrolysis of Zr^{4+} and Hf^{4+} are between the pH ranges for hydrolysis of Fe^{3+} and Al^{3+}. However, some anions can act as ligands in the solution, which was ignored in this representation. In the absence of SO$_4^{2-}$, the hydrolysis of Zr^{4+} can be initiated even in high acidity. If complexation in the aqueous solution is considered, the pH ranges for precipitation will increase.

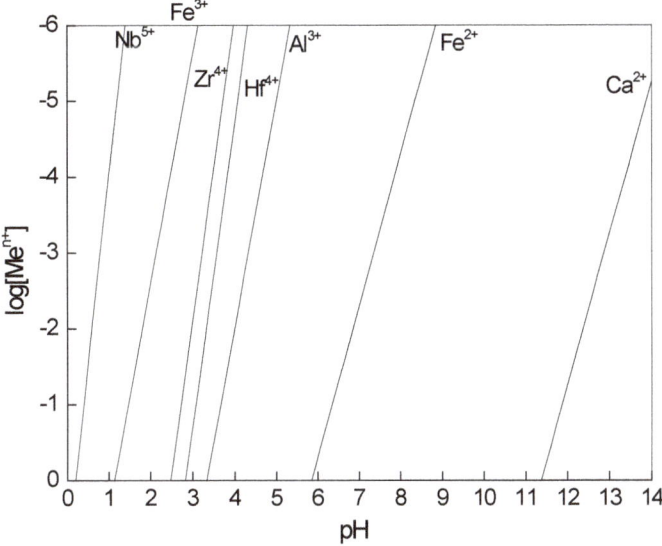

Figure 1. log[Me^{n+}]-pH diagram based on the solubility product constants (K$_{sp}$) of metals hydroxide at 298.15 K (K$_{sp}$ data obtained from the literature [22]).

For the test solution in this study, a complex reaction with SO$_4^{2-}$ inhibits the hydrolysis of Zr^{4+} and Hf^{4+} when the pH is increased, because the use of SO$_4^{2-}$ produces competition between SO$_4^{2-}$ and OH$^-$ ions to react with Zr^{4+} and Hf^{4+}. The complex reaction of Zr^{4+} and Hf^{4+} in the acidic sulfate solution can be represented by Equation (1), and the corresponding stability constants (β_i) are listed in Table 2 [23]. In other words, the formation of complexes with SO$_4^{2-}$ can expand the stable region of the dissolved Zr and Hf. In addition, NbO(SO$_4$)$_2^-$ is also exist in the acidic sulfate solution, and the relevant equation is represented by Equation (3) [24]. While sulfate ions have little effect on other metal ions in the sulfuric acid leach solution.

$$\mathrm{Me^{4+}_{(aq)} + i\, HSO_4^- = Me(SO_4)_i^{4-2i}{}_{(aq)} + i\, H^+} \tag{1}$$

$$\beta_i = \frac{\left[\mathrm{M(SO_4)_i^{4-2i}}\right]}{\left[\mathrm{M^{4+}}\right]\left[\mathrm{SO_4^{2-}}\right]^i} \tag{2}$$

$$\mathrm{Nb(OH)_5 + 3H^+ + 2SO_4^{2-} = NbO(SO_4)_2^- + 4H_2O} \ (\log K = 7.24) \tag{3}$$

Table 2. Stability constants of sulphate complexes of zirconium and hafnium.

Stability Constant	Zr	Hf
β_1	466	130
β_2	3.48×10^3	2.1×10^3
β_3	3.92×10^6	3.02×10^6

Figure 2 shows the fraction of ions of Nb and Zr in the sulfate aqueous solution at different pH values. As can be seen, the pH for Nb(V) precipitation via hydrolysis was as low as the pH in the absence of SO_4^{2-}. However the predominant forms of Zr and Hf in the acidic sulfate solution with pH > 0 are $Zr(SO_4)_3^{2-}$ and $Hf(SO_4)_3^{2-}$ and the pH range for hydrolysis increased. These conditions are consistent with those reported in the literature. The selective recovery of Zr(IV) from sulfuric acid leach solution via ordinary neutralization is not feasible due to the co-precipitation of Fe(III) and Al(III) [6].

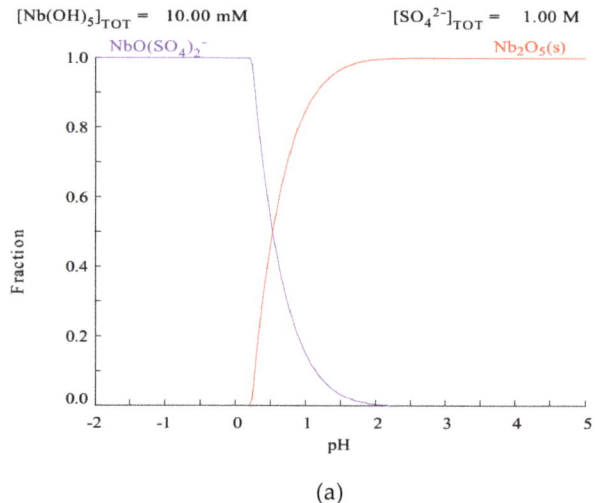

(a)

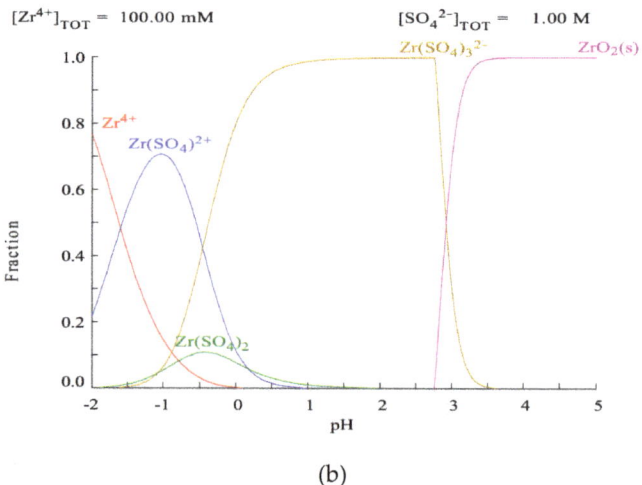

(b)

Figure 2. Cont.

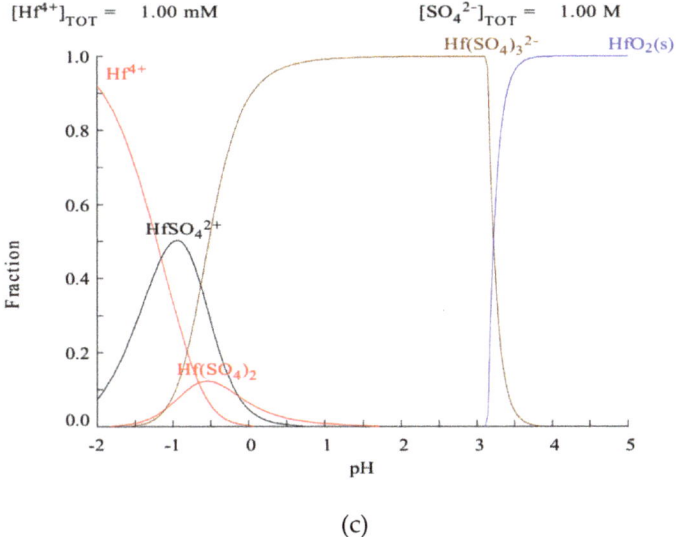

(c)

Figure 2. Predicted fraction diagrams of (**a**) Nb, (**b**) Zr, and (**c**) Hf in aqueous solutions using MEDUSA software (Make Equilibrium Diagrams Using Sophisticated Algorithms, 32 bit version 2010, Royal Institute of Technology, Stockholm, Sweden).

Precipitation using basic zirconium sulfate ($Zr_5O_8(SO_4)_2 \cdot xH_2O$) in place of $Zr(OH)_4$ is another alternative for producing zirconate. According to the literature [25,26], basic zirconium sulfate can be obtained when zirconium chloride solution is mixed with sulfuric acid solution at a set temperature (60–90 °C) and pH (1.2–2) with a zirconium to sulfate ratio of 5:2. This reaction can be expressed by Equation (4).

$$5Zr^{4+} + 2HSO_4^- + (8 + x)H_2O \rightarrow Zr_5O_8(SO_4)_2 \cdot xH_2O\ (s) + 18H^+ \quad (4)$$

Since this reaction occurs at a pH < 2.0, co-precipitation of Fe(III) and Al(III) will be low.

A flow chart for the recovery of Zr using the basic sulfate precipitation method is shown in Figure 3. As shown, first, Nb is preferentially precipitated by neutralization. After separating the Nb from the solution, $CaCl_2$ is added to remove some SO_4^{2-}, and then zirconium is selectively precipitated using basic sulfate zirconium at low pH without the co-precipitation of Fe(III) and Al(III). A small amount of Hf in the solution would follow Zr in the precipitation process, since it has chemical properties that are very similar to Zr. In the process, Na_2CO_3 and $CaCl_2$ was added in solid form little by little.

In order to prepare a Zr product, the next stage of the process can be connected to the conventional technology in the Zr production from zircon [27]. The prepared basic zirconium sulfate was dissolved with 4 mol/L HCl. Evaporative crystallization was then used to prepare $ZrOCl_2 \cdot 8H_2O$. After obtaining the $ZrOCl_2 \cdot 8H_2O$, the ZrO_2 can be obtained by calcination at 600 °C in a muffle furnace (Thermo Scientific™ M110) for 2 h.

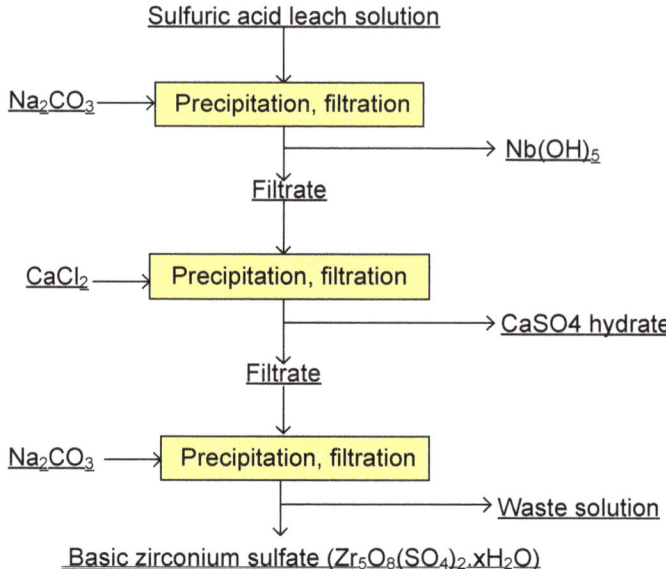

Figure 3. Proposed flowchart for recovery of Zr using precipitation methods.

3. Results and Discussion

3.1. Effect of pH on the Precipitation of Metal Ions by Neutralization

Figure 4 shows the effect of pH on the precipitation of the main elements in the leach solution. Based on the aqueous solution chemistry analysis and precipitation behavior, Nb, Zr, Hf, Al, and Fe were precipitated from the solution by hydrolysis at a suitable pH. The quantity of dissolved Ca^{2+} in the sulfate solution was very low, and it also partially precipitated in the form of $CaSO_4 \cdot 2H_2O$ when the pH was increased. Zr and Hf precipitated until the pH was increased to 3.5, but Fe and Al precipitated simultaneously, so that it was not feasible to selectively recover Zr and Hf via the ordinary precipitation method from the sulfuric acid leach solution. Figure 5 shows a precipitate in a difficult state of filtration at pH 4.0.

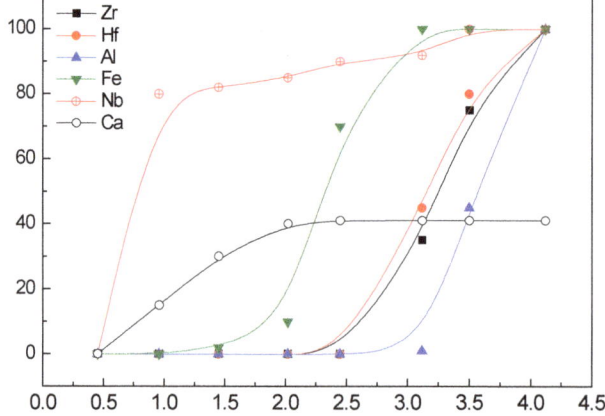

Figure 4. Effect of pH on the precipitation of elements from the leach solution.

Figure 5. The precipitate in a difficult state of filtration at pH 4.0 after adding Na_2CO_3.

The precipitation method for the selective recovery of Zr, Hf, and Nb included precipitation of Nb by hydrolysis and precipitation of Zr and Hf using basic zirconium sulfate. The optimum pH for the precipitation of Nb was 1.0 based on the results shown in Figure 4. Table 3 shows the composition of the resulting precipitate. As can be seen, the content of Nb was 40.1%, and the concentration of other impurities was low, but a further purification was necessary to obtain the Nb product. After Nb precipitation, Zr and Hf were recovered using basic zirconium sulfate precipitation or ion exchange.

Table 3. Composition of Nb precipitate.

Element	Nb	Zr	Hf	Al	Fe
Content (wt%)	40.1	1.1	<0.1	3.7	5.7

3.2. Effect of the Quantity of $CaCl_2$ on the Basic Sulfate Precipitation

In order to control a suitable zirconium to sulfate ratio for the Zr precipitation process, an amount of $CaCl_2$ was added to remove a portion of the sulfate ions from the Zr-bearing solution after the Nb recovery. As can be seen in Figure 6, the Zr precipitate yield increased with the addition of $CaCl_2$ from 100 to 120 g per 1 L solution, and further increases in the amount of $CaCl_2$ resulted in a decrease in the yield of Zr precipitation. The Hf precipitation yield exhibited the same trend, but the precipitate yields of Fe(III) and Al(III) were always very low. Hence, 120 g/L of $CaCl_2$ was chosen as the optimum quantity for removing the sulfate ions. After calculation, the concentration of sulfate in the solution was decreased from 112.0 g/L to 12.4 g/L, and the suitable molar ratio of zirconium to sulfate was about 2.5:1.

3.3. Effect of pH on the Basic Sulfate Precipiation

During the basic zirconium sulfate precipitation process, Na_2CO_3 was used to adjust the solution pH, and the effect of varying the final pH was also examined. As shown in Figure 7, the Zr precipitation yield increased from 62.1% to 96.1% when the final pH was increased from 1.2 to 1.6, and above 1.6, the precipitation yield of Fe(III) and Al(III) noticeably increased. Thus, the optimum final pH for the basic zirconium precipitation process was 1.6.

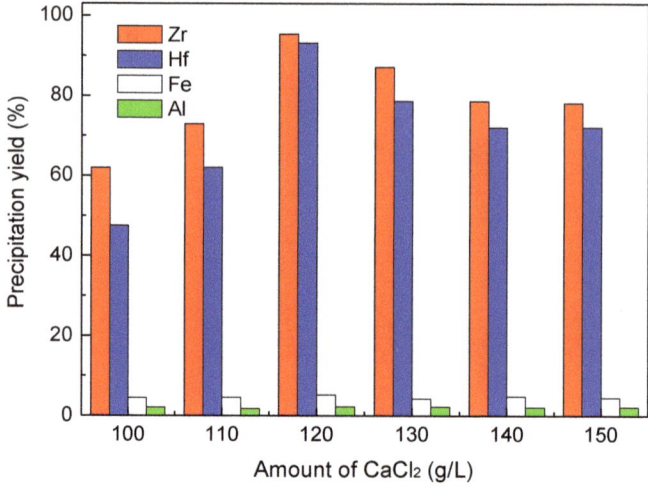

Figure 6. Effect of CaCl$_2$ quantity on the Zr precipitation yield, Zr precipitation conditions: 75 °C, final pH ~1.6, 60 min.

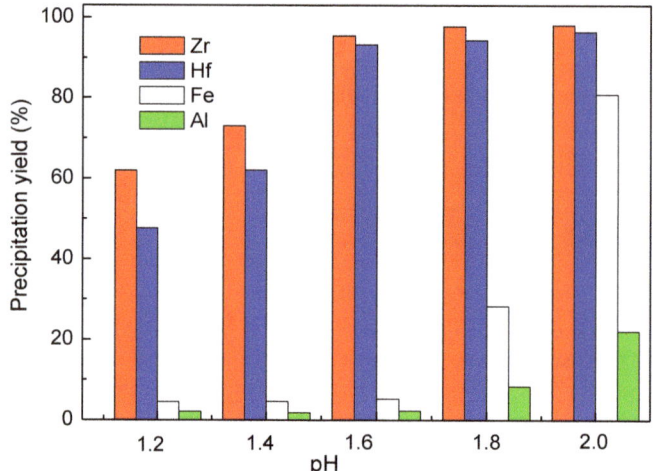

Figure 7. Effect of pH on the Zr precipitation yield, Zr precipitation conditions: 75 °C, C(SO$_4^{2-}$) = 12.4 g/L, 60 min.

3.4. Effect of Temperature on the Basic Sulfate Precipiation

The effect of temperature on the precipitation yields of metals was also determined experimentally. It can be seen from the results in Figure 8 that the precipitation yields of Zr and Hf both increased with the increase in temperature, and 75 °C was the best for the recovery of Zr and Hf. Most of Fe and Al could be kept in the solution by controlling the pH at ~1.6.

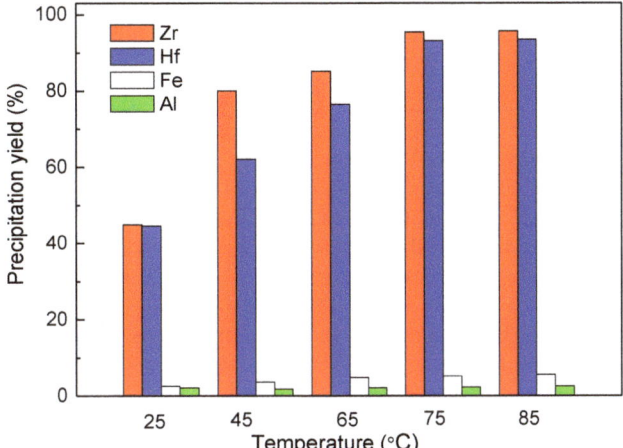

Figure 8. Effect of temperature on the Zr precipitation yield, Zr precipitation conditions: final pH ~1.6, $C(SO_4^{2-})$ = 12.4 g/L, 60 min.

3.5. Effect of Time on the Basic Sulfate Precipiation

The experimental results shown in Figure 9 indicate that the precipitation process was very fast. The precipitation yields of Zr and Hf reached 96.1% and 91.5% when the precipitation time was 60 min. No further increase in yield was noted with longer reaction time. Thus, 60 min was considered to be the optimum reaction time for the basic zirconium precipitation here. It may be considered to shorten the residence time when this method is applied in industry, because the solution after precipitation with a small amount of Zr could be returned to the previous process.

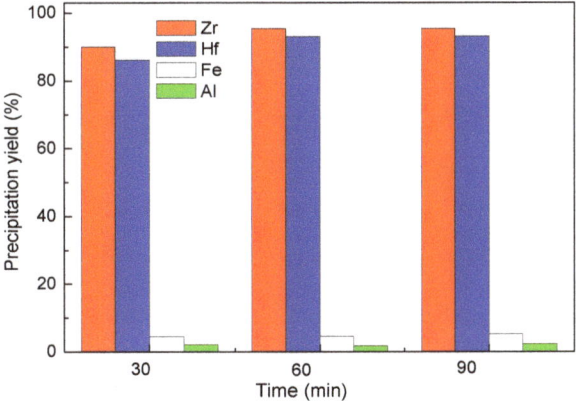

Figure 9. Effect of time on Zr precipitation yield, Zr precipitation conditions: final pH ~1.6, 75 °C, $C(SO_4^{2-})$ = 12.4 g/L.

3.6. Characterization of Precipitates

Table 4 shows the chemical composition of calcium sulfate precipitate as a result of the addition of 120 g/L $CaCl_2$. As can be seen, the loss of Zr in the SO_4^{2-} removal step was low. After washing, the calcium sulfate was a byproduct.

Table 4. Chemical composition of calcium sulfate precipitate.

Element	Zr	Hf	Nb	Al	Fe	O	S	Ca
Content (wt%)	0.11	0.05	0.08	0.09	0.10	55.7	19.10	23.26

Table 5 shows the chemical composition of the basic zirconium sulfate precipitate prepared using the optimized conditions. As can be seen, the contents of Fe and Al in the precipitate were very low. The Zr basic sulfate prepared by this method was easily dissolved in acids, thus it could be used in conventional techniques to produce a final product, such as $ZrOCl_2$ and ZrO_2. Figure 10 shows the XRD pattern of the basic zirconium sulfate precipitate. Qual-X software was used to analyze the data. It was observed that some of the peaks matched those of the calcium sulfate anhydride (bassanite, $CaSO_4 \cdot 0.5H_2O$) phase, and the basic zirconium sulfate was amorphous. The presence of bassanite can be explained by the fact that a few Ca^{2+} in the sulfate media were not stable and precipitated when the temperature or time was increased.

Table 5. Chemical composition of the basic zirconium sulfate precipitate.

Element	Zr	Hf	Nb	Al	Fe	Si	S	Ca
Content (wt%)	33.77	0.59	0.13	0.33	0.11	0.05	10.30	6.28

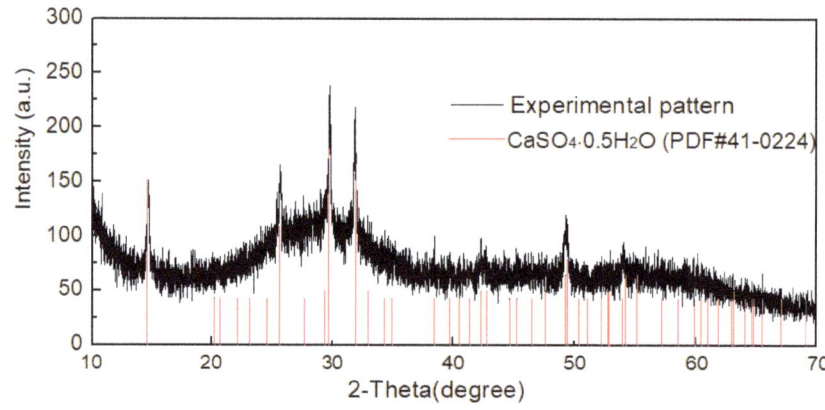

Figure 10. XRD pattern of the basic zirconium sulfate precipitate.

Fourier-transform infrared spectroscopy analysis (FT-IR) was used to study the structure of basic zirconium sulfate precipitate further, and the FT-IR spectra of the resulting precipitate, $Zr(SO_4)_2 \cdot 4H_2O$, and $ZrOSO_4 \cdot 4H_2O$ are shown in Figure 11. The FT-IR spectra of the precipitate agreed well with those of $ZrOSO_4 \cdot 4H_2O$, which confirmed that they had a same structure. It was speculated that the adsorption bands at 1000 cm^{-1} and 1220 cm^{-1} corresponded to Zr-O-S stretching vibrations [26].

3.7. Flowchart with Metal Balance

Figure 12 shows the proposed flowchart for the recovery of Zr, Hf, and Nb from the sulfuric acid leach solution using selective precipitation, and it also reveals the directions of the different metals. As shown, although the basic zirconium sulfate precipitation could achieve selective recovery of Zr and Hf, the concentration of sulfate in the solution needed to be controlled, and the resulting precipitate still contained some impurities. Further purification of the product was needed for Zr production.

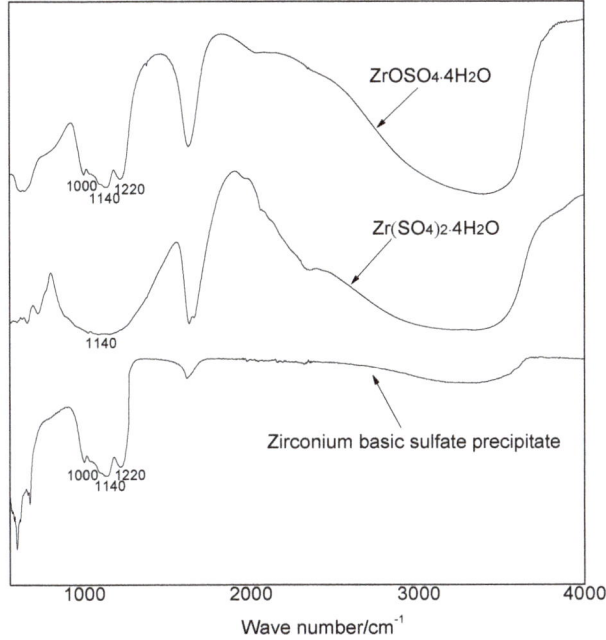

Figure 11. IR spectra of the basic zirconium sulfate precipitate and pure zirconium sulfate salts.

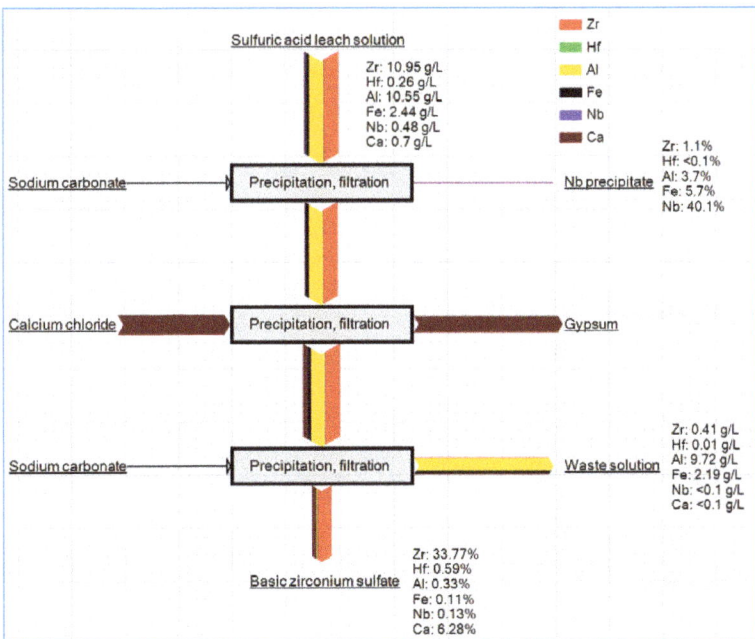

Figure 12. Proposed flowchart for the recovery of Zr, Hf, and Nb from the sulfuric acid leach solution using selective precipitation.

3.8. ZrO₂ Preprared from Basic Zirconium Sulfate

XRD results in Figure 13 confirmed that the powder prepared by the basic zirconium sulfate precipitate was ZrO$_2$.

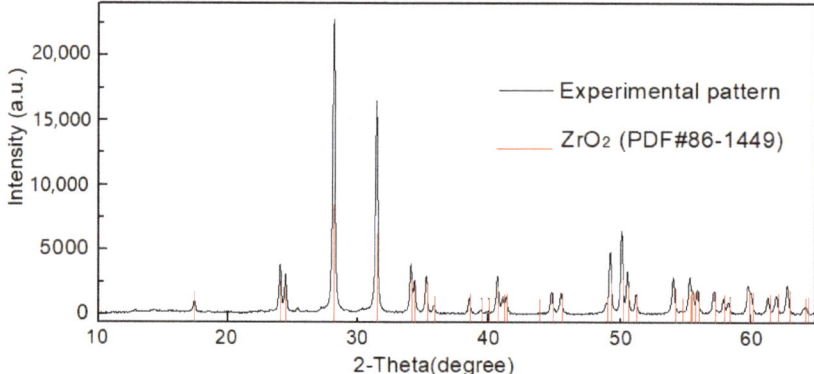

Figure 13. XRD pattern of the ZrO$_2$ prepared from the basic zirconium sulfate precipitate.

Table 6 shows the chemical composition of ZrO$_2$ prepared after selective precipitation. As can be seen, the contents of ZrO$_2$ + HfO$_2$ in the product were higher than 99%, which met the requirements of industrial ZrO$_2$.

Table 6. Chemical composition of ZrO$_2$ prepared from the basic zirconium sulfate precipitate.

Composition	ZrO$_2$	HfO$_2$	Al$_2$O$_3$	Fe$_2$O$_3$	SiO$_2$	CaO	Na$_2$O	MgO
Content (wt%)	97.22	1.88	0.4	0.1	0.03	0.2	0.05	<0.01

4. Conclusions

In this study, the treatment of Zr-bearing sulfuric acid leach solution to yield ZrO$_2$ was successfully carried out. After Nb was preferentially precipitated by adjusting pH to around 1.0, selective precipitation via basic zirconium sulfate (Zr$_5$O$_8$(SO$_4$)$_2$·xH$_2$O) was the novel method used to selectively recover Zr. This method achieved removal of the main impurities, such as Fe and Al, and enrichment of Zr from the sulfuric acid leach solution. After partial removal of SO$_4^{2-}$ by adding 120g CaCl$_2$ per 1L solution, 96.18% Zr precipitation yield was obtained by adjusting the pH to ~1.6 and keeping the temperature at 75 °C for 60 min. The resulting precipitate contained 33.77% Zr and 0.59% Hf with low Fe and Al. A high-quality product of ZrO$_2$ can be obtained from the basic sulfate precipitate. The present study showed a promising process to recover Zr from sulfuric acid leach solution.

Author Contributions: Y.M. performed the experiments and wrote the paper; X.W. and S.S. contributed the reagents/materials/analysis tools; Y.M., K.F., X.W., and B.F. analyzed the data. All authors have read and agreed to the published version of the manuscript.

Funding: This research received no external funding.

Acknowledgments: One of the authors (Yiqian Ma) is grateful to the Chinese Government for providing a scholarship. Authors are thankful to the EURARE (www.eurare.eu) project for providing the initial material.

Conflicts of Interest: The authors declare no conflict of interest.

References

1. Nielsen, R.H.; Schlewitz, J.H.; Nielsen, H. Zirconium and Zirconium Compounds. In *Kirk-Othmer Encyclopedia of Chemical Technology*; Kirk-Othmer, Ed.; Wiley online Library, John Wiley & Sons, Inc.: New York, NY, USA, 2000; Volume 1, pp. 1–46.
2. Zirconium in Steels. Available online: http://ispatguru.com/zirconium-in-steels/ (accessed on 5 March 2018).
3. Biswas, R.K.; Habib, M.A.; Karmakar, A.K.; Islam, M.R. A novel method for processing of Bangladeshi zircon: Part I: Baking, and fusion with NaOH. *Hydrometallurgy* **2010**, *10*, 124–129. [CrossRef]
4. Johnsen, O.; Ferraris, G.; Gault, R.A.; Grice, J.D.; Kampf, A.R.; Pekov, I.V. The nomenclature of eudialyte-group minerals. *Can. Mineral.* **2003**, *4*, 785–794. [CrossRef]
5. Rastsvetaeva, R.K. Structural Mineralogy of the Eudialyte Group: A Review. *Crystallogr. Rep.* **2007**, *52*, 47–64. [CrossRef]
6. Lebedev, V.N. Sulfuric acid technology for processing of eudialyte concentrate. *Russ. J. Appl. Chem.* **2003**, *76*, 1559–1563. [CrossRef]
7. Davris, P.; Stopic, S.; Balomenos, E.; Panias, D.; Paspaliaris, I.; Friedrich, B. Leaching of rare earth elements from Eudialyte concentrate by suppressing silicon dissolution. *Miner. Eng.* **2017**, *108*, 115–122. [CrossRef]
8. Voßenkaul, D.; Birich, A.; Müller, N.; Stoltz, N.; Friedrich, B. Hydrometallurgical processing of eudialyte bearing concentrates to recover rare earth elements via low-temperature dry digestion to prevent the silica gel formation. *J. Sustain. Met.* **2017**, *3*, 79–89.
9. Balomenos, E.; Davris, P.; Deady, E.; Yang, J.; Panias, D.; Friedrich, B.; Binnemans, K.; Seisenbaeva, G.A.; Dittrich, C.; Kalvig, P.; et al. The EURARE Project: Development of a Sustainable Exploitation Scheme for Europe's Rare Earth Ore Deposits. *Johns. Matthey Technol. Rev.* **2017**, *61*, 142–153. [CrossRef]
10. Ma, Y.; Stopic, S.; Gronen, L.; Friedrich, B. Recovery of Zr, Hf, Nb from eudialyte residue by sulfuric acid dry digestion and water leaching with H_2O_2 as a promoter. *Hydrometallurgy* **2018**, *181*, 206–214. [CrossRef]
11. Ma, Y.; Stopic, S.; Friedrich, B. Hydrometallurgical Treatment of a Eudialyte Concentrate for Preparation of Rare Earth Carbonate. *Johns. Matthey Technol. Rev.* **2019**, *63*, 2–13. [CrossRef]
12. Zakharov, V.I.; Maiorov, D.V.; Alishkin, A.R.; Matveev, V.A. Causes of insufficient recovery of zirconium during acidic processing of lovozero eudialyte concentrate. *Russ. J. Non Ferr. Met.* **2011**, *52*, 423–428. [CrossRef]
13. Dibrov, I.A.; Chirkst, D.E.; Litvinova, T.E. Experimental study of zirconium(IV) extraction from fluoride-containing acid solutions. *Russ. J. Appl. Chem.* **2002**, *75*, 195–199. [CrossRef]
14. Chanturiya, V.A.; Minenko, V.G.; Samusev, A.L.; Chanturia, E.L.; Koporulina, E.V.; Bunin, I.Z.; Ryazantseva, M.V. The Effect of Energy Impacts on the Acid Leaching of Eudialyte Concentrate. *Min. Proc. Ext. Met. Rev.* **2020**. [CrossRef]
15. Lebedev, V.N.; Schur, T.E.; Maiorov, D.V.; Popova, L.A.; Serkova, R.P. Specific Features of Acid Decomposition of Eudialyte and Certain Rare-Metal Concentrates from Kola Peninsula. *Russ. J. Appl. Chem.* **2003**, *76*, 1191–1196. [CrossRef]
16. Ma, Y.; Stopic, S.; Huang, Z.; Friedrich, B. Selective recovery and separation of Zr and Hf from sulfuric acid leach solution using anion exchange resin. *Hydrometallurgy* **2019**, *189*, 105143. [CrossRef]
17. Gates, P.A.; Horlacher, C.F.; Reed, G. *Preliminary Economic Assessment NI 43-101 Technical Report for the Norra Kärr (REE-Y-Zr) Deposit Gränna*; Tasman Metals Limited: Norra Kärr, Sweden, 2012.
18. Davidson, T.; Thompson, J.; Short, M.; Moseley, G.; Mounde, M.; La Digges Touche, G. *Norra Kärr Project PFS Prefeasibility Study-NI 43-101-Technical Report for the Norra Kärr Rare Earth Element Deposite*; GBM Minerals Engineering Consultants Limited: London, UK, 2015.
19. Kozak, C.M.; Mountford, P. *Encyclopedia of Inorganic Chemistry, Zirconium & Hafnium: Inorganic—Coordination Chemistry*; John Wiley & Sons: Oxford, UK, 2006.
20. Wang, L.Y.; Lee, M.S. A review on the aqueous chemistry of Zr(IV) and Hf(IV) and their separation by solvent extraction. *J. Ind. Eng. Chem.* **2016**, *39*, 1–9. [CrossRef]
21. Duan, Z.; Yang, H.; Satoh, Y.; Murakami, K.; Kano, S.; Zhao, Z.; Shen, J.; Abe, H. Current status of materials development of nuclear fuel cladding tubes for light water reactors. *Nucl. Eng. Des.* **2017**, *316*, 131–150. [CrossRef]
22. Dean, J. *Langes's Handbook of Chemistry*, 13th ed.; McGraw-Hill, Inc.: New York, NY, USA, 1985.

23. Nielsen, R.H.; Govro, R.V. *Zirconium Purification: Using a Basic Sulfate Precipitation*; U.S. Dept. of the Interior, Bureau of Mines: Washington, DC, USA, 1956.
24. Ryabchikov, D.I.; Marov, I.N.; Ermakov, A.N.; Belyaeva, V.K. Stability of some inorganic and organic complex compound of zirconium and hafnium. *J. Inorg. Nucl. Chem.* **1964**, *26*, 965–980. [CrossRef]
25. Berg, R.W. Progress in niobium and tantalum coordination chemistry. *Coord. Chem. Rev.* **1992**, *113*, 1–130. [CrossRef]
26. Chatterjee, M.; Ray, J.; Chatterjee, A.; Ganguli, D. Characterization of basic zirconium sulphate, a precursor for zirconia. *J. Mater. Sci. Lett.* **1989**, *8*, 548–553. [CrossRef]
27. Xiong, B.; Wen, W.; Yang, X.; Li, H.; Luo, F.; Zhang, W. *Zirconium and Hafnium Metallurgy*; Metallurgical Industry Press: Beijing, China, 2002. (In Chinese)

 © 2020 by the authors. Licensee MDPI, Basel, Switzerland. This article is an open access article distributed under the terms and conditions of the Creative Commons Attribution (CC BY) license (http://creativecommons.org/licenses/by/4.0/).

Article

Electrorefining Process of the Non-Commercial Copper Anodes

Radmila Markovic [1,*], Vesna Krstic [1], Bernd Friedrich [2], Srecko Stopic [2], Jasmina Stevanovic [3], Zoran Stevanovic [1] and Vesna Marjanovic [1]

1. Mining and Metallurgy Institute Bor, Zeleni Bulevar 35, 19210 Bor, Serbia; vesna.krstic@irmbor.co.rs (V.K.); zoran.stevanovic@irmbor.co.rs (Z.S.); vesna.marjanovic@irmbor.co.rs (V.M.)
2. IME Process Metallurgy and Metal Recycling, Intzestraße 3, 52056 Aachen, Germany; bfriedrich@ime-aachen.de (B.F.); sstopic@metallurgie.rwth-aachen.de (S.S.)
3. Institute of Chemistry, Technology and Metallurgy, National Institute of the Republic of Serbia, University of Belgrade, Njegoševa 12, 11000 Belgrade, Serbia; jaca@tmf.bg.ac.rs
* Correspondence: radmila.markovic@irmbor.co.rs; Tel.: +381-62-450762

Citation: Markovic, R.; Krstic, V.; Friedrich, B.; Stopic, S.; Stevanovic, J.; Stevanovic, Z.; Marjanovic, V. Electrorefining Process of the Non-Commercial Copper Anodes. *Metals* **2021**, *11*, 1187. https://doi.org/10.3390/met11081187

Academic Editors: Geoffrey Brooks and Antoni Roca

Received: 15 June 2021
Accepted: 23 July 2021
Published: 26 July 2021

Publisher's Note: MDPI stays neutral with regard to jurisdictional claims in published maps and institutional affiliations.

Copyright: © 2021 by the authors. Licensee MDPI, Basel, Switzerland. This article is an open access article distributed under the terms and conditions of the Creative Commons Attribution (CC BY) license (https://creativecommons.org/licenses/by/4.0/).

Abstract: The electrorefining process of the non-commercial Cu anodes was tested on the enlarged laboratory equipment over 72 h. Cu anodes with Ni content of 5 or 10 wt.% and total content of Pb, Sn, and Sb of about 1.5 wt.% were used for the tests. The real waste solution of sulfuric acid character was a working electrolyte of different temperatures (T1 = 63 ± 2 °C and T2 = 73 ± 2 °C). The current density of 250 A/m^2 was the same as in the commercial process. Tests were confirmed that those anodes can be used in the commercial copper electrorefining process based on the fact that the elements from anodes were dissolved, the total anode passivation did not occur, and copper is deposited onto cathodes. The masses of cathode deposits confirmed that the Cu ions from the electrolyte were also deposited onto cathodes. The concentration of Cu, As, and Sb ions in the electrolyte was decreased. At the same time, the concentration of Ni ions was increased by a maximum of up to 129.27 wt.%. The major crystalline phases in the obtained anode slime, detected by the X-ray diffraction analyses, were $PbSO_4$, Cu_3As, $SbAsO_4$, Cu_2O, As_2O_3, PbO, SnO, and Sb_2O_3.

Keywords: electrorefining; non-commercial copper anode; waste solution; high content; Ni; Pb; Sn; Sb; passivation; anode slime

1. Introduction

Electrorefining of the anode materials is a purification process in order to remove the ingredients that may have a negative impact on the physical, chemical, and mechanical properties of the base materials. Almost all metals can be purified by this process but based on the production data of approximately 24 Mt in 2019, copper exceeds all other metals [1,2]. Electrolytes and other process parameters must be selected so that the anode dissolution and metal deposition take place with a high degree of efficiency and a transfer degree of ingredients from anode onto cathode is minimized. Additionally, the process parameters should prevent the passivation of the anode and enable the precipitation of appropriate physical and chemical characteristics. The values of characteristic technological parameters for the electrorefining process of some metals are different for different metals (copper, nickel, cobalt, lead, and tin) [3]. If it is necessary, the additives are added to the electrolyte to ensure a proper operation of both electrodes. The addition of thiourea, gelatin, and chlorine ions during the electrorefining process had a positive effect on minimizing the nodules', porosities', and dendrites' appearance [4]. Deposited copper surface roughness was decreased with the increase in the concentration of thiourea and gelatin [5].

Anodic passivation is one of the basic problems that occur in the industrial electrorefining process and could be explained as a copper sulfate layer formation on an anode surface [6]. Cu_2O form was registered in the anodes with oxygen content in the range of 0.1–0.3%. This form was also registered in the slime present on the anode surface. Through the reaction of copper (I) oxide form and sulfuric acid, a partial dissolution of copper oxide form occurs, whereby a salt of copper sulfate and elemental copper is formed. As the chemical dissolution of this form is more intense than the electrochemical one, the copper concentration in the electrolyte increases, which is a characteristic of the commercial process of anode copper electrorefining [7].

Raw materials for smelting are becoming more and more complex so that the effect of ingredients on the anode copper electrorefining process is constantly changing. Anodes obtained from the secondary raw materials are generally rich in nickel, lead, antimony, and tin, and a low content of selenium, tellurium, and silver has been reported [8]. During solidification, Ni is enriched in the solid phase, but some elements pass into the solid solution in copper crystals (Sb, Sn, Pb, As, and Bi) and in Cu-Pb-As-Sb-Bi oxide phases [9]. The so-called "mineralogy" of the anode directly affects the passivity of the anode, formation of suspended slime, deposition of slime, as well as the possibility of separating the useful and high-value components from the obtained slime [10]. The amount of inclusion phases, primarily oxides, is a direct consequence of the content of elements in the anode. Only Cu_2O and NiO are formed during the primary crystallization, while the other forms are formed during the secondary crystallization, causing a local accumulation.

The behavior of ingredients and phases during the electrolytic refining of anodes is in principle the same as for anodes obtained from the primary raw materials, but the higher content of ingredients can cause some problems during the electrorefining process such as the increase in a cell voltage or even anode passivation [11]. The mostly inhomogeneous distribution of phases leads to a different dissolution percentage in a contact with the working electrolyte.

During the electrorefining process, the ingredients from anodes could be distributed into three groups. The first group consists of ingredients with a dissolution potential that is much more negative than a dissolution potential of copper: Ni, Fe, Zn, and Co. During the process, those elements are accumulated in the electrolyte. In the nickel-rich copper anodes, the following oxide phases can be formed: NiO, Cu-Sb-Ni, and Cu-Sn-Ni [12]. Anodes obtained from the copper-based secondary materials are characterized by the nickel content of more than 0.3% wt.%. A negative impact of Ni on the refining process is reflected in the fact that NiO accelerates the process of passivation of the anode, reduces the solubility of Cu from the anode, and generates a large amount of nickel in the anode slime [13]. The second group consists of ingredients with a dissolution potential that is much more positive than the copper dissolution potential, which prevents their dissolution in the electrolyte, but they directly pass into the anode slime: Au, Ag, Pt, Se, and Te. This group also includes the insoluble salts such as $PbSO_4$ and $Sn(OH)_2SO_4$, which pass from electrolyte into anode slime. The various Pb oxide inclusions are the major Pb carriers in anode material. Pb dissolves from anodes along with Cu and immediately precipitates as the $PbSO_4$ insoluble salt. In a copper anode, rich in Ni, Sb, and Sn content, Sn could be found in different forms of Cu-Sn-Ni oxide, such as Cu_2NiSnO_5 and Cu-Sb-Ni oxide, as the "kupferglimmer" form $Cu_3Ni_{2-x}SbO_{6-x}$ where x is in the range from 0.1 to 0.2. During the electrolysis process, Sn reacts with electrolyte and precipitates in anode slime [14]. The third group consists of ingredients with a dissolution potential close to a dissolution potential of copper: As, Sb, and Bi. They are dissolved during the anode refining, and under certain conditions can be precipitated (high concentration of these elements, low concentration of Cu). Arsenic (AsO_4^{-3}) can also react with Sb and Bi to form an insoluble compound of As-Sb-Bi known as the "floating slime" with the predicted composition of $BiAsO_4$, $SbAsO_4$, Sb_2O_3, and Bi_2O_3 [15]. Based on the data of the Jafari et al. [15], the predicted composition of "floating slime" is $BiAsO_4$, $SbAsO_4$, Sb_2O_3, and Bi_2O_3. Additionally, the operational As/(Sb + Bi) molar ratios could be maintained at or above 2 aim to reduce the presence of floating slimes, but this ratio is lower than 2 in about 26% of electrolysis plants [16].

The aim of this work was to investigate the electrorefining process of copper anodes with increased Ni, Pb, Sn, and Sb content in comparison with the content of those elements in the commercial copper anode electrorefining process. The two types of copper anodes are proposed for testing regarding the Ni, Pb, Sn, and Sb content as the ingredients. Copper content was also different for different anode types. Real waste sulfate solution was used as a working electrolyte. Furthermore, the effect of two different working electrolyte temperatures, where one was the same as in the commercial copper electrorefining process (T1 = 63 ± 2 °C) and the second was 10 °C higher (T2 = 73 ± 2 °C), was tested. The anode elements dissolving during each test over the duration of 72 h was defined on the basis of values of a cell voltage. The mass of dissolved anodes and the mass of anode slime were used for calculating the mass percentage of anode slime in accordance with the mass of dissolved anodes. Each calculation was performed for different working conditions. Electrolyte composition is checked on each 24 h for the main elements. The phases in anode slime samples, obtained for different experimental conditions, are defined by the X-ray diffraction (XRD) analysis.

2. Materials and Methods
2.1. Materials
2.1.1. Non-Commercial Copper Anodes

Two types of non-commercial copper anodes regarding the chemical content of Ni, Pb, Sn, and Sb are prepared for the electrorefining tests. Oxygen content was under 200 ppm. The main difference between the anodes was in Ni and Cu content. In the anodes with sample code An-1, Ni content was approximately 5 wt.%, and in the anodes with sample code An-2, Ni content was approximately 10 wt.%. The total content of Pb, Sn, and Sb for both types of anodes was approximately 1.5 wt.%. The content of each element (Pb, Sn, and Sb) was about 0.5 wt.%. Copper content is calculated on the basis of values for the content of Ni, Pb, Sn, and Sb as a difference up to 100 wt.%. Cathode copper and pure metals (nickel, lead, tin, and antimony) were used for the non-commercial copper anode preparation. Preparation of a suitable mixture for melting was made in an induction furnace. All other metals were added into the furnace after reaching the copper melt temperature of 1300 °C. The content of oxygen is controlled in the melt (concentration range was from 1 to 12,000 ppm) and in a copper sample. In the case that the oxygen content was under 200 ppm, the melt was cast into the suitable steel molds. After the self-cooling process, the mechanical processing was used later, as it aimed to prepare anodes for the electrorefining process. The mass of each anode was approximately 7 kg. A detailed procedure of anode preparation was presented previously [17].

2.1.2. Working Electrolyte

The waste sulfate solution from the commercial copper electrolysis AURUBIS AG, Hamburg, Germany (earlier Norddeutsche Affinerie AG), of the following chemical composition (g/L): Cu—32.5; Ni—20.5; As—4; Sb—0.3; Sn—0.001; Pb—0.004, was used as a working electrolyte. Based on the values of concentration of the same elements in the commercial electrorefining process, Ni concentrations were increased and Cu concentration was decreased [18]. The aim was to avoid a sludge precipitation during the electrolyte self-cooling. The samples for analysis were specially prepared, explained in the previous paper [17].

2.1.3. Surfactants

Thiourea and gelatin were used as a surfactant. Thiourea, CH_4N_2S, purity grade (purchased from Sigma Aldrich, Company: Merck KGaA, Frankfurter Str. 250, D-64271 DARMSTADT), with a molecular weight of 76.12 g/mol, water solubility of 137 g/L at 20 °C, was used for the minimization of the nodules, porosities, and dendrites' appearance. The solution was prepared with the 18 MΩ cm deionized water.

Gelatin, purity grade, (purchased from Sigma Aldrich, Company: Merck KGaA, Frankfurter Str. 250, D-64271 DARMSTADT) was also used for minimizing the nodules, porosities, and dendrites' appearance to decrease the copper surface roughness. The preparation of the appropriate solution was made with 18 MΩ cm deionized water at a temperature of about 40 °C.

2.1.4. Chemicals for Preliminary Testing the Electrolyte Mixing System

Potassium iodide (purchased from Company: Merck KGaA, Frankfurter Str. 250, D-64271 Darmstadt, Germany) was used for the preparation of 10 wt.% aqueous solution (KI) that was used as electrolyte during preliminary testing of the system for electrolyte mixing.

Sodium thiosulfate ($Na_2S_2O_3$.), purchased from Carl Roth GmbH + Co KG Schoemperlenstr. 3–5, D-76185 Karlsruhe, Germany), was used for the preparation of the aqueous solution for the decolorization of the KI solution.

The preparation of the appropriate solutions was performed with 18 MΩ cm deionized water.

2.2. Methods

2.2.1. Experimental Set-Up and Procedure

Electrorefining tests were conducted at the IME Process Metallurgy and Metal Recycling, Aachen, Germany, on the enlarged laboratory equipment, under the constant galvanostatic pulse conditions with the applied current density of 250 A/m^2 [19–21]. An external source with characteristics of 50 A and 10 V (model HEINZINGER TNB-10-500, Heinzinger Electronic GmbH the Power Supply Company, Rosenheim, Germany) was used as a direct current supplier. The starting cathode sheets were made of stainless steel. Cathode copper (99.95 wt.% Cu) was used as a reference electrode.

A rectangular electrolytic cell made of polypropylene (PP) was used for the electrorefining process. The electrode arrangement in the cell was cathode–anode–cathode, and the electrolyte working volume was 5.85 dm^3. The bus bar from copper provided a direct current supply. The cell was current connected with the system for automatic measurement and data processing. A stainless steel water tank with recalculated hot water was used to maintain the required electrolyte temperature in a cell inserted in this tank. The water tank was insulated with styrofoam/aluminum material and covered with a portable cover with openings for the cells. A plastic hose with a manual valve was used for regulation of the water recirculation and circulation speed. The overflow was made through the overflow box at the top of the water tank. Thermostat, HAAKE B7—PHOENIX 2 (Thermo Fisher Scientific, Waltham, MA, USA) was used to maintain the heating water operating temperature. The flow pump was used for water recirculation. Two pumps, BVP Standard (No. ISM 444 ISMATEC, Cole-Parmer GmbH, Wertheim, Germany), were used for dosing an aqueous solution of surfactants, as well as for dosing the deionized water for evaporation in cells. A more detailed description of the experimental setup has been presented previously [17]. The consumption of thiourea and gelatin (50 mg per 1 t of cathode copper, respectively) was in accordance with the literature data, where the best results were obtained when the ratio of those two components was in the range of 0.8 to 1.7 [4]. The aqueous solution of surfactants and deionized water were dosed into a cell continually over 72 h.

The nitrogen distribution system was used for electrolyte mixing. The introduction of nitrogen into a cell was achieved through a glass tube with a diameter of Ø 2 mm, which was placed in the cell, and mixing of electrolytes was achieved by bubbles of solution through a tube with a diameter of Ø 8 mm. Preliminary tests were conducted with two types of nitrogen inlet. In a case (a), a tube with a smaller diameter of 360 mm was placed inside the tube of larger diameter, and in case (b), the tube of smaller diameter was outside and that of the length of 20 mm entered through a lower hole of a tube of larger diameter to the tube itself. In both cases, nitrogen moves from the bottom up through a tube of larger diameter, ensuring the movement of solution in the same direction. The flow of nitrogen is regulated by the existing rotameters. The preliminary tests showed that mixing the electrolyte by introducing nitrogen, as shown in Case (b). gave satisfactory results. The preliminary tests were performed in an existing electrochemical cell with an anode made of PVC material and aluminum cathodes. Synthetic aqueous KI solution was used as an electrolyte. The decolorization course of an aqueous KI solution was tested with the addition of 0.1 mL of $Na_2S_2O_3$. At the same time, nitrogen was introduced through a tube to mix the solution. The color of KI aqueous solution before the addition of $Na_2S_2O_3$

solution was blue. The change in color of the solution from the top to the bottom was observed after the addition of $Na_2S_2O_3$ solution and the introduction of nitrogen in order to mix the solution. The chemical reaction was completed after complete decolorization of the solution. The test results confirmed that the proposed method of introduction of nitrogen into the cell enables the circulation and mixing of the solution.

2.2.2. Characterization Methods

The Electro-Nite system (model HERAEUS, Heraeus Electro-Nite International N.V. Centrum Zuid Houthalen, Belgium) was used to control the oxygen content in the copper mixture melt.

The oxygen content in the copper mixture samples was controlled using the Juwe Onmat 8500 instrument (Ströhlein ON-mat 8500, JUWE Laborgeräte GmbH, Germany instrument).

Chemical analysis of copper anodes was done by the Spector Xepos Energy Dispersive X-ray Fluorescence Spectroscope (ED-XRF, SPECTRO, Kleve, Germany). The X-ray anode material was Au. The major and minor elements were determined via the fusion method (1000 °C for 1 h with a mixture of $Li_2B_4O_7/KNO_3$ followed by direct dissolution in 10% HNO_3 solution). Samples of copper anodes were obtained by cutting a part of the anode before polishing from the top, middle, and bottom.

The atomic emission spectrometry with inductively-coupled plasma technique (ICP-AES) (model Spectro Ciros Vision, SPECTRO Analytical Instruments GmbH, Kleve, Germany) was used for the determination of the chemical composition of electrolyte samples.

The inductively coupled plasma mass spectrometry (ICP-MS) (model Agilent 7700, Agilent Technologies, Inc., Tokyo 192-8510 Japan) was used for the determination of the element content in anode slime.

Each chemical analysis was carried out in duplicate and accompanied by a quality control (blank and certified reference materials (CRM) analysis).

The X-ray diffraction analysis (XRD) (model Explorer, G.N.R., Via Torino, Italy) was used for the anode slime phase analysis.

3. Results and Discussion

3.1. Anode Chemical Composition

The results of standard chemical analysis of copper anodes were carried out on 26 elements (Table 1). Samples for the analysis were taken from the bottom, middle, and top of anodes aimed to check the uniformity of anode materials.

Table 1. Chemical composition of copper anodes with Ni, Pb, Sn, and Sb non-standard chemical content.

Element	Position of Anode Sampling							
	Bottom	Middle	Top	Average	Bottom	Middle	Top	Average
	Anode with Samples Code An-1				Anode with Samples Code An-2			
	Content, wt.%							
Ni	4.66	4.68	4.67	4.67	10.04	10.11	9.96	10.04
Pb	0.417	0.426	0.419	0.421	0.392	0.363	0.400	0.385
Sn	0.443	0.445	0.444	0.444	0.412	0.395	0.419	0.41
Sb	0.443	0.450	0.449	0.447	0.382	0.369	0.394	0.382
Zn	<0.0015	<0.0015	<0.0015	<0.0015	0.0055	0.0058	0.0060	0.0058
P	0.0052	0.0052	0.0051	0.0052	0.0065	0.0064	0.0064	0.0065
Fe	0.016	0.016	0.014	0.0153	0.014	0.016	0.015	0.015
Si	0.0047	0.0056	0.0024	0.0042	0.010	0.0031	0.0027	0.0012
Cr	<0.0003	<0.0003	<0.0003	<0.0003	0.0005	0.0006	0.0006	0.0006
Te	0.0064	0.0075	0.0075	0.007	0.0020	0.0024	0.0022	0.0022
As	0.028	0.029	0.029	0.0287	0.021	0.020	0.020	0.020
Cd	0.0019	0.0019	0.0018	0.0019	0.0010	0.0011	0.0011	0.0011
Bi	0.0038	0.0036	0.0035	0.0036	0.0027	0.0028	0.0028	0.0028
Ag	0.068	0.066	0.067	0.067	0.053	0.055	0.053	0.054

Table 1. Cont.

Element	Position of Anode Sampling							
	Bottom	Middle	Top	Average	Bottom	Middle	Top	Average
	Anode with Samples Code An-1				Anode with Samples Code An-2			
	Content, wt.%							
S	0.0031	0.0032	0.0031	0.0031	0.0048	0.0048	0.0049	0.0049
Au	<0.0005	<0.0005	<0.0005	<0.0005	0.0027	0.0018	0.0018	0.0018
C	0.020	0.011	0.020	0.0300	0.025	0.032	0.031	0.029
Ti	0.0017	0.0017	0.0017	0.0017	0.0023	0.0027	0.0025	0.0025
Se	0.0060	0.0061	0.0060	0.0060	0.0078	0.0079	0.0079	0.0079

Close values for the content of elements from different sampling positions confirmed the efficiency of the anode preparation process. Copper content was calculated as a difference of Ni, Pb, Sn, and Sb average values up to 100 wt.%. The calculated value of Cu content in the anodes with sample codes An-1 and An-2 was 94.02 wt.% and 88.78 wt.%, respectively. Additionally, Table 1 presents the average values for all elements obtained on the basis of mathematical calculation for three values for each element. The results of chemical analyses have shown that the content of Zn, Mn, Mg, Cr, Co, Al, Be, Zr, Au, and B in the anode with sample code An-1 was below the sensitivity limit of the used analytical method. For the anode with sample code An-2, the content of Zn, Cr, and Au had the values over the sensitivity limit of the used analytical method, but the content of Mn, Mg, Co, Al, Be, Zr, and B was below the sensitivity limit of the used analytical method for the anode with sample code An-1. The content of P, Fe, Si, Cd, Bi, Ag, S, C, Ti, and Se in both anode types was within the limits in industry practice [16]. The obtained results for oxygen content in anode An-1 and An-2 had values of 120 and 98 ppm, respectively. Those values are in accordance with the investigation realized by the authors [22] that confirmed the aim to decrease the content of oxide and "kupferglimmer" forms in anodes with oxygen content to be lower than 200 ppm.

3.2. Anode Dissolution and Cathode Deposit

The mass of anodes was measured at the beginning and end of the process. Dissolved anode mass was calculated as the difference between those two values. Table 2 presents the values for the mass of dissolved elements from anodes during 72 h of the experiments at different electrolyte temperatures.

Table 2. Mass of dissolved copper anodes and cathode deposit mass.

Anode with Samples Code An-1		Anode with Samples Code An-2	
Electrolyte working temperature			
T1 = 63 ± 2 °C	T2 = 73 ± 2 °C	T1 = 63 ± 2 °C	T2 = 73 ± 2 °C
Dissolved anode mass, g			
1688	1779	1367	1442
Cathode correspond to anode An-1		Cathode correspond to anode An-2	
Cathode deposit mass, g			
1708	1789	1371	1526

Cathode mass as a cumulative mass of starting cathode sheet and deposited material after 72 h was measured at the end of the process. The difference between the mass of the starting cathode sheet and the cumulative cathode mass at the end of the process was the mass of material deposited during the process (Table 2).

Data from Table 2 show that the mass of the cathode deposit is higher than the dissolved mass of corresponded anode. It could be explained by the electrowinning process of Cu ions from the working electrolyte (Table 3). By the electrowinning process, copper is deposited directly from the working electrolyte onto a cathode sheet under the action of the direct current. General electrochemical reactions for this process are shown by the individual reactions on the cathode and anode as well as by the cumulative reaction in the system [23]:

Table 3. Electrolyte chemical composition changing.

Element	Anode with Samples Code An-1					
	T1 = 63 ± 2 °C			T2 = 73 ± 2 °C		
	Process Duration, h					
	24	48	72	24	48	72
	Element Concentration, g/L					
Cu	23.8	18.5	10.9	22.5	17.6	2.2
Ni	25.6	29.4	32.7	25.5	31	36.5
As	3.7	2.7	1.9	3.5	2.6	1
Sb	0.292	0.290	0.217	0.267	0.237	0.156
	Anode with Samples Code An-2					
	T1 = 63 ± 2 °C			T2 = 73 ± 2 °C		
	Process Duration, h					
	24	48	72	24	48	72
	Element Concentration, g/L					
Cu	21	10.4	6.8	18	12.5	1.5
Ni	28.5	33.5	43.6	31	41	47
As	3.6	2.8	0.85	3.5	2.5	0.8
Sb	0.295	0.292	0.24	0.294	0.288	0.188

Cathode reaction:
$$Cu^{2+} + 2e \rightarrow Cu^0, \ E^0 = +0.34 \ V \quad (1)$$

Anode reaction:
$$H_2O \rightarrow H^+ + (OH)^- \rightarrow \frac{1}{2}O_2 + 2H^+ + 2e, \ E^0 = +1.23 \ V \quad (2)$$

Cumulative reaction:
$$Cu^{2+} + SO_4^{2-} + H_2O \rightarrow Cu^0 + \frac{1}{2}O_2 + 2H^+ + SO_4^{2-}, \ E^0 = -0.89 \ V \quad (3)$$

The dissolved anode mass had the lowest value for the anode with the sample code An-2 after electrorefining at the working electrolyte temperature of T1 = 63 ± 2 °C (Table 2). Masses of dissolved anodes had higher values for anodes with lower content of ingredients. In the case when the value for the average total content of Ni, Pb, Sn, and Sb was 5.982 wt.% (anode An-1), the mass of the dissolved anode was higher for 321 g in comparison with the dissolved mass of anode An-2 (average value for the total content of ingredients was 11.217 wt.%) for the same process condition. When the Ni content in anodes is lower than 3 wt.%, Ni dissolution in the electrolyte is near 100%. In a case that the Ni content is higher than 3 wt.%, the ingredients are present in the working electrolyte as the Ni and Cu-Ni-Sb oxide forms and $NiSO_4$. Ni salt ($NiSO_4$) participates in a transmission of electricity and creates the unfavorable conditions for the discharge of copper ions, reducing its concentration in the vicinity of the cathode. In addition, increasing the concentration of ballast salts in the electrolyte reduces its electrical conductivity and the dissolution of elements.

Cell Voltage Changing

During the electrorefining process, the cell voltage was measured every 10 s. The cell voltage changing for different anode types and different working electrolyte temperatures is defined on the basis of about 25,000 data (Figures 1 and 2).

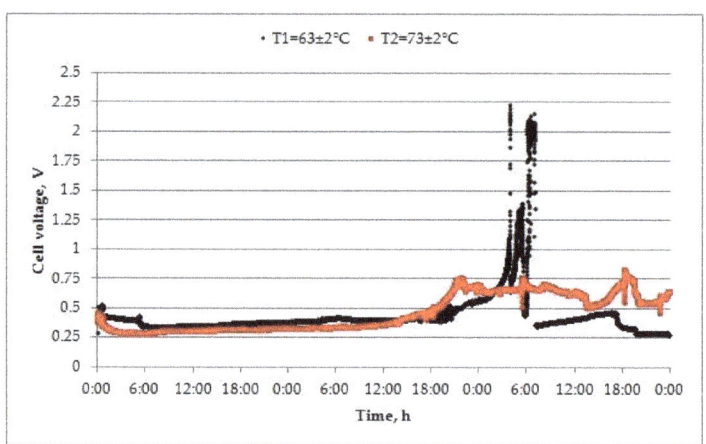

Figure 1. Cell voltage changing during the electrorefining of anode with sample code An-1, for different working electrolyte temperatures, 72 h duration.

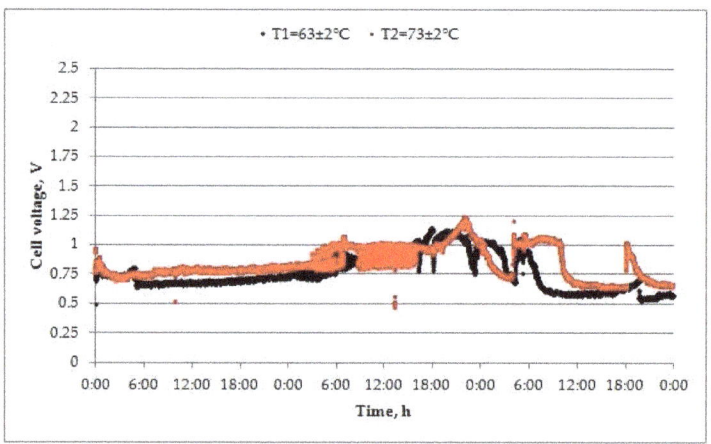

Figure 2. Cell voltage changing during the electrorefining of the anode with sample code An-2, for different working electrolyte temperatures, 72 h duration.

The changing of the cell voltage during each test (Figures 1 and 2) was discussed on the basis of several characteristic changes of cell voltage during the non-commercial electrorefining process [24]. Further explanation of data from Figures 1 and 2 will be appointed for the first appearance points of each phase.

From Figure 1 which that corresponds to the anode with sample code An-1, the time of active dissolution was 42:19 h (T1 = 63 ± 2 °C). During that time period, the cell voltage oscillation was in the range of 0.2 V. After the end of the first phase period, which is called the "active dissolution phase", the value for cell voltage oscillation was more than 0.2 V, and this phase was called the "oscillation phase". The duration of the first appearance of the oscillation phase was approximately 2 h for the test on T1 = 63 ± 2 °C. The phase when the cell voltage suddenly was increased is called the "occurrence of full passivation". The first appearance of the suddenly increased cell voltage (2.22 V, Figure 1) is registered after 51:49 h (anode An-1) on T1 = 63 ± 2 °C. The cell voltage changing at a higher working electrolyte temperature (T2 = 73 ± 2 °C) had asimilar trend but without the appearance of the passivation peak. It is confirmed by a higher value of dissolved anode mass (Table 2).

It is characteristic that the total anode passivation time for the anode with sample code An-1 was a few minutes, and the lower value of the dissolved anode mass at the working electrolyte temperature (T1 = 63 ± 2 °C) can be explained by a layer of anode slime on the anode surface. Therefore, for the test at higher temperature (T2 = 73 ± 2 °C), the cell voltage oscillation of more than 0.2 V was not registered, but an increase of cell voltage has appeared at the same time when the passivation peak at a lower working electrolyte temperature (T1 = 63 ± 2 °C) appeared. The electrorefining process continues on the increased value of cell voltage (0.75 V). This value of cell voltage could be explained as a consequence of particular anode passivation (Figure 1). The "full-time passivation phase" could be explained as a phase with a constant high cell voltage value, and this phase does not appear in Figure 1.

The results from Figure 2 have indicated almost identical behavior of the anode with sample code An-2 during the electrorefining process at different working electrolyte temperatures. Electrorefining processes are carried out on the cell voltage from approximately 0.6 to 1.26 V (Figure 2). As a consequence of particular anode passivation, the electrowinning of Cu ions from the electrolyte at both working electrolyte temperatures occurred (Table 3). Additionally, the dissolved anode mass was lower than for the anode with sample code An-1. The cell voltage for the stable phase also had similar values as the anode with sample code An-1. For the test realized at T1 = 63 ± 2 °C, this value was stable at about 25 h, and for the test at T2 = 73 ± 2 °C, the duration of the stable phase was about 26 h (Figure 2). Additionally, the mass of anode slime and Cu content in anode slime confirmed no evident difference during the electrorefining process of the anode with sample code An-2 at a different working electrolyte temperature (Table 4).

Table 4. Anode slime chemical composition, mass of anode slime, and anode slime mass percentage in relation to the dissolved anode mass.

Anode Type	Anode with Sample Code An-1		Anode with Sample Code An-2	
Working Electrolyte Temperature	T1 = 63 ± 2 °C	T2 = 73 ± 2 °C	T1 = 63 ± 2 °C	T2 = 73 ± 2 °C
Element	Element content, wt.%			
Cu	13.1	14.7	59.0	59.8
Ni	3	3.2	1.2	1.4
Pb	29.2	31.9	17.5	16.9
Sn	2.8	3.1	1.3	1.4
Sb	19.7	17.6	6.3	6.4
As	6.2	5.8	2.5	2.1
	Total Mass of Anode Slime, g			
Total Mass of Slime from the Anode Surface and Cell Bottom	59.9	63.06	88.57	72.24
	Anode Slime Mass Percentage in Relation to the Dissolved Anode Mass, wt.%			
Total Anode Slime Mass Percentage	3.55	3.58	6.47	5.01

In relation to the kinetics, the cell voltage and current are theoretically connected over the Tafel approximation of the electrochemical kinetic equation (Butler–Volmer). Once the exchange current density and Tafel slope is known for copper dissolution/deposition, the overpotentials on the cathodic and anodic side can be readily calculated at the known temperature of the refining process. Being of a large exchange current density and low Tafel slope, the copper electrochemical reaction is of rather small overpotentials of several hundreds of mV. However, the hydrodynamics also affects the overpotential, which is hard to calculate for a given cell geometry and can be only gained empirically by the independent measurements, e.g., current vs. flow. In addition, the cell voltage also comprises the ohmic drop within the electrolyte, which can change as the process goes on, and the voltage drops at electric contacts and within the electrodes, which are to be arbitrarily set to reach the measured cell voltage. It follows that any measured cell voltage can be easily compared to the theoretically calculated by assigning the hardly measurable drops (electrolyte, contacts, electrodes, etc.) to the difference between measured and calculated values, which has nothing to do with the refining process itself or to the type of dissolving anode as the scopes of this manuscript. Comparison is important if some specific cell geometry and construction have to be chosen, but in this case, these parameters were fixed in order to investigate the anode dissolution.

3.3. Working Electrolyte Chemical Composition Changing

A real sulfuric waste solution of the following chemical composition (g/L): Cu—32.5; Ni—20.5; As—4; Sb—0.3; Sn—0.001; Pb—0.004 was used as the working electrolyte. Each sample analyzed at 24, 48, and 72 h was taken from the middle of the cells. Sludge precipitation has occurred in samples that were analyzed without any previous preparation. Further, the samples were prepared by the next procedure: 10 mL of the electrolyte from the cell medium, plus 10 mL of concentrated HCl and deionized water up to 50 mL. The appropriated dissolution was used for a higher concentration of the analyzed elements. Monitoring the change in the concentration of Cu, Ni, As, and Sb is presented in Table 3.

Data from Table 3 were used for calculating the changes in Cu, Ni, As, and Sb concentration. The concentrations of those elements in the starting electrolyte were the basic values (100%).

The results presented in Figure 3 have confirmed that the concentration of Cu, As, and Sb ions was decreased during the tests at different temperatures. On the other hand, Ni concentration is increased during the tests. Cu ions' concentration decrease had similar values for both anodes in the same operational conditions (Figure 3a,b). The highest value of Cu ions decreasing is registered for an anode with the sample code An-2 (with Cu content in the anode of 88.78 wt.%). At the same time, the increase in Ni ions' concentration has the highest value of 129.27 wt.% at the end of electrorefining for the same anode. The decrease in Cu ions' concentration in the working electrolyte directly indicates the fact that the electrowinning process of copper ions has occurred at the same time. The identical phenomenon is characteristic for the electrorefining test with anode with the sample code An-1, where Cu content is 94.02 wt.%. At the increased working electrolyte temperature, the values for decreasing Cu, As, and Sb ions concentration were higher in comparison with the values at lower temperature for both anode types.

Literature data have confirmed the negative impact of Ni on the electrorefining process [13,25,26]. Nickel ions' concentration in the working electrolyte up to 25 g/L is a concentration that has no harmful effect on cathode precipitate. However, a high Ni ion concentration in the electrolyte reduces the copper solubility at the anode/electrolyte interface and leads to partial or full anode passivation and interruption of the electrorefining process (Figure 2) [17,24].

The highest decrease in Cu ions' concentration (95.38 wt.%) has been achieved by the electrorefining of the anode with sample code An-2 at the working electrolyte temperature of T2 = 73 ± 2 °C (Figure 3). By comparison of the values for decreasing the Cu ions; concentration at the end of the process for different anode types, it could be seen that the Cu ions' concentration decrease is higher by about 30% during the electrorefining process of the Cu anode with additional 10 wt.% Ni. It is the same for both working electrolyte temperatures.

The same behavior is characteristic for changing the As ions' concentration. The highest decrease in As ions' concentration (80 wt.%) was also registered during the electrorefining process of anode with the sample code An-2 at the working electrolyte temperature of T2 = 73 ± 2 °C (Figure 3).

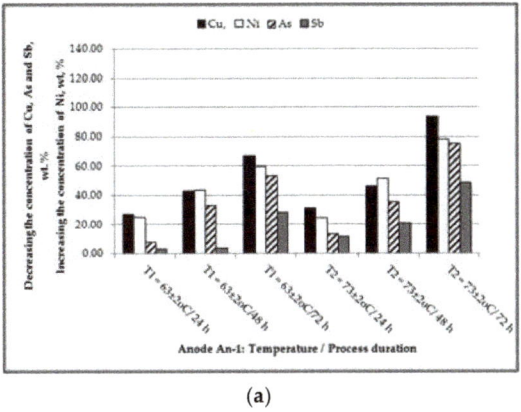 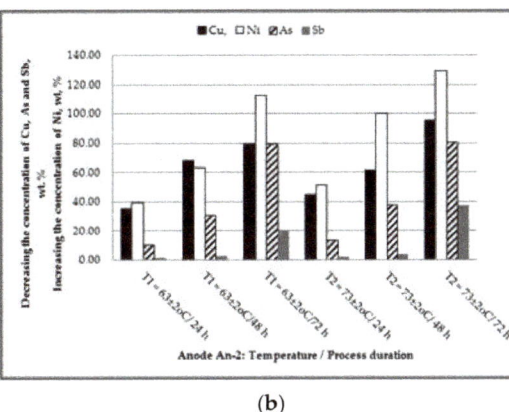

Figure 3. Cu, Ni, As, and Sb concentration changing during the electrorefining process for anodes: (**a**) with sample code An-1 and (**b**) with sample code An-2.

The decrease in Sb ions' concentration is also registered for a higher temperature of the working electrolyte (T2 = 73 ± 2 °C), and this value had the maximum value for the anode with sample code An-1 of 48 wt.%.

The increased concentration of Ni ions, as a consequence of Ni electrolytic dissolution from anodes, had the highest value for the anode with sample code An-2 (129.27%). This value is registered at the end of the electrorefining process that was carried out at a working electrolyte temperature of T2 = 73 ± 2 °C. This value is about 30% higher than the value of Ni ions, increasing during the process that was carried out at a lower working electrolyte temperature. In the case of changing the Cu ions concentration, it could be a consequence of the increased dissolution of copper and nickel salts at an elevated temperature. However, a high concentration of nickel in the electrolyte reduces the solubility of copper at the anode–electrolyte interface and leads to the passivation of the anode and interruption of the electrolysis process [27].

3.4. Anode Slime

The slime formed during the electrorefining process of Cu anodes with non-commercial chemical composition is a result of electrochemical and chemical processes onto the anodes and in the working electrolytes. The slime originates from the anode surface and from the bottom of the electrolytic cell. At the end of each test, the anode slime was stripped from the anode surfaces, separately filtered, and after washing and drying, individually measured in order to compare the mass of slime present in the working electrolyte. The sample used for chemical characterization was a composite sample from the anode surface and electrolytic cell. The anode slime sample of 0.5 g was dissolved with aqua regia, transferred into a volumetric flask of 50 mL, and completed with deionized water. The appropriated dissolution was used for a higher concentration of analyzed elements.

The results from Table 4 show that the anode slime makes up 3.55–6.47 wt.% of dissolved anodes mass. Those values are higher than in the commercial electrorefining copper anodes process, where the anode slime makes up 0.2–0.8 wt.% [28].

The Cu content has a higher value in the anode slime obtained during the electrorefining of the anode with sample code An-2. Decreasing the Cu ions' concentration in the working electrolyte has a negative impact on the morphology of the cathode deposit. Additionally, the elementary copper fell off from a cathode sheet as a consequence of the non-compact cathode deposit (Figure 2 and data from Table 3) [17]. Values for the cell voltage during the electrorefining tests had a similar value in both cases (Figure 2). The oscillation phase had similar values, and cell voltage had a higher value than the values in the standard copper electrorefining process [29]. The maximum reduction of Cu ions' concentration in the working electrolyte is registered for the anode with the sample code An-2 during the electrorefining process at higher electrolyte temperature (Table 3). The results from Table 3 confirmed that the electrowinning of Cu ions from the working electrolyte is also carried out during 72 h of tests. The cathode deposit has different characteristics in comparison with the standard Cu electrorefining process, and a part of cathode copper was dropped into anode slime in a form of copper (I) oxide or as copper powder. The copper content in the anode with the sample code An-1 has a lower value, and it is in accordance with changing the Cu ions concentration in the working electrolyte during each test. The presence of nickel in anode slime is explained as a consequence of electrolyte inclusion in anode slime. Sn content has similar values for both anode types. The Sb/As ratio is about 3/1 for each anode slime sample.

XRD Analysis of Anode Slime

The X-ray diffraction (XRD) (model Explorer, G.N.R. srl, Via Torino, Italy) was used for the anode slime phase analysis. The device uses CuKα X-ray radiation of wavelength 1.54 Å. The X-ray tube voltage is 40 kV, and current consumption is 30 mA. The detector is a scintillation counter, and the geometry of the device is θ-θ. Each analysis lasted 40 min. Match!, Version 2 and "Crystallography Open Database" COD REV44788 databases were used for the phase identification from the powder diffraction. The Match! identifies a phase in a sample by comparing its powder diffraction pattern to the reference patterns of known phases. Hence, it needs a so-called "reference database" in which these reference patterns are provided. The Match! is extremely flexible in this context, with several options for obtaining/using a reference database. A reference database based on the COD is installed automatically along with Match!. This reference database contains the powder diffraction patterns calculated from the crystal structure data taken from the COD, which itself provides the crystal structure data published by the IUCr journals, the "American Mineralogist Cristal Structure Databases" (AMCSD), and various other sources. All entries taken from the COD reference database contain the atomic coordinates, based on which the corresponding powder diffraction patterns have been calculated. Besides this, the I/Ic-values have been calculated for all entries, so that a semi-quantitative analysis can be carried out. Each sample is recorded in the interval from 10° to 70° 2θ with a step of 0.05° 2θ. The anode slime samples were prepared in a powder form of a particle size of 5 to 10 microns in an agate mortar.

The quantitative results are the results of a software evaluation that is automatically calculated as each new phase is entered into the analysis results. The Match! software package, version 2.0 with the COD reference database was used for identifying the present phases and determining their content in the anode slime obtained during the electrorefining process of the anode with the sample code An-1 and An-2. The phase identification was performed by comparing the obtained data to data from a standard database. The content of the phases was calculated using the software and as such can be used as the relative data to compare the number of phases.

The results of each tested anode slime sample (from anode with the sample code An-1 and An-2), the corresponding patterns (Figures 4–7), and identified phases with formulas and names, expressed in the weight percent (wt.%) (Table 5), are presented. Peaks that belong to one phase are denoted by the same letter symbol.

It can be seen from Figure 4a that the major crystalline phases in the homogenized sample of anode slime were Anglesite ($PbSO_4$), Copper(I) arsenide Domeykite low (Cu_3As), Antimony arsenate ($SbAsO_4$), Copper(I) oxide Cuprite (Cu_2O), Claudetite (As_2O_3), Litharge (PbO), and Senarmontite (Sb_2O_3).

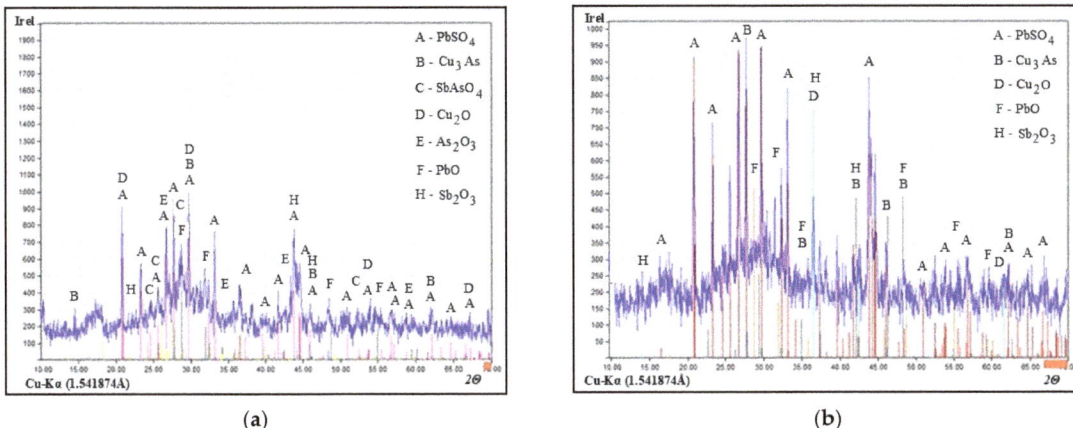

Figure 4. X-ray diffraction patterns of homogenized slime after electrorefining of anode with the sample code An-1: (**a**) working electrolyte temperature of T1 = 63 ± 2 °C and (**b**) working electrolyte temperature of T2 = 73 ± 2 °C.

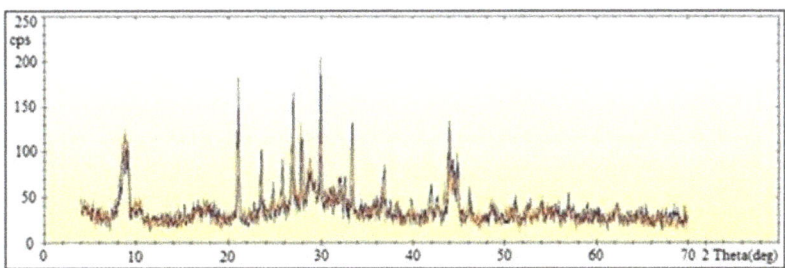

Figure 5. Comparative presentation of anode slime phase analyzes. Red represents the patterns at a temperature of T1 = 63 ± 2 °C, and blue represents the patterns at a temperature of T2 = 73 ± 2 °C for the anode with sample code An-1.

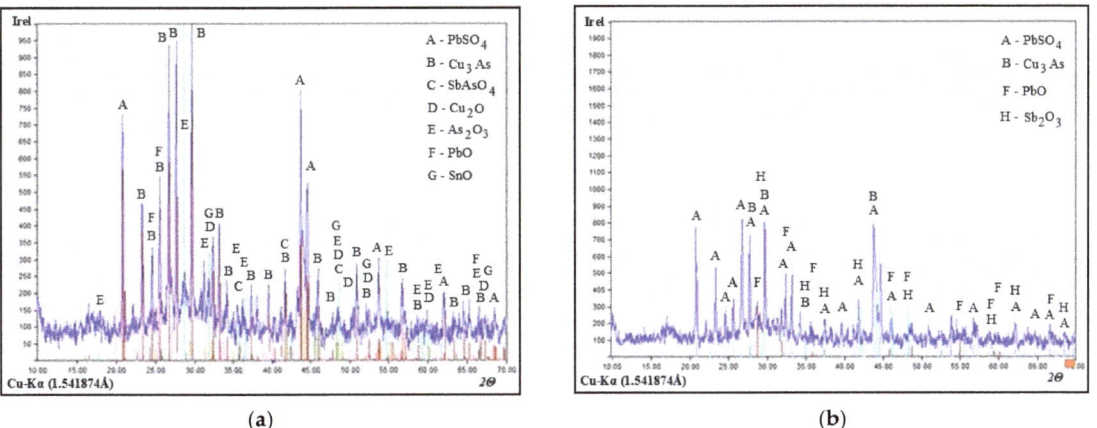

Figure 6. X-ray diffraction patterns of homogenized slime after the electrorefining of the anode with sample code An-2: (**a**) working electrolyte temperature of T1 = 63 ± 2 °C and (**b**) working electrolyte temperature of T2 = 73 ± 2 °C.

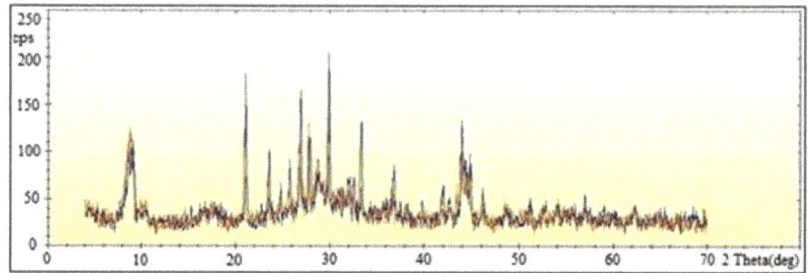

Figure 7. Comparative presentation of anode slime phase analyses. Red represents the patterns at a temperature of T1 = 63 ± 2 °C, and blue represents the patterns at a temperature of T2 = 73 ± 2 °C for the anode with sample code An-2.

With the increase in the working electrolyte temperature, the major crystalline phases were reduced on $PbSO_4$, Cu_3As, Cu_2O, PbO, and Sb_2O_3. PbO peaks intensity is decreased in the anode slime obtained by the electrorefining process at higher working electrolyte temperature. From data presented in Table 5, the phase As_2O_3 is present at a value of 30.4 wt.% and $SbAsO_4$ in value of 7.5 wt.% for anode slime obtained at a temperature of T1 = 63 ± 2 °C. Those phases are not registered in slime obtained from the same anode at temperature of T2 = 73 ± 2 °C. The chemical composition of anode slime confirms that the amount of As and Sb at a higher temperature is lower compared to the values at lower working electrolyte temperature (Table 4).

Based on the patterns from Figure 4, $PbSO_4$ has a more crystalline form in anode slime obtained at a lower temperature. Clusters at lower temperatures tend to reach the crystalline form for copper. Cu (I) in the form of Cu_2O and Cu_3As is more present in anode slime obtained at higher working electrolyte temperature based on the fact that the copper clusters achieved a more difficult crystalline form at lower temperatures (Table 5) and based on the fact that the concentration of Cu(I) ions in the electrolyte increases with the increase in the working electrolyte temperature [30]. Because of that, the more complete crystallization of Cu (I) ions was achieved at a temperature of T2 = 73 ± 2 °C. This is also confirmed by the values of the estimated mass percentage of these compounds given in Table 5.

Table 5. Estimated percentage composition of the registered phases in anode slime from different types of electrorefining processes.

Symbol	Pattern Numbers	Phase Name	Chemical Formula	Anode with Samples Code An-1, (wt.%)		Anode with Samples Code An-2, (wt.%)	
				T1 = 63 ± 2 °C	T2 = 73 ± 2 °C	T1 = 63 ± 2 °C	T2 = 73 ± 2 °C
A	[96-900-0653] [96-900-5525]	Anglesite	$PbSO_4$	45.0	48.4	55.0	75.9
B	[96-101-0976]	Copper(I) arsenide Domeykite low	Cu_3As	4.9	31.2	20.2	20.7
C	[96-901-1215]	* Antimony Arsenate	$SbAsO_4$	7.5	-	16.1	-
D	[96-101-0927] [96-101-0964]	Copper(I) oxide Cuprite	Cu_2O	4.2	15.4	6.5	-
E	[96-900-9692] [96-900-9694]	Claudetite	As_2O_3	30.4	-	1.2	-
F	[96-901-2701] [96-901-2700]	Litharge	PbO	6.5	3.9	0.9	2.4
G	[96-110-1036]	Cassiterite	SnO	-	-	0.2	-
H	[96-900-9748]	Senarmontite	Sb_2O_3	1.2	1.2	-	1.0

* The existence of $SbAsO_4$ form was emphasized in the papers [15,31] as presented in those papers.

The presence of Ni phases was not observed. This indicates that the Ni content in anode slime is very low, which is confirmed by the results of anode slime chemical composition given in Table 4.

Figure 5 presents a comparative presentation of the phase analyses of anode slime obtained during the electrorefining processes of the anode with sample code An-1 at different working electrolyte temperatures.

Overlapping patterns confirmed that the phases are better crystallized at a higher temperature. The appearance of higher concentrations of Cu_3As and Cu_2O gave the higher peak intensities at a higher temperature of $T2 = 73 \pm 2$ °C, and this was confirmed with the results given in Table 5.

Figure 6 shows a phase diagram of anode slime obtained by the electrorefining processes of the anode with sample code An-2 at temperatures (a) $T1 = 63 \pm 2$ °C and (b) $T2 = 73 \pm 2$ °C. Characteristic for those patterns is a large number of peaks and overlapping the peaks of some phases. $PbSO_4$, As_2O_3, $SbAsO_5$, Cu_3As, Cu_2O, PbO, SnO, and Sb_2O_3 are detected in anode slime and presented by the X-ray diffraction patterns in Figure 6a. The next phases, $PbSO_4$, Cu_3As, PbO, and Sb_2O_3, are detected in the anode slime sample obtained at $T2 = 73 \pm 2$ °C (Figure 6b). $SbAsO_4$, Cu_2O, and As_2O_3 are presented in the mass percentage of 16.1, 6.5, and 1.2, respectively, only in the anode slime obtained at a temperature of $T1 = 63 \pm 2$ °C (Table 5, Figure 6b).

The main characteristic for patterns in Figures 4 and 6 is that $SbAsO_4$ and As_2O_3 are not registered in the anode slime samples obtained during the electrorefining of anodes with sample code An-1 and An-2 at higher working electrolyte temperature ($T2 = 73 \pm 2$ °C).

Figure 7 gives a comparative presentation of the phase analyzes of anode slime obtained during the electrorefining processes of the anode with sample code An-2 at different working electrolyte temperatures.

Overlapping patterns indicated that the phases are better crystallized at a higher temperature. The appearance of higher concentrations of $PbSO_4$, Cu_3As, and PbO gave the higher peak intensities at a higher temperature, and it is confirmed by the results given in Table 5. The results shown in Table 5 are semi-quantitative, and those data were not the exact data. Those data were used in order to compare the characteristics of anode slime obtained from the same non-standard copper anode at different temperatures.

The amorphous part of the material obtained during the electrorefining processes of anodes could not be discussed, although a form of X-ray patterns confirmed the presence of amorphous material, so that the percentage composition of crystallized phase is not a precise figure, but estimates by the software automatically calculate and apply only for the crystalline phase. The formation of the floating slimes with the amorphous characteristics was associated with oxidation of Sb(III) to Sb(V) [32,33]. For future investigation, it is possible to estimate the amorphous phase content, for example, by adding a known amount of an internal standard sample (e.g., silicon). The amount of amorphous phase can be estimated from the difference between the exact amount of silicon added and the amount of silicon determined by the XRD.

4. Conclusions

Copper anodes of non-commercial chemical composition regarding the Ni, Pb, Sn, and Sb content were prepared to test the electrorefining process. Ni content was 5 or 10 wt.%, and the content of the main ingredients (Pb, Sn, and Sb) was 0.5 wt.% for each one. The real waste solution of sulfuric acid character that was used as a working electrolyte had the Cu ions of decreased concentration in relation to the commercial value. It was confirmed that the electrorefining process of those anodes can be carried out for a limited number of hours. The decrease in the Cu ions' concentration (max. 95.38 wt.%) in the working electrolyte is a limited parameter for electrorefining the non-commercial copper anodes. The concentration of Ni ions increased, and the highest value of 129.27 wt.% was for the anode with a higher Ni content in the anode. The concentration of As ions in the working electrolyte decreased by up to 80 wt.%, and Sb ions' concentration decreased by a maximum of 48 wt.%. It was also observed that the full passivation did not occur. The masses of cathode deposits confirmed that the Cu ions from the electrolyte were deposited onto cathodes. The mass of anode slime makes up 3.55–6.47 wt.% of the dissolved anode mass. The major crystalline phases in anode slime, detected by the X-ray diffraction analyses, were $PbSO_4$, Cu_3As, $SbAsO_4$, Cu_2O, As_2O_3, PbO, SnO, and Sb_2O_3. A larger number of crystalline phases was registered at a temperature of T1 = 63 ± 2 °C.

Author Contributions: Individual contributions of authors are as follows: conceptualization, R.M., J.S. and S.S.; methodology, R.M. and B.F.; XRD analysis and software, V.K.; validation, V.M. and Z.S.; formal analysis, V.M. and R.M.; investigation, R.M.; resources, B.F., S.S. and R.M.; data curation, R.M. and V.M.; writing—original draft preparation, R.M., V.K. and V.M.; writing—review and editing, R.M., S.S. and V.M.; visualization, Z.S.; supervision, J.S.; project administration, R.M. All authors have read and agreed to the published version of the manuscript.

Funding: Financial support for this study was partly provided by the Ministry of Education, Science and Technological Development of the Republic of Serbia (Contract No. 451-03-9/2021-14/200052).

Institutional Review Board Statement: Not applicable.

Informed Consent Statement: Not applicable.

Data Availability Statement: The data presented in this study are available on request from the corresponding author.

Conflicts of Interest: The authors declare no conflict of interest.

References

1. International Copper Study Group, Lisbon, Portugal. Available online: http://www.icsg.org/index.php/component/jdownloads/finish/170/2876 (accessed on 26 July 2021).
2. Garside, M. Copper Refinery Production Worldwide 2000–2019. Available online: https://www.statista.com/statistics/254917/total-global-copper-production-since-2006/ (accessed on 26 July 2021).
3. Srinivasan, S.; Bommaraju, T. Chapter 3: Electrochemical Technologies and Applications. In *Fuel Cells, from Fundamentals to Application*; Srinivasan, S., Ed.; Springer: Boston, MA, USA, 2006; pp. 93–186.
4. Veilleux, B.; Lafront, A.M.; Ghali, E. Effect of thiourea on nodulation during copper electrorefining using scaled industrial cells. *Can. Metall. Q.* **2001**, *40*, 343–354. [CrossRef]
5. Suzuki, A.; Oue, S.; Kobayashi, S.; Nakano, H. Effects of Additives on the Surface Roughness and Throwing Power of Copper Deposited from Electrorefining Solutions. *Mater. Trans.* **2017**, *58*, 1538–1545. [CrossRef]
6. Abe, S.; Burrows, B.W.; Ettel, V.A. Anode passivation in copper refining. *Can. Metall. Q.* **1980**, *19*, 289–296. [CrossRef]
7. Palaniappa, M.; Jayalakshmi, M.; Prasad, P.M.; Balasubramanian, K. Chronopotentiometric studies on the passivation of industrial copper anode at varying current densities and electrolyte concentrations. *Int. J. Electrochem. Sci.* **2008**, *3*, 452–461.
8. Chen, T.T.; Dutrizac, J.E. *A Mineralogical Study of the Deportment of Impurities during the Electrorefining of Secondary Copper Anodes*; Copper 99-Cobre; Pointe Hilton South Mountain Resort: Phoenix, AZ, USA, 1999; pp. 437–460.
9. Chen, T.T.; Dutrizac, J.E. A Mineralogical Study of the Effect of the Lead Content of Copper Anodes on the Dissolution of Arsenic, Antimony and Bismuth during Copper Electrorefining. *Can. Metall. Q.* **2003**, *42*, 421–432. [CrossRef]
10. Robinson, T.; Quinn, J.; Davenport, W.G.; Karcas, G. Electrolytic Copper Refining–2003 World Tankhouse Operating Data. *Proc. Copp.* **2003**, *5*, 3–66.
11. Forsén, O.; Aromaa, J.; Lundström, M. Primary Copper Smelter and Refinery as a Recycling Plant—A System Integrated Approach to Estimate Secondary Raw Material Tolerance. *Recycling* **2017**, *2*, 19. [CrossRef]
12. Wenzl, C. Structure and Casting Technology of Anodes in Copper Metallurgy. Ph.D. Dissertation, University of Leoben, Leoben, Austria, 2008.
13. Anzinger, A.; Wallner, J.; Wobking, H. Uber die Bedeutung des Elementes Nickel fur die Kupferraffinationselektrolyse einer Sekundarhutte. *BHM-Berg Und Huttenmann. Mon.* **1998**, *143*, 82–85.
14. Swanson, C.E.; Shaw, M.F. *The Interface of Anode Refining with Electrorefining. Converting, Fire Refining and Casting*; McCain, J.D., Floyd, J.M., Eds.; Metals and Materials Society: Pittsburgh, PA, USA, 1993; pp. 269–284.
15. Jafari, S.; Kiviluoma, M.; Kalliomäki, T.; Klindtworth, E.; Aji, A.T.; Aromaa, J.; Lundström, M. Effect of typical impurities for the formation of floating slimes in copper electrorefining. *Int. J. Miner. Process.* **2017**, *168*, 109–115. [CrossRef]
16. Moats, M.; Robinson, T.; Wang, S.; Filzwieser, A.; Siegmund, A.; Davenport, W. Global survey of copper electrorefining operations and practices. *Copper* **2013**, *5*, 307–318.
17. Marković, R.; Friedrich, B.; Stajić–Trošić, J.; Jordović, B.; Jugović, B.; Gvozdenović, M.; Stevanović, J. Behaviour of non-standard composition copper bearing anodes from the copper refining process. *J. Hazard. Mater.* **2010**, *182*, 55–63. [CrossRef] [PubMed]

18. Moats, M.; Robinson, T.; Davenport, W.; Karcas, D.; Demetrio, S. Electrolytic Copper Refining 2007 World Tankhouse Operating Data. *Copper* **2007**, *5*, 195–242.
19. Christian Hermann Domingo Hecker Cartes Process for Optimizing the Process of Copper Electro-Winning and Electro-Refining by Superimposing a Sinusoidal Current over a Continuous Current. Available online: https://patents.google.com/patent/US20110024301A1/en (accessed on 26 July 2021).
20. Zheng, Z. Fundamental Studies of the Anodic Behavior of Thiourea in Copper Electrorefining. Ph.D. Dissertation, University of British Columbia, Vancouver, BC, Canada, 2001.
21. Maatgi, M.K.; Al-Zubi, O.M.; Bheej, H.A. The Influence of Different Parameters on the Electro Refining of Copper. *Int. J. Eng. Inf. Technol.* **2018**, *4*, 120–124.
22. Mubarok, Z.; Antrekowisch, H.; Mori, G. Proceedings of the Sixth International Copper-Cobre Conference, Toronto, ON, Canada, 25–30 August 2007. Available online: https://www.worldcat.org/title/proceedings-of-the-sixth-international-copper-cobre-conference-august-25-30-2007-toronto-ontario-canada/oclc/233786354 (accessed on 26 July 2021).
23. Schlesinger, M.E.; Sole, K.C.; Davenport, W.G. *Extractive Metallurgy of Copper*, 5th ed.; Elsevier: Amsterdam, The Netherlands, 2011; pp. 529–538.
24. Marković, R.; Stevanović, J.S.; Gvozdenović, M.M.; Jugović, B.; Grujić, A.; Nedeljković, D.; Stajić, T.J. Treatment of Waste Copper Electrolytes Using Insoluble and Soluble Anodes. *Int. J. Electrochem. Sci.* **2013**, *8*, 7357–7370.
25. Antrekowitsch, H.; Hein, K.; Paschen, P.; Zavialov, D. Einfluss der struktur und elementverteilung auf das auflosungsverhalten von kupferanoden. *Erzmetall* **1999**, *52*, 337–345.
26. Abe, S.; Takasawa, Y. Prevention of floating slimes precipitation in copper electrorefining. *Electrorefin. Win. Copp.* **1987**, *15*, 87–98.
27. Hoffman, J.E. The Purification of Copper Refinery Electrolytes. *JOM* **2004**, *56*, 30–33. [CrossRef]
28. Hait, J.; Jana, R.K.; Kumar, V.; Sanyal, S.K. Some studies on sulfuric acid leaching of anode slime with additives. *Ind. Eng. Chem. Res.* **2002**, *41*, 6593–6599. [CrossRef]
29. Aromaa, J. Electrochemical Engineering. In *Encyclopedia of Electrochemistry*; Wiley-VHC Verlag GmbH & Co. KGaA: Weinheim, Germany, 2007; pp. 161–196.
30. Bajmakov, V.; Žurin, A. *Elektroliz v Gidrometallurgii*; Metallurgija: Moskva, Russia, 1977; pp. 77–83.
31. Aguilera, E.M.; Vera, M.C.H.; Viñals, J.; Seguel, T.G. Characterization of raw and decopperized anode slimes from a Chilean refinery. *Metall. Mater. Trans. B* **2016**, *47*, 1315–1324. [CrossRef]
32. Hiskey, J.B. Mechanism and thermodynamics of floating slimes formation. In *TT Chen Honorary Symposium on Hydrometallurgy, Electrometallurgy and Materials Characterization*; TMS, John Wiley & Sons, Inc.: Hoboken, NJ, USA, 2012; pp. 101–112.
33. Xue-Wen, W.; Qi-Yuan, C.; Zhou-Lan, Y.; Lian-Sheng, X. Identification of arsenato antimonates in copper anode slimes. *Hydrometallurgy* **2006**, *84*, 211–217. [CrossRef]

Article

Aluminium Recycling in Single- and Multiple-Capillary Laboratory Electrolysis Cells

Andrey Yasinskiy [1,2], Sai Krishna Padamata [1,*], Ilya Moiseenko [1], Srecko Stopic [2], Dominic Feldhaus [2], Bernd Friedrich [2] and Peter Polyakov [1]

[1] Laboratory of Physics and Chemistry of Metallurgical Processes and Materials, Siberian Federal University, Krasnoiarskii rabochii 95, 600025 Krasnoyarsk, Russia; ayasinskiykrsk@gmail.com (A.Y.); ilya9.97@mail.ru (I.M.); P.v.polyakov@mail.ru (P.P.)

[2] IME Process Metallurgy and Metal Recycling, RWTH Aachen University, Intzestraße 3, 52056 Aachen, Germany; sstopic@ime-aachen.de (S.S.); DFeldhaus@metallurgie.rwth-aachen.de (D.F.); bfriedrich@ime-aachen.de (B.F.)

* Correspondence: saikrishnapadamata17@gmail.com

Abstract: This work is a contribution to the approach for Al purification and extraction from scrap using the thin-layer multiple-capillary molten salt electrochemical system. The single- and multiple-capillary cells were designed and used to study the kinetics of aluminium reduction in LiF–AlF$_3$ and equimolar NaCl–KCl with 10 wt.% AlF$_3$ addition at 720–850 °C. The cathodic process on the vertical liquid aluminium electrode in NaCl–KCl (+10 wt.% AlF$_3$) in the 2.5 mm length capillary had mixed kinetics with signs of both diffusion and chemical reaction control. The apparent mass transport coefficient changed from $5.6 \cdot 10^{-3}$ cm.s^{-1} to $13.1 \cdot 10^{-3}$ cm.s^{-1} in the mentioned temperature range. The dependence between the mass transport coefficient and temperature follows an Arrhenius-type behaviour with an activation energy equal to 60.5 kJ.mol^{-1}. In the multiple-capillary laboratory electrolysis cell, galvanostatic electrolysis in a 64LiF–36AlF$_3$ melt showed that the electrochemical refinery can be performed at a current density of 1 A.cm^{-2} or higher with a total voltage drop of around 2.0 V and specific energy consumption of about 6–7 kWh.kg^{-1}. The resistance fluctuated between 0.9 and 1.4 Ω during the electrolysis depending on the current density. Thin-layer aluminium recycling and refinery seems to be a promising approach capable of producing high-purity aluminium with low specific energy consumption.

Keywords: aluminium; thin-layer electrolysis; molten salts; halides; capillary cell

1. Introduction

Aluminium is the second most utilised metal in the world, only outranked by steel, due to its outstanding mechanical and metallurgical properties. Aluminium and its alloys are extensively used in aerospace, electronics, household utensils, construction, etc. The demand for this metal is only growing, and sufficient supply would require a considerable amount of secondary aluminium production [1]. Through primary aluminium production, residues such as salt slag, aluminium dross, and red mud are generated. The environmentally hazardous residues that are generated can be drastically reduced if secondary aluminium productivity increases. Moreover, 93% of CO_2 emissions can be reduced using secondary aluminium production [2]. Processes such as remelting [3], electrolysis [4], and fractional solidification [5] are used for secondary aluminium production. The main disadvantage of aluminium recycling by remelting is that the recovered metal contains a significant number of impurities. Fractional solidification can produce highly pure metal, but it is difficult to accomplish on a large scale. Demand for high-purity aluminium is growing due to its wide range of applications including high-tech areas. It is used for anode foils for aluminium electrolytic capacitors, hard-disk substrates, sputtering targets, and wiring materials for semiconductor devices and liquid crystal display panels. In the near future, new demands for such applications as compact self-ballasted fluorescent lamps, LED bulbs, solar power generation units, and wind-power generation units are expected [6]. The reasons for extensive use of high-purity aluminium are as follows [7]:

- oxide layers having high permittivity and insulation properties can be obtained through surface treatment;
- high-purity aluminium contains only a small number of impurity elements, precipitates, and inclusions;
- it exhibits high electrical and thermal conductivities.

Ultrapure aluminium can be produced using the Hoopes process, but the process demands a high energy consumption (18 kWh kg^{-1}) [8]. An ideal recycling process should be environmentally friendly, have a high metal yield with low impurities, and low energy consumption. A comparison made in [9] shows that the average energy requirement for the remelting process is 2.2 kWh/kg, while the theoretically minimal value is 510 kWh/kg. For primary aluminium production, the average and the minimum values are 26 and 10.2 kWh/kg, respectively.

A recent study suggests that platinum group metals (PGMs) can be electrochemically reduced along with aluminium from the spent catalysts [10]. The γ-Al_2O_3 catalyst carrier dissolves in fluoride melts and is electrochemically decomposed to reduce aluminium [11]. These aluminium-PGM alloys can be separated through electrochemical or pyrometallurgical processes to recover PGMs as well as the liquid aluminium [12]. Molten salt electrolysis in horizontally placed electrodes is considered one of the most promising processes for metal recovery [11]. The NaCl–KCl–AlF_3 (or Na_3AlF_6) melt with the addition of BaF_2 in the horizontal-electrode cell is used at 690–850 °C with an interelectrode distance of ~10 cm in the traditional three-layer refinery; the main drawback of this process is an extremely high energy consumption [11]. Chloride-based molten salts have also shown some promising results in aluminium extraction from aluminium alloys [13]. Yet, it still faces some problems [14] such as:

- extremely high hygroscopicity of $AlCl_3$;
- significant volatility of $AlCl_3$.

These two factors lead to the strong evaporation of electrolytes and hydrolysis of aluminium chloride in a gaseous phase. Another problem that comes from high hygroscopicity is the low corrosion resistance of cell compartments in the presence of water vapour in the molten salt.

This work is a continuation of previously attempted research [15,16], where the electrochemical reduction of aluminium was performed through a thin layer refinery. In [15,16], two different types of single-capillary cells were used to study electrode processes, and the results were comprehensively compared to those obtained in a traditional (non-capillary) cell. It was found that the addition of 10 wt.% of AlF_3 to an equimolar NaCl–KCl melt gives the best results among all tested compositions and allows operating the cell at a current density of 1.4 A.cm^{-2}. A significant problem raised from the study was that the resistance fluctuated from 0.7 to 2.5 Ω. Using a LiF–AlF_3 melt with relatively high electric conductivity was proposed as a means to reduce the resistance in the capillary [16]. The distance between the two liquid electrodes (aluminium alloy and pure aluminium) was reduced in an attempt to drastically minimize the Ohmic voltage drop and to increase the reaction rates due to overlapping diffusion layers. This can be achieved by the introduction of corrosion- and heat-resistant porous ceramic soaked with the electrolyte, which acts as a physical barrier between two aluminium electrodes. Cathodic processes in chloride [17,18] and fluoride [19,20] melts were studied to understand the electroreduction of aluminium. The results show the promising performance of these molten salts as they possess high electrical conductivity. The NaCl–KCl–AlF_3 and LiF–AlF_3 melts were chosen for this study based on their good performance in previous experiments [15,16]. The LiF–AlF_3 melts have a low liquidus temperature and high electrical conductivity. Chloride-based melts have also been considered as low-temperature electrolytes for both aluminium reduction and refinery processes [21]. Alumina is generally not dissolved in chloride melts, although introducing a small proportion of AlF_3 into the melt can improve the solubility and dissolution rate of alumina.

The recycling of aluminium with capillary molten salt electrolysis can be a good alternative to other recycling processes, such as simple remelting, due to several factors. While remelting has a very low specific energy consumption, capillary thin-layer electrolysis is performed in an electrochemical cell (Al–Me)Al | Al^{3+} | Al with EMF equal (or close) to zero, which has been shown in previous work [15], and which makes it possible to produce pure aluminium with much smaller energy consumption than that required for primary production. The EMF depends on the activity of aluminium in Al–Me alloy (where Me are common impurity elements such as Si, Fe, Cu), which were extensively studied [22,23]. The activity coefficient γ of Al in Al–Si is given by the equation [23]:

$$\gamma_{Al} = \frac{1}{X_{Al}} \exp\left(-\frac{\Delta G_T^0}{RT}\right) \frac{a_{Al_2O_3}^{\frac{13}{8}} \cdot a_{Si}^{\frac{3}{4}}}{a_{Al_6Si_{12}O_{13}}^{\frac{3}{8}}} \tag{1}$$

where X_{Al} is the molar fraction of Al in the alloy, a is the activity, T is the temperature, R is the gas constant equal to 8.314 J.(mol.K)$^{-1}$, ΔG_T^0 is the standard Gibbs energy change, and ΔG_T^0 = 138 600–23.89 T (J/mol). In the case of equilibration of Al–Si alloys with their fluorides, the equation should be changed accordingly. Another factor is the absence of carbon in the process and, therefore, the absence of CO_2, CO, and CFx emissions, which take place in primary aluminium production.

In this paper, the effects of temperature, molten salt composition, and the number of capillaries (single vs. multiple) on the electrochemical behaviour of liquid Al electrodes were discussed. Electrochemical characterization of the single- and multiple-capillary systems using equimolar NaCl–KCl with 10 wt.% of AlF_3 and eutectic LiF–AlF_3 at 720–850 °C was performed. The results are intended to exhibit the complexity of the electrode process in the capillary electrolysis cells and to contribute to the development of the aluminium refinery technology in thin layers of molten halides. The previously used single-capillary cell [15] was improved in terms of simple manufacturing, and the pilot multiple-capillary laboratory cell was designed based on previous data.

2. Experiment

The NaCl–KCl–AlF_3 and LiF–AlF_3 melts were prepared from individual LiF, KCl, NaCl, and AlF_3 salts of reagent grade. Initially, anhydrous salts were dried at 400 °C for 4 h. Before the state of the experiments, the electrolytes were heated to operating temperature, then purified using the graphite electrode at 0.2V (vs. the Al^{3+}/Al potential) for 2 h. The undesirable residues were electrodeposited on the graphite surface. Two- and three-electrode cells with both electrodes being liquid aluminium were used for electrochemical measurements. The single- and multiple-capillary cells were used to perform aluminium anodic dissolution and cathodic reduction. Copper was not added to the anode, as is done in the Hoopes process, as there was no need to increase the density of the anode. The single-capillary cell consisted of a boron nitride (BN) two-electrode (working electrode/WE and reference electrode/RE) set-up placed into the graphite crucible (65 mm inner diameter, 70 mm height) filled with aluminium (counter electrode/CE) and molten NaCl–KCl–AlF_3 electrolyte, as shown in Figure 1.

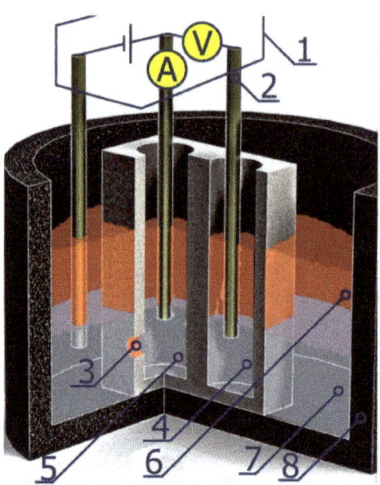

Figure 1. Schematic representation of a single capillary cell: 1—PGSTAT Autolab 302n, 2—tungsten current leads, 3—capillary (0.6 mm diameter, 2.5 mm length) drilled in BN two-electrode set-up, 4—Al/AlF$_3$ reference electrode, 5—Al working electrode, 6—NaCl–KCl (1:1 molar ratio) + 10 wt.%AlF$_3$, 7—Al counter electrode, 8—graphite crucible.

The BN two-electrode setup was a block (length × width × height: 30 × 15 × 72 mm^3) with two closed-end channels (10 mm diameter, 60 mm depth) filled with aluminium and molten salt. The tungsten rods (2 mm diameter) were immersed into aluminium in both channels to serve as current leads. The capillary was drilled out through the wall of the channel, which acted as a working electrode. The other channel was liquid-tight. However, due to the porosity of the BN block, molten salt soaked the walls and made it possible to establish an electrolytic contact between RE and WE. Due to the difference in the surface tension and wettability of the materials by aluminium and fused salt, the capillary contained only salt, with aluminium being kept outside. The apparent surface area of the electrodes in the capillary was 0.003 cm^2. The length of the capillary was 0.25 cm with a diameter of 0.6 mm.

In the multiple-capillary cell, a BN crucible (30 mm inner diameter and 50 mm height) with a perforated wall (10 × 20 mm^2) was immersed into a graphite crucible (65 mm inner diameter, 70 mm height) as shown in Figure 2. The number of capillaries was 840. Each one had a diameter of 0.04 cm and a length of 0.25 cm. The total apparent electrode surface area was 1.055 cm^2. Both crucibles were filled with molten aluminium and electrolytes. Two tungsten 2 mm rods were immersed into the liquid aluminium in both crucibles and acted as current leads.

The PGSTAT302n potentiostat (MetrOhm Autolab B.V., Utrecht, the Netherlands) with the 20 A booster and Nova 2.1.2 software (MetrOhm Autolab B.V., Utrecht, the Netherlands) was used to implement the studies under galvanostatic conditions. The temperature of the furnace was maintained constant by using a USB-TC01 thermocouple module (National Instruments, Austin, TX, USA) and measured using a k-type thermocouple (not shown in figures). Stationary polarization curves were obtained with current densities applied in the range of 0.07 to 15.38 A.cm^{-2}. The Ohmic voltage drop was determined via the I-interrupt technique. The polarization duration before the current interrupt was 120 s, and the interruption duration was 20 s. The duration between potential measurements during interruption was about 100 μs. The value of potential was taken after a rapid drop in the voltage. The experiments were conducted in an air atmosphere with temperature fluctuations of no more than ±3 °C.

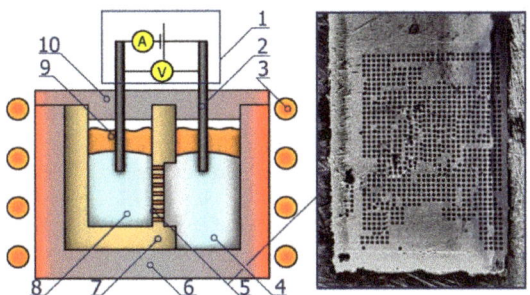

Figure 2. Schematic representation of a multiple-capillary cell: 1—PGSTAT Autolab 302n with BOOSTER20A, 2—tungsten current leads, 3—electric heater, 4—90Al–10Cu (wt.%) working electrode, 5—multiple-capillary wall (0.4 mm diameter, 2.5 mm length, 840 pieces), 6—outer graphite crucible, 7—inner boron nitride crucible, 8—Al counter electrode, 9—64LiF–36AlF$_3$ (molar ratio), 10—boron nitride cover.

3. Results and Discussion
3.1. Single-Capillary Electrolysis

From the previous experiments [12], it was found that an equimolar NaCl–KCl melt with 10 wt.% of AlF$_3$ had a good performance in terms of aluminium reduction kinetics. The same melt was examined to study the effects of temperature on the limiting current density. The results obtained during stationary polarization are shown in Figure 3.

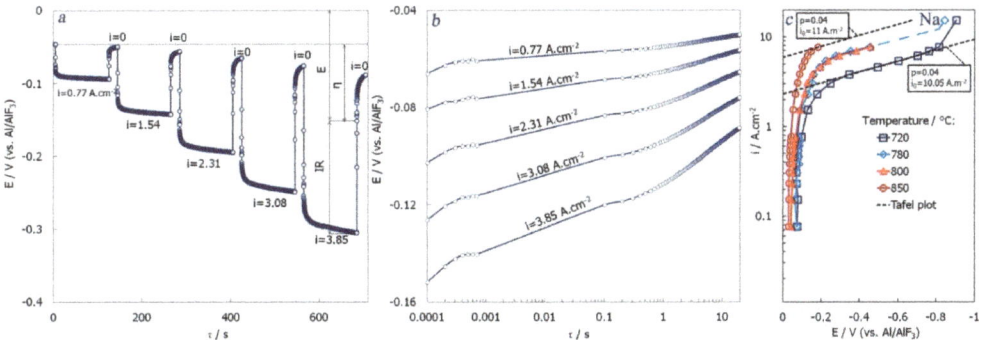

Figure 3. Results obtained in the single-capillary cell with the NaCl–KCl (1:1 molar ratio) + 10 wt.% AlF$_3$ melt: (**a**)—typical potential vs. time dependence obtained at 800 °C, (**b**)—potential relaxation during 20 s current interrupt, (**c**)—stationary galvanostatic polarization curves obtained at 720–850 °C.

The electrode potential was naturally shifted negative during cathodic polarization. The reversible potential was about −0.04 V vs. the Al/AlF$_3$ reference electrode. The stationary polarization curves (Figure 3c) fit the Tafel-type behaviour as they have a linear part in a wide range of current densities and potentials, which is not typical for diffusion-controlled processes. The mixed kinetics is observed due to the co-reduction of Al and alkali metals (Na or K). Co-reduction onset current density relates to the limiting current density i$_l$, which appears in the equation for diffusion-controlled processes [24]:

$$\eta_{conc} = \frac{RT}{zF} \ln\left(1 - \frac{i}{i_l}\right) \quad (2)$$

where η_{conc} is the concentration overvoltage, z is the number of electrons transferred per one atom of Al reduced, F is the Faraday's constant equal to 96485 C.mol^{-1}, and i is the current density.

Since sodium is reduced along with aluminium, Na dissolves in Al and the activity of Na in Al increases. It slightly shifts the electrode potential to more negative values which can be seen in Figure 3a,b. To avoid this situation, the current density for the multiple-capillary electrolysis should be chosen below 2 A.cm^{-2} at 850 °C (and less for lower temperatures), or electrolyte choice should be revised.

Another possible explanation of the linear part of the polarization curve is the activation overvoltage appearance. It is barely possible that charge transfer can be the rate-determining step at high temperatures between 720 and 850 °C. The chemical reaction control seems to be more realistic. If this is the case, then the overvoltage is governed by the equation [24]:

$$\eta_{act} = \frac{RT}{pzF}\ln i_0 + \frac{RT}{pzF}\ln i \quad (3)$$

where η_{act} is the activation overvoltage, p is the reaction order, and i_0 is the exchange current density.

The theoretical curves were plotted in Figure 3c. The reaction order of 0.04 fitted the experimental data for all the temperatures, and the exchange current density increased from 10 to 11 A.m^{-2} with an increase in the temperature.

The kinetic parameters obtained from the stationary polarization curves are summarized in Table 1 where θ is the temperature in °C (while T is used for the absolute temperature in K), $C_{Al^{3+}}$ is the concentration of electroactive particles, OCP is the equilibrium open circuit potential, i_l is the limiting current density found by extrapolation of the linear part of the polarization curves to the OCP value, and K_m is the apparent mass transport coefficient calculated according to the known relation:

$$K_m = \frac{i_l}{zFC} \quad (4)$$

Table 1. Kinetic parameters of aluminium reduction in the single-capillary cell.

θ, °C	$C_{Al^{3+}}$ mol.cm^3	OCP, V	i_l, A.cm^{-2}	$K_m \cdot 10^3$, cm.s^{-1}
720	0.001791	−0.075	2.9	5.6
780	0.001753	−0.073	4.7	9.3
800	0.001740	−0.043	4.9	9.7
850	0.001709	−0.031	6.5	13.1

The change in the concentration is due to a change in the molar volume of the molten salt with the temperature. The apparent mass transport coefficient vs. the temperature dependence was found to follow an Arrhenius-type relation as shown in Figure 4:

$$\left(\frac{d\ln K_m}{dT}\right) = \frac{E_A}{RT^2} \quad (5)$$

where E_A is the activation energy for diffusion.

The dependence between $\ln K_m$ and T^{-1} is close to linear, which allows calculating the diffusion activation energy that equals 60.5 kJ.mol^{-1}.

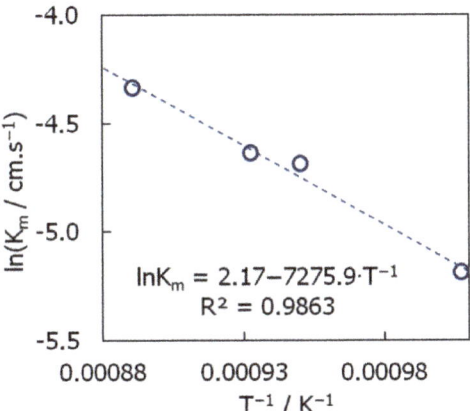

Figure 4. The change in the logarithm of the apparent mass transport coefficient vs. the inversed temperature.

3.2. Multiple-Capillary Electrolysis

The final goal was to establish the electrolysis in a multiple-capillary system to perform aluminium refinery or recycling with high productivity and low energy consumption. Keeping this in mind, the multiple-capillary laboratory cell was designed to perform small-scale short electrolysis tests. The electrolytic system was revised after the performance of the single-capillary electrolysis. Among several tested systems, which namely are NaCl–KCl–AlF$_3$, KF–AlF$_3$ [15], and LiF–AlF$_3$ [10,16], the 64LiF–34AlF$_3$ (mol. %) melt was chosen because of the huge potential window between Al and Li, low liquidus temperature, high electrical conductivity and good performance in the previous experiments [10]. The set of various current densities in the range from 0.01 to 4.74 A.cm^{-2} was applied during each 180 s with current interrupts between the runs to estimate the resistance and the back EMF, which can be found from the equation:

$$\text{Back EMF} = U - IR = \text{EMF} + \eta_a + \eta_c \tag{6}$$

where U is the total voltage, I is the current applied, R is the resistance, EMF is electromotive force (or decomposition voltage), and η_a and η_c are the anodic and the cathodic overvoltages, respectively. The results of the electrolysis runs are presented in Figure 5.

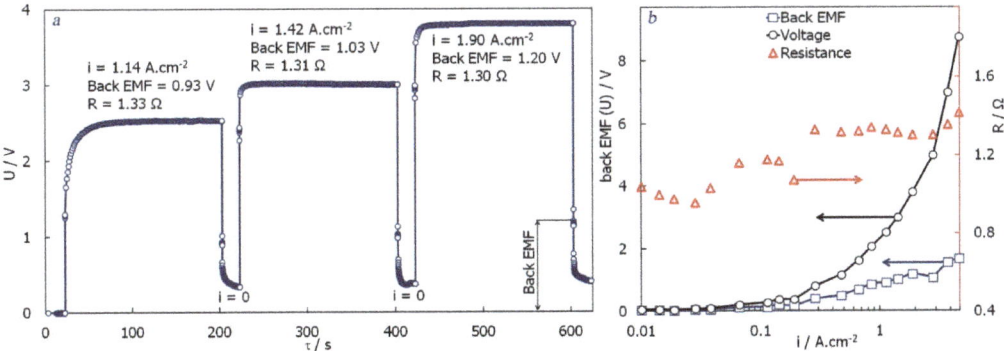

Figure 5. Results obtained in the multiple-capillary cell with the 64LiF–36AlF$_3$ (molar ratio) melt at 800 °C: (**a**)—typical potential vs. time dependence during galvanostatic polarization with current interrupts, (**b**)—quasi-stationary resistance, voltage, and back EMF as a function of applied current density.

The total voltage and the back EMF naturally increased with current density. The resistance changed stepwise. In one series of runs, it slightly decreased probably due to the local increase in the temperature with the current applied. There were a few leaps up in the resistance between the series of runs, which may be caused by an unexpected change in the capillary parameters due to local salt crystallization in separate capillaries during the whole experiment. From Figure 5b, at a current density of 0.9–1.1 A.cm^{-2}, one can expect a total voltage drop of around 2 V, which may result in a specific energy consumption of 6–7 kWh/kg at high current efficiency values above 85%. The energy requirements of this method are three times higher than those of the remelting process, which is 2.2 kWh/kg. However, the metal purity that can be obtained is much higher. While primary aluminium production requires about 26 kWh/kg, and the three-layer refinery process needs 18 kWh/kg, the capillary electrolysis process seems a promising alternative that yields high-purity aluminium with much lower specific energy consumption, which depends mainly on the specific electrical conductivity of the electrolyte and the length of capillaries, while back EMF can be rather low.

The resistance at 0.9–1.1 A.cm^{-2} was about 1.3 Ω, which agrees with values obtained previously in the single-capillary cell [16]; however, it is still rather high. This value can be further reduced by optimizing the capillary parameters (the length and the diameter) and the electrolyte composition.

4. Conclusions

Thin-layer aluminium recycling and refinery seems to be a promising approach capable of producing high-purity aluminium with low specific energy consumption. The single-capillary, three-electrode cell was improved after previous work. The multiple-capillary laboratory electrolysis cell was first present. The main findings from the single- and multiple-capillary electrolysis are:

- the cathodic process on a vertical liquid-aluminium electrode in the NaCl–KCl (+10 wt.% AlF$_3$) in the 2.5 mm length capillary had mixed kinetics with signs of both diffusion and chemical reaction control;
- the apparent mass transport coefficient changed from 5.6×10^{-3} cm.s^{-1} to 13.1×10^{-3} cm.s^{-1}, which is at least 10 times higher than usually observed in traditional molten salt cells;
- the dependence between the mass transport coefficient and the temperature follows an Arrhenius-type behaviour with the activation energy being 60.5 kJ.mol^{-1};
- the presence of sodium or potassium in the electrolyte leads to the co-reduction of these metals with aluminium at relatively low current densities. For the refinery process, it is reasonable to keep the current density below 1 A.cm^{-2} or consider revising the electrolyte (the LiF-AlF$_3$ was tested as a promising candidate);
- the galvanostatic electrolysis in the multiple-capillary cell with the 64LiF–36AlF$_3$ melt showed that the electrochemical refinery can be performed at a current density of 1 A.cm^{-2}, or higher, with the total voltage around 2.0 V and the specific energy consumption about 6–7 kWh.kg^{-1};
- the resistance fluctuated between 0.9 and 1.4 Ω during the electrolysis depending on the current density.

Further efforts should be directed to the study of the effect of electrolysis conditions and capillary parameters on the extraction degree, current efficiency, and aluminium purity. The Ohmic voltage drop should also be reduced to enable refinery with a specific energy consumption of 5 kWh.kg^{-1} or lower.

Author Contributions: Conceptualization, A.Y. and P.P.; funding acquisition, A.Y; investigation, I.M. and A.Y.; methodology, A.Y. and D.F.; supervision, P.P. and B.F.; writing—original draft, A.Y, S.K.P. and S.S. All authors have read and agreed to the published version of the manuscript.

Funding: The presented study was performed with the financial support of the Russian Science Foundation (Grant No. 19-79-00004).

Institutional Review Board Statement: Not applicable.

Informed Consent Statement: Not applicable.

Data Availability Statement: The data presented in this study are available on request from the corresponding author.

Conflicts of Interest: The authors declare no conflict of interest.

References

1. Soo, V.K.; Peeters, J.; Paraskevas, D.; Compston, P.; Doolan, M.; Duflou, J.R. Sustainable aluminium recycling of end-of-life products: A joining techniques perspective. *J. Clean. Prod.* **2018**, *178*, 119–132. [CrossRef]
2. Bureau of International Recycling (BIR). Annual Report 2019. Available online: https://www.bir.org/publications/annual-reports/download/648/1000000235/36?method=view (accessed on 5 May 2021).
3. Eggen, S.; Sandaunet, K.; Kolbeinsen, L.; Kvithyld, A. *Recycling of Aluminium from Mixed Household Waste, Light Metals*; Tomsett, A., Ed.; TMS: San Diego, CA, USA, 2020; p. 1091.
4. Huan, S.; Wang, Y.; Liu, K.; Peng, J.; Di, Y. Impurity Behavior in Aluminum Extraction by Low-Temperature Molten Salt Electrolysis. *J. Electrochem. Soc.* **2020**, *167*, 103503. [CrossRef]
5. Venditti, S.; Eskin, D.; Jacot, A. *Fractional Solidification for Purification of Recycled Aluminium Alloys, Light Metals*; Tomsett, A., Ed.; TMS: San Diego, CA, USA, 2020; p. 1110.
6. Mikubo, S. The Latest Refining Technologies of Segregation Process to Produce High Purity Aluminum. In Proceedings of the 12th International Conference on Aluminium Alloys, Yokohama, Japan, 5–9 September 2010; p. 224.
7. Hoshikawa, H.; Tanaka, I.; Megumi, T. Refining Technology and Low Temperature Properties for High Purity Aluminium, Translated from R&D Report, "SUMITOMO KAGAKU", vol. 2013. Available online: https://www.sumitomo-chem.co.jp/english/rd/report/files/docs/2013E_2.pdf (accessed on 14 May 2021).
8. Kondo, M.; Maeda, H.; Mizuguchi, M. The production of high purity aluminum in Japan. *JOM* **1990**, *42*, 36–37. [CrossRef]
9. Das, S.K.; Long, W.J., III; Hayden, H.W.; Green, J.A.S.; Hunt, W.H., Jr. Energy implications of the changing world of aluminum metal supply. *JOM* **2003**, *56*, 14–17. [CrossRef]
10. Padamata, S.K.; Yasinskiy, A.S.; Polyakov, P.V. The cathodic behavior of aluminum from Pt/Al2O3 catalysts in molten LiF-AlF3-CaF2 and implications for metal recovery from spent catalysts. *J. Electrochem. Soc.* **2021**, *168*, 013505. [CrossRef]
11. Yasinskiy, A.; Polyakov, P.; Varyukhin, D.Y.; Padamata, S.K. *Liquid Bipolar Electrode for Extraction of Aluminium and PGM Concentrate from Spent Catalysts, 150th Annual Meeting & Exhibition Supplemental Proceedings*; TMS: Orlando, FL, USA, 2021; p. 812.
12. Padamata, S.K.; Yasinskiy, A.S.; Polyakov, P.V.; Pavlov, E.A.; Varyukhin, D.Y. Recovery of noble metals from spent catalysts: A review. *Metall. Mater. Trans. B* **2020**, *51*, 2413–2435. [CrossRef]
13. Xu, J.; Zhang, J.; Shi, Z. Extracting aluminum from aluminum alloys in AlCl3-NaCl molten salts. *High Temp. Mater. Proc.* **2013**, *32*, 367–373. [CrossRef]
14. Grjotheim, K.; Matiasovsky, K. Some Problems Concerning Aluminium Electro-plating in Molten Salts. *Acta Chem. Scand. A* **1980**, *34*, 666–670. [CrossRef]
15. Yasinskiy, A.; Polyakov, P.; Yang, Y.; Wang, Z.; Suzdaltsev, A.; Moiseenko, I.; Padamata, S.K. Electrochemical reduction and dissolution of liquid aluminium in thin layers of molten halides. *Electrochim. Acta* **2021**, *366*, 137436. [CrossRef]
16. Yasinskiy, A.; Polyakov, P.; Moiseenko, I.; Padamata, S.K. *Electrochemical Reduction and Dissolution of Aluminium in a Thin-Layer Refinery Process, Light Metals*; TMS: Orlando, FL, USA, 2021; p. 519.
17. Kan, H.; Wang, Z.; Wang, X.; Zhang, N. Electrochemical deposition of aluminum on W electrode from AlCl3-NaCl melts. *Trans. Nonferrous Met. Soc. China.* **2010**, *20*, 158–164. [CrossRef]
18. Huan, S.; Wang, Y.; Peng, J.; Di, Y.; Li, B.; Zhang, L. Recovery of aluminum from waste aluminum alloy by low-temperature molten salt electrolysis. *Miner. Eng.* **2020**, *154*, 106386. [CrossRef]
19. Nikolaev, A.Y.; Suzdaltsev, A.V.; Zaikov, Y.P. Cathode Process in the KF-AlF3-Al2O3 Melts. *J. Electrochem. Soc.* **2019**, *166*, D784. [CrossRef]
20. Suzdaltsev, A.V.; Nikolaev, A.Y.; Zaikov, Y.P. Towards the Stability of Low-Temperature Aluminum Electrolysis. *J. Electrochem. Soc.* **2021**, *168*, 046521. [CrossRef]
21. Thonstad, J.; Fellner, P.; Haarberg, G.M.; Híveš, J.; Kvande, H.; Sterten, Â. *Aluminium Electrolysis: Fundamentals of the Hall-Hérout Process*, 3rd ed.; Aluminium: Dusseldorf, Germany, 2001; p. 359.
22. Murray, J.L.; McAlister, A.J. The Al-Si (Aluminum-Silicon) System. *Bull. Alloy Phase Diagr.* **1984**, *5*, 74–84. [CrossRef]
23. Miki, T.; Morita, K.; Sano, N. Thermodynamic Properties of Aluminum, Magnesium, and Calcium in Molten Silicon. *Metall. Mater. Trans. B* **1998**, *29*, 1043–1049. [CrossRef]
24. Bard, A.J.; Faulkner, L.R. *Electrochemical Methods. Fundamentals and Applications*, 2nd ed.; John Wiley & Sons, Inc.: New York, NY, USA, 2001; p. 833.

Article

One Step Production of Silver-Copper (AgCu) Nanoparticles

Münevver Köroğlu [1], Burçak Ebin [2], Srecko Stopic [3,*], Sebahattin Gürmen [1] and Bernd Friedrich [3]

- [1] Department of Metallurgical & Materials Engineering, Istanbul Technical University, Istanbul 34469, Turkey; koroglu@itu.edu.tr (M.K.); gurmen@itu.edu.tr (S.G.)
- [2] Department of Nuclear Chemistry and Industrial Materials Recycling, Chalmers University of Technology, Kemivagen 4, 41296 Gothenburg, Sweden; burcak@chalmers.se
- [3] IME Process Metallurgy and Metal Recycling, RWTH Aachen University, 52056 Aachen, Germany; bfriedrich@ime-aachen.de
- * Correspondence: sstopic@ime-aachen.de; Tel.: +49-176-7826-1674

Abstract: AgCu nanoparticles were prepared through hydrogen-reduction-assisted Ultrasonic Spray Pyrolysis (USP) and the Hydrogen Reduction (HR) method. The changes in the morphology and crystal structure of nanoparticles were studied using different concentrated precursors. The structure and morphology of the mixed crystalline particles were characterized through X-ray diffraction analysis (XRD), scanning electron microscopy (FEG-SEM), transmission electron microscopy (TEM) and Energy-dispersive X-ray spectroscopy (EDS). The average particle size decreased from 364 nm to 224 nm by reducing the initial solution concentration from 0.05 M to 0.4 M. These results indicate that the increase in concentration also increases the grain size. Antibacterial properties of nanoparticles against Escherichia coli were investigated. The obtained results indicate that produced particles show antibacterial activity (100%). The AgCu nanoparticles have the usage potential in different areas of the industry.

Keywords: silver; copper; nanoparticles; ultrasonic spray pyrolysis; antibacterial

Citation: Köroğlu, M.; Ebin, B.; Stopic, S.; Gürmen, S.; Friedrich, B. One Step Production of Silver-Copper (AgCu) Nanoparticles. *Metals* **2021**, *11*, 1466. https://doi.org/10.3390/met11091466

Academic Editor: Leonid M. Kustov

Received: 7 August 2021
Accepted: 14 September 2021
Published: 16 September 2021

Publisher's Note: MDPI stays neutral with regard to jurisdictional claims in published maps and institutional affiliations.

Copyright: © 2021 by the authors. Licensee MDPI, Basel, Switzerland. This article is an open access article distributed under the terms and conditions of the Creative Commons Attribution (CC BY) license (https://creativecommons.org/licenses/by/4.0/).

Highlights

AgCu nanoparticles have been produced in one step by ultrasonic spray pyrolysis.
AgCu particle size was controlled by changing the concentration of the solution.
AgCu nanosized particles exhibit improved antibacterial activity.

1. Introduction

In recent years, there has been increasing interest in bimetallic nanoparticles because of their potential applications to magnetism, catalysis and optics. These nanoparticles are often called nanoalloys [1]. Bimetallic nanoparticles, either as alloys or as core–shell structures, exhibit unique electronic, optical and catalytic properties compared to monometallic nanoparticles [2,3]. Several bimetallic nanoparticles have been recommended for use in a catalytic system [4–6]. A series of bimetallic catalysts, such as Cu-Au, Cu-Pd, Cu-In and Cu-Sn, have been introduced to exhibit improved surface activities toward CO. Cu-Pt alloy or Cu-modified Pt electrocatalysts still could show the capability of Cu to reduce CO_2 into hydrocarbon products [7].

Although a lot of work has been done on the preparation of noble metal alloys, there are only a few reports on bimetallic particles of copper, especially with silver. Lattice constants of Ag and Cu are 0.409 nm and 0.361 nm, respectively, and this large difference in the lattice constants of Cu and Ag makes the preparation of their alloy difficult. Additionally, it is difficult to control the simultaneous reduction of Cu and Ag because of the difference in redox potential, and the instability of Cu in an aqueous medium is an added difficulty [2]. The fact that copper (Cu) is an important metal used in modern technologies increases its attention [8]. Nanospheric Cu particles are more attractive than other metals because of their advantages, such as being cheap, easy to find and their wide range of uses [8,9]. Based on these advantages, Cu nanoparticles, capacitor material, catalyst, conductive coating, ink-jet printing technology, conductive paste, insulating material, oil additive and sintering additives can be used [10,11]. In particular, Cu has gained more interest because of its capability to reduce CO_2 into hydrocarbon fuels [12]. It has long been known that silver (Ag) is a very strong antibacterial material in both metallic form and compound forms, and its inhibitory effect on bacteria has been studied by many researchers [13–15]. Ag nanoparticles are a gold standard bacteriostatic agent [16]. Ag is known to be used in antibacterial applications since ancient times, and Ag nanoparticles are often preferred in biosensor applications [17]. Ag nanoparticles play an important role in increasing the sensitivity of biosensors because of their ability to accelerate the transfer of electrons [16–19]. The seriousness of problems with energy supplies and environmental pollution is creating greater interest in fuel cells and lithium batteries [20–22]. Fuel cells produce electricity by electrochemically converting hydrogen and oxygen into water, and noble metals, such as Pt, are used as a catalyst for the oxygen reduction reaction. However, the high cost of Pt has sparked a search for a Pt substitute or new ways of reducing the quantity of Pt required. AgCu bimetallic nanoparticles have proper adsorption strength and become a good catalyst for the oxygen reduction reaction. Moreover, Ag and Cu are considerably less expensive than Pt or Pd [6].

Various methods have been proposed to synthesize metallic nanoparticles, including wet chemical reduction, electrochemical, laser ablation and solution combustion [23–27]. Among them, the ultrasonic spray pyrolysis (USP) technique has been rarely used for this purpose. The USP technique was preferred for its low cost and especially for its simplicity for fabricating oxides with good qualities [28]. The wet chemical synthesis is based on the reduction of metals salts by a reducing agent. It consists of many steps to obtain products, and controlling the process is challenging compared to the USP method. Strong reducing agents are necessary for producing metallic nanoparticles, such as sodium borohydride [29], hydrazine [30] and sodium hypophosphite [31]. USP is a process in which solid particles are produced by evaporation, drying and thermal decomposition/reduction processes in a controlled atmosphere, starting from droplets obtained from ultrasonic frequency from metal salt precursors. Single-step and atmospheric pressure droplet to particle conversion and particle collection processes in USP results in spherical, needle-like, plate, flower-like, diagonal and micro- or nanosized metal, as well as oxide, ceramic, carbon-based or nanocomposite-agglomerated materials with a narrow size distribution [32–35].

In this study, we aimed to produce nanoalloy particles, which can be used in energy supplies (fuel cells, lithium batteries) and antibacterial products. Since there is no previous study that has been reported on the synthesis of the AgCu nanoalloy by the USP-HR method, the antibacterial particles were prepared with the one-step method with a controlled Ag content and particle morphology, which is the original aspect presented in this study. In comparison to the previous synthesis of single nanosized particles of copper and silver, this USP synthesis from mixed precursors will offer the improved characteristics of the final AgCu particles [36–40].

2. Experimental

AgCu nanosized particles were synthesized using the aqueous solution of silver nitrate ($AgNO_3$) and copper nitrate ($CuNO_3)_2 \cdot 3H_2O$) under 1 L/min H_2 flow rate at an 800 °C reduction temperature. The nitrate salts (all from Merck, Darmstadt, Germany) were dissolved in deionized water and stirred with a magnetic stirrer for 30 min. The metal concentration in the precursor was between 0.05 mol/L and 0.4 mol/L. The precursor solution was atomized using an ultrasonic atomizer with a resonant frequency of 1.3 MHz (RBI-Instrumentation, Meylan, France). The reduction of aerosol droplets occurred at 800 °C in the electrically heated furnace with the heating zone of 0.25 m and the diameter of the quartz tube of 0.02 m (Nabertherm, Germany). The details of the experimental parameters for the synthesis of AgCu nanosized particles are given in Table 1.

Table 1. Experimental parameters.

AgNO$_3$ (mol/L)	Cu(NO$_3$)$_2$ (mol/L)	Temperature (°C)	H$_2$ Flow Rate (L/min)	N$_2$ Flow Rate (L/min)	Ultrasonic Frequency (MHz)
0.05	0.05	800	1.0	0.5	1.3
0.1	0.1	800	1.0	0.5	1.3
0.2	0.2	800	1.0	0.5	1.3
0.4	0.4	800	1.0	0.5	1.3

X-ray diffraction patterns were obtained for the crystal structure determination of alloy particles by the Philips-1700 X-ray diffractometer (Philips, Eindhoven, The Netherlands) employing Cu-Kα radiation. The chemical compositions of particles were analyzed by energy dispersive spectroscopy (EDS). The particle size and morphology of the samples were investigated by field emission scanning electron microscopy (FEG-SEM, JEOL JSM 700F, Tokyo, Japan) and transmission electron microscopy (FEI Tecnai G^{2}F20 S-TWIN-TEM, Hillsboro, OR, USA). The temperature behavior of silver nitrate salt and copper nitrate salts was investigated by using a differential scanning calorimeter and thermal gravimetry (DSC-TG SDT Q600, TA Instrument, New Castle, DE, USA). Moreover, the antibacterial activities of AgCu nanosized particles were evaluated according to the American Society for Testing and Materials (ASTM) E 2149-01 standard test method.

3. Results and Discussion

3.1. Thermodynamic Analysis of Ag and Cu Nitrate Salts

In order to understand the reaction mechanism in the production of nanosized AgCu alloy particles with USP and HR, the breakdown of the Ag and Cu nitrate salts was investigated using thermochemical analysis. For this purpose, the thermal behavior of the Ag and Cu nitrate solutions were investigated, respectively, then the free energy changes of the nitrate salts in the nitrogen and hydrogen atmosphere under the heat dissolution/reduction reactions were investigated using the enthalpy (H), entropy (S) and heat capacity (C) HSC (Outotec, Espoo, Finland) program. The thermal behavior of the nitrate salts used as starting material in experimental studies was carried out by using Differential Scanning Calorimeter-Thermal Gravimetry (DSC-TG), (TA Instrument, New Castle, DE, USA) between the room temperature and 1000 °C using a heating rate of 10 °C/min in a nitrogen atmosphere. Figures 1 and 2 show the thermal behavior of the AgNO$_3$ and Cu(NO$_3$)$_2$.3H$_2$O salts, respectively.

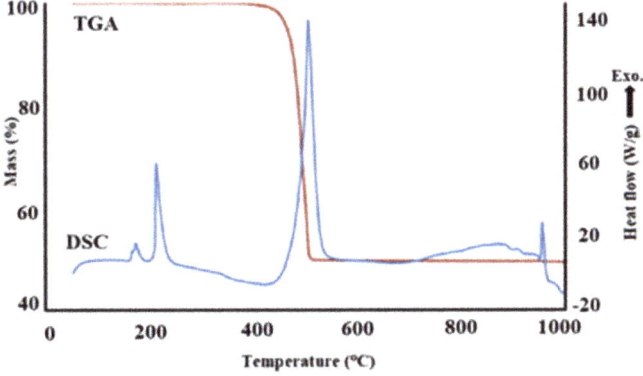

Figure 1. DSC-TGA analysis of AgNO$_3$ salt.

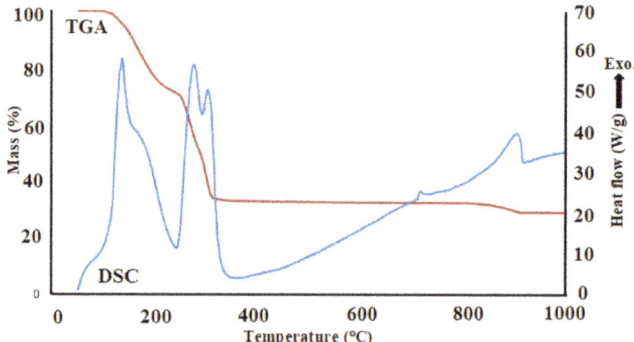

Figure 2. DSC-TGA analysis of (CuNO$_3$)$_2$·3H$_2$O) salt.

The expected reaction of the thermal decomposition ware proposed below with Equations (1) and (2):

$$2AgNO_3 = 2Ag + 2NO_2(g) + O_2(g) \quad (1)$$

$$2Cu(NO_3)_2 = 2CuO + 4NO_2(g) + O_2(g) \quad (2)$$

The DSC curve shown in Figure 1 gives a peak indicating that there is an endothermic reaction at a temperature of about 200 °C, while the TG curve indicates some loss of mass at this temperature. The loss of this mass is due to the small amount of crystal water present in the AgNO$_3$ salt. The TG curve indicates a significant loss of mass (about 40%) in the structure around 400 °C, and this mass loss continues to a temperature of 500 °C. After 500 °C, it is observed that there is no mass loss in the structure, and the structure maintains its stability. Considering this and the endothermic reaction peak shown by the DSC curve at the same temperature, it can be said that the AgNO$_3$ salt of this mass loss is subjected to thermal breakdown and the NO$_2$ gas is away from the structure. The stable structure resulting from this thermal decomposition (after 500 °C) is silver and indicates the melting point of endothermic peak silver at about 950 °C.

The DSC curve shown in Figure 2 shows that there is a mass loss of up to about 250 °C. This mass loss is caused by the removal of the crystal water in the Cu(NO$_3$)$_2$ salt. The DSC curve seen in the temperature range of this mass loss gives a peak indicating the endothermic reaction. The TG curve indicates a significant loss of mass (approximately 50%) in the structure around 250 °C, and this mass loss continues to a temperature of about 310 °C. After 310 °C, it is observed that there is no mass loss in the structure, and the structure maintains its stability. Considering this and the endothermic reaction peak of the DSC curve at the same temperature, it can be said that this mass loss is caused by the thermal breakdown of Cu(NO$_3$)$_2$ salt. The stable structure resulting from this thermal decomposition (after 310 °C) is Cu and indicates the melting point of the endothermic peak copper formed at about 880 °C.

In particle production by the USP-HR method, the hydrogenation temperature of the aerosols obtained by atomizing the high purity metal salt is of great importance. For this purpose, the HSC program was used in the investigation of nitrogen and hydrogen gases and thermal decomposition thermodynamics of AgNO$_3$ and Cu(NO$_3$)$_2$ salts, which we used in our experiments.

Figure 3 shows the graph of the temperature-free energy change obtained by the FactSage program (FactSage, Montreal, QC, Canada and Aachen, Germany) for decomposition of AgNO$_3$ and Cu(NO$_3$)$_2$. The thermodynamic reaction for hydrogen reduction of AgNO$_3$ and Cu(NO$_3$)$_2$ can be described as in Equations (3) and (4).

$$AgNO_3 + H_2 = Ag + NO_2(g) + H_2O(g) \quad (3)$$

$$Cu(NO_3)_2 + 2H_2(g) \rightarrow Cu + 2NO_2(g) + 2H_2O \quad (4)$$

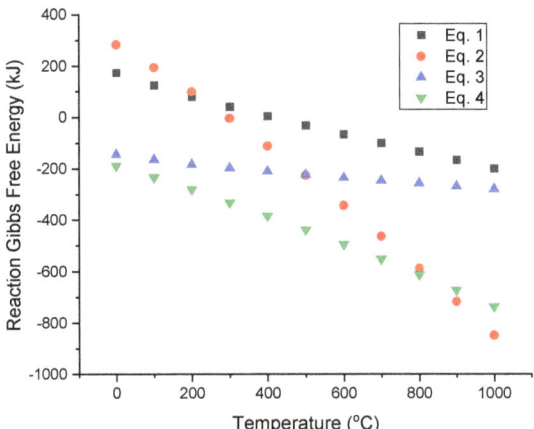

Figure 3. The change of the Gibbs Free Energy value with temperature in the reaction of the $AgNO_3$ salt with hydrogen and nitrogen.

As seen from Figure 3, the reduction of $AgNO_3$ salt with hydrogen is thermodynamically possible at room temperature, where Gibbs free energy is negative, and a decrease in free energy is observed with increasing temperature. In the analysis for the thermal decomposition of $AgNO_3$ in a nitrogen atmosphere, it is seen that the Gibbs free energy change of the reaction remained in the positive zone at low temperatures and decreased with increasing temperature. It is also seen that the thermal decomposition reaction in the nitrogen atmosphere of $AgNO_3$ will begin to occur at 420 °C, where the thermodynamically Gibbs free energy change passes to the negative region. The reduction of $Cu(NO_3)_2$ salt by hydrogen is thermodynamically possible even at 0 °C, where Gibbs free energy is negative. It is also seen that the thermal decomposition reaction in the nitrogen atmosphere of $Cu(NO_3)_2$ begins to occur at 390 °C, where the thermodynamically Gibbs free energy exchange passes to the negative region Equations (1)–(4) and Figure 3 proved that silver and copper could be formed through the hydrogen reduction of silver and copper nitrates.

3.2. Structural Characterization of AgCu Particles

X-ray diffraction patterns of the AgCu alloy nanoparticles produced at an 800 °C reduction temperature using the solutions with different concentrations are given in Figure 4.

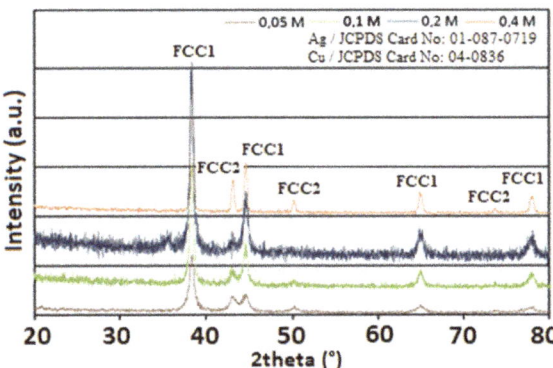

Figure 4. XRD patterns of AgCu nanoparticles (800 °C, 1.0 L/min H_2 flow rate and frequency of 1.3 MHz).

In Figure 4, the peaks at 38°, 44°, 64° and 77° according to 2θ values are assigned to the (111), (200), (220) and (311) reflection lines, and it confirms the formation of a face-centered cubic structure of Ag (JCPDS Card No: 01-087-0719). The face-centered cubic Cu phase at 2θ = 43°, 50° and 74° coincide with (111), (200) and (220) (JCPDS Card No: 04-0836). According to the XRD results, the alloy consisted of FCC1 (α) and FCC2 (β). The diffraction peaks for the stable Ag-rich (α) and Cu-rich (β) phase were observed in Figure 5. In addition, Cu_2O nanoparticles phases were found in the cubic structure at 36° in 2θ values (111) at only a 0.2 M concentration. When the initial solution concentration decreases, the peaks' expansion and the peak's intensity decrease, which can be explained by the decrease in crystalline and particle size. In addition, crystallite sizes were calculated using the Scherrer Equation (5) from the diffraction pattern of the X-ray diffractogram in Figure 5 (see Figure 6).

$$D = \frac{K * \lambda}{B * \cos \theta} \quad (5)$$

where D is the average crystalline size, B is the broadening of the diffraction line measured at half of the maximum intensity, λ is the wavelength (Cu-Kα = 1.541874 Å), θ is the Bragg angle for a given diffraction, and K is a constant, which is a value ranging from 0.85 to 0.9 for powders. Figure 5 shows the average crystalline size of nanosized particles depending on the concentration of the precursor.

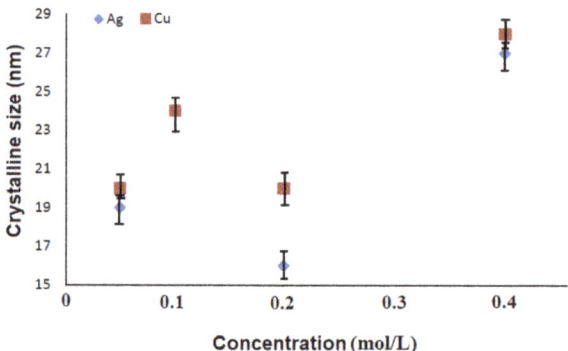

Figure 5. Relationship between concentration and crystalline size (800 °C, 1.0 L/min H_2 flow rate and 1.3 MHz).

Figure 6. SEM analyses of AgCu nanosized particles, (**a**) 0.05 mol/L, (**b**) 0.1 mol/L, (**c**) 0.2 mol/L and (**d**) 0.4 mol/L.

Figure 5 shows the crystalline size calculated using the Scherrer Equation of the nanoparticles produced from the initial solutions with different concentrations, respectively, from 16 nm to 26 nm for silver and from 20 nm to 28 nm for copper. The nanosized particles obtained using 0.2 M solution exhibit the lowest crystallite size.

3.3. Morphological Characterization of AgCu Nanocomposite Particles

SEM images of the particles obtained by increasing solution concentrations (0.05, 0.1, 0.2 and 0.4 mol/L) at 800 °C are given in Figure 6. All samples exhibit spherical shape morphology and almost smooth surfaces. With the reduction of the solution concentration, it is seen that particles that have a finer particle size and generally a narrower particle size distribution are produced. It was observed that large-grained particles together with very fine grains were present in samples produced from solutions with different concentrations. Furthermore, these produced particles show a tendency to cluster. Growing silver-copper nanoparticles showed a tendency to cluster more with a decrease in concentration. The differences in the agglomeration of the particles produced in the environment where all the conditions except the concentration are the same are explained by the surface area and activity of the particles.

EDS analyses of the nanosized particles are given in Table 2. The presence of Ag and Cu was affirmed by EDS analysis. Any possible impurities, such as nitrogen due to undecomposed reactants, were not detected in the EDS spectrums. However, as a result of EDS analysis of the solution with a concentration of 0.05 M, the presence of oxygen here comes from the possible oxide structure in the sample preparation and is not seen as an impurity in the produced particles.

Table 2. EDS results of AgCu nanoparticles produced at different concentrations at 800 °C.

Concentration (mol/L)	Element (%)		
	Ag	Cu	O
0.05	32.2	46.4	21.4
0.1	48.1	51.9	-
0.2	52.1	47.9	-
0.4	49.1	50.9	-

In order to measure the sizes of particles obtained from the initial solutions of different concentrations, the mean particle sizes were calculated by measuring the dimensions of the particles seen on the SEM images with the help of the ImageJ program (NIH, Maryland, USA) (Figure 7).

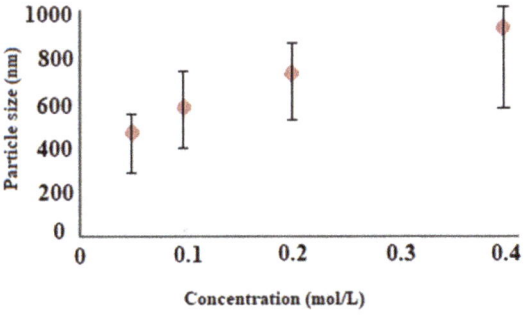

Figure 7. Variation in particle size depending on the initial concentration (800 °C, 1.0 L/min H$_2$ flow and 1.3 MHz).

When the particle size obtained from 0.05 M silver nitrate-copper nitrate was approximately 224 nm in size, the particles were grown with increasing concentrations, and particles the size of 364 nm were formed in the 0.4 M concentration. TEM analyses of AgCu nanoparticles produced from silver nitrate and copper nitrate starting solutions with 0.05 M–0.4 M concentrations by the USP-HR technique were performed using the FEI Tecnai G2 F20 S—TWIN 200 kV STEM/TEM device. Particles stored in ethanol were kept in the ultrasonic homogenizer, and the possible agglomerations were removed from the structure. Then, it was covered with copper grids by the immersion method. The results of these characterization studies are given in Figure 8.

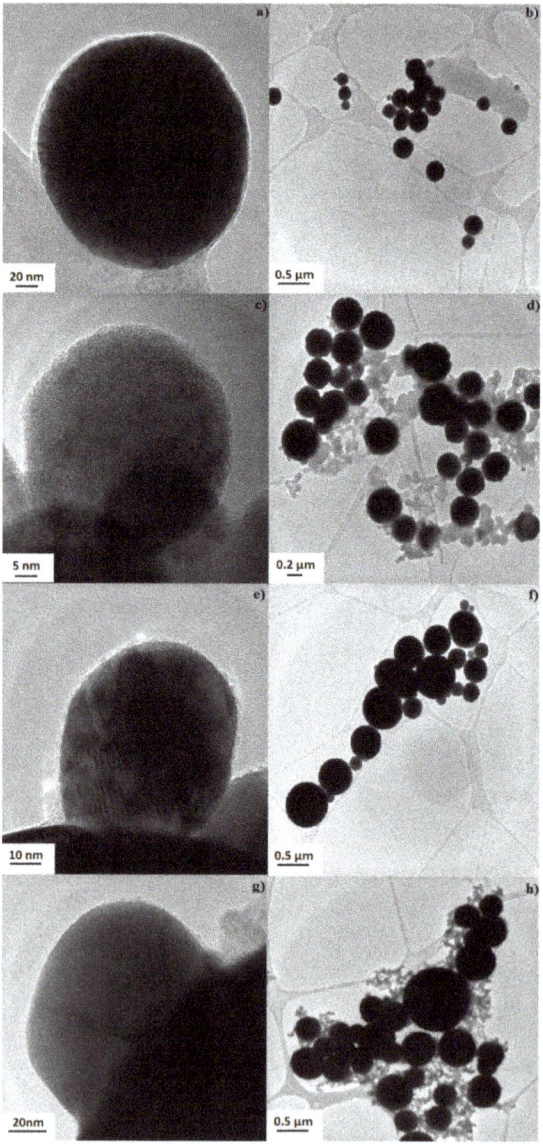

Figure 8. TEM images of AgCu nanosized particles, (**a,b**) 0.05 mol/L (**c,d**) 0.1 mol/L (**e,f**) 0.2 mol/L and (**g,h**) 0.4 mol/L.

When Figure 8 is examined, primary particles with a particle size less than 100 nm from the TEM images have been clearly identified (Figure 8a,c,e,g). In addition, secondary particles (≥100 nm) formed by the incorporation of primary particles are clearly seen (Figure 8b,d,f,h). It was determined that the hollow structure was replaced by the dense particle due to sintering at a high temperature.

3.4. Antibacterial Properties of AgCu Nanosized Particles

The measurement of antibacterial activity was made with the AgCu alloy nanoparticles shown in Table 1. The antibacterial activity of the nanosized particles defined in Table 1 was assessed against E. coli (gram-negative) bacteria (American Type Culture Collection (ATCC) 35218) via planting the bacteria in the agar medium according to the American Society for Testing and Materials (ASTM) E 2149-01 standard under dynamic contact conditions. The measurement was made after incubation at 37 °C for 24 h. This test standard is suitable for particles that do not have migrations property. The antibacterial activity of these particles against E. coli bacteria was given in Table 3.

Table 3. The value of antibacterial activity of the nanosized particles against Escherichia coli bacteria after 24 h.

The Sample of Specimen	Bacterial Reduction (%)
Untreated reference sample	+140.00
0.05 mol/L	−100.00
0.1 mol/L	−100.00
0.2 mol/L	−100.00
0.4 mol/L	−100.00

As shown in Table 3, the improved antibacterial (ASTM E 2149-01) properties of AgCu nanoparticles were observed via a decrease in bacterial reduction (approx. 100%)

4. Conclusions

AgCu nanoparticles were successfully synthesized via the hydrogen-reduction-assisted ultrasonic spray pyrolysis method in one step at 800 °C using an aqueous solution of silver/copper nitrates as a precursor. The USP-HR method was used for the production of AgCu nanoparticles in the desired size and morphology by using nitrogen as the inert gas and hydrogen as the carrier/reducing gas. The effects of various precursor concentrations on the morphology and crystal structure of the AgCu nanoparticles were investigated. The average particle size decreased from 364 nm to 224 nm by reducing the initial solution concentration from 0.4 mol/L to 0.05 mol/L. The size range of the AgCu nanoparticles produced in experimental studies is 20 nm–100 nm for the solution where the concentration is 0.05 mol/L, 100–450 nm for 0.1 mol/L, 180–1100 nm for 0.2 mol/L and 50–1050 nm for 4 mol/L. These results indicate that the increase in concentration also increases the grain size. In XRD analysis, silver and copper particles were determined to be cubic structures. The particle crystalline sizes calculated for the concentrations of 0.05 mol/L, 0.1 mol/L, 0.2 mol/L and 0.4 mol/L according to the Scherrer equation, respectively, for silver, 19.92 nm, 24.3 nm, 15.91 nm and 27 nm, and for copper, 20.84 nm, 24.05 nm, 20.4 nm and 28.7 nm for 34 nm. In the TEM images of nanosized AgCu particles, primary particles of smaller than 100 nm and larger secondary particles were observed. Furthermore, the elimination of 100% bacteria was achieved by all synthesized AgCu nanoparticles. Improved antibacterial (ASTM E 2149-01) properties of AgCu nanoparticles demonstrated that these nanoparticles could be used as antibacterial agents in various areas.

Author Contributions: Conceptualization, M.K., B.E. and S.S.; funding acquisition, S.G.; investigation, M.K. and B.E.; methodology, M.K. and S.G.; supervision, S.G. and B.F.; writing—original draft, M.K., S.G. and S.S. All authors have read and agreed to the published version of the manuscript.

Funding: This research received no external funding.

Institutional Review Board Statement: Not applicable.

Informed Consent Statement: Not applicable.

Data Availability Statement: Not applicable.

Acknowledgments: Authors thank Gultekin Goller, Onur Balci, Ikbal Isik and Technician Huseyin Sezer for SEM, TEM and XRD characterizations and for the antibacterial studies.

Conflicts of Interest: The authors declare no conflict of interest.

References

1. Laasonen, K.; Panizon, E.; Bochicchio, D.; Ferrando, R. Competition between Icosahedral Motifs in AgCu, AgNi, and AgCo Nanoalloys: A Combined Atomistic−DFT Study. *J. Phys. Chem. C* **2013**, *117*, 26405–26413. [CrossRef]
2. Valodkar, M.; Modi, S.; Pal, A.; Thakore, S. Synthesis and anti-bacterial activity of Cu, Ag and Cu–Ag alloy nanoparticles: A green approach. *Mater. Res. Bull.* **2011**, *46*, 384–389. [CrossRef]
3. Jabbareh, M.A.; Monji, F. Thermodynamic modeling of Ag-Cu nanoalloy phase diagram. *Calphad* **2018**, *60*, 208–213. [CrossRef]
4. Khan, N.A.; Uhl, A.; Shaikhutdinov, S.; Freund, H.J. Alumina supported model Pd–Ag catalysts: A combined STM, XPS, TPD and IRAS study. *Surf. Sci.* **2006**, *600*, 1849–1853. [CrossRef]
5. Gao, Y.; Shao, N.; Bulusu, S.; Zeng, X.C. Effective CO Oxidation on Endohedral Gold-Cage Nanoclusters. *J. Phys. Chem. C* **2008**, *112*, 8234–8238. [CrossRef]
6. Shin, K.; Kim, D.H.; Yeo, S.C.; Lee, H.M. Structural stability of AgCu bimetallic nanoparticles and their application as a catalyst: A DFT study. *Catal. Today* **2012**, *185*, 94–98. [CrossRef]
7. Chang, Z.; Huo, S.; Zhang, W.; Fang, J.; Wang, H. The Tunable and Highly Selective Reduction Products on Ag@Cu Bimetallic Catalysts toward CO_2 Electrochemical Reduction Reaction. *J. Phys. Chem. C* **2017**, *121*, 11368–11379. [CrossRef]
8. Ciacci, L.; Vassura, I.; Passarini, F. Urban Mines of Copper: Size and Potential for Recycling in the EU. *Resources* **2017**, *6*, 6. [CrossRef]
9. Shahcheraghi, S.H.; Schaffie, M.; Ranjbar, M. Development of an electrochemical process for production of nano-copper oxides: Agglomeration kinetics modeling. *Ultrason. Sonochem.* **2018**, *44*, 162–170. [CrossRef]
10. Zhu, W.; Zhang, L.; Yang, P.; Chang, X.; Dong, H.; Li, A.; Hu, C.; Huang, Z.; Zhao, Z.J.; Gong, J. Morphological and Compositional Design of Pd-Cu Bimetallic Nanocatalysts with Controllable Product Selectivity toward CO2 Electroreduction. *Nano-Micro Small* **2018**, *14*, 1703314. [CrossRef]
11. Pajor-Swierzy, A.; Farraj, Y.; Kamyshny, A.; Magdassi, S. Effect of carboxylic acids on conductivity of metallic films formed by inks based on copper@silver core-shell particles. *Colloids Surf. A* **2017**, *522*, 320–327. [CrossRef]
12. Costentin, C.; Robert, M.; Saveant, J.M. Catalysis of the Electrochemical Reduction of Carbon Dioxide. *Chem. Soc. Rev.* **2013**, *42*, 2423–2436. [CrossRef]
13. Kawashita, M.; Tsuneyama, S.; Miyaji, F.; Kokubo, T.; Kozuka, H.; Yamamoto, K. Antibacterial silver-containing silica glass prepared by sol-gel method. *Biomaterialia* **2000**, *21*, 393–398. [CrossRef]
14. Qin, D.; Yang, G.; Wang, Y.; Zhou, Y.; Zhang, L. Green synthesis of biocompatible trypsin-conjugated Ag nanocomposite with antibacterial activity. *Appl. Surf. Sci.* **2019**, *469*, 528–536. [CrossRef]
15. Sirelkhatim, A.; Mahmud, S.; Seeni, A.; Kaus, N.H.; Ann, L.C.; Bakhori, S.K.; Hasan, H.; Mohamad, D. Review on zinc oxide nanoparticles: Antibacterial activity and toxicity mechanism. *Nano-Micro Lett.* **2015**, *7*, 219–242. [CrossRef]
16. Hong, H.; Cao, G.; Qu, J.; Deng, Y.; Tang, J. Antibacterial activity of Cu_2O and Ag co-modified rice grains-like ZnO nanocomposites. *J. Mater. Sci. Technol.* **2018**, *34*, 2359–2367. [CrossRef]
17. Ebraiminezhad, A.; Raee, M.; Manafi, Z.; Sotoodeh Jahromi, A.; Ghasemi, Y. Ancient and Novel Forms of Silver in Medicine and Biomedicine. *J. Adv. Med. Sci. Appl. Technol.* **2016**, *2*, 122–128. [CrossRef]
18. Ren, X.; Mena, X.; Chen, D.; Tang, F.; Jiao, J. Using silver nanoparticle to enhance current response of biosensor. *Biosens. Bioelectron.* **2005**, *21*, 433–437. [CrossRef]
19. Zuo, F.; Zhang, C.; Zhang, H.; Tan, X.; Chen, S.; Yuan, R.A. solid-state electrochemiluminescence biosensor for Con A detection based on CeO2@Ag nanoparticles modified graphene quantum dots as signal probe. *Electrochim. Acta* **2019**, *294*, 76–83. [CrossRef]
20. Amiri, A.; Tang, S.; Steinberger-Wilckens, R.; Tade, M.O. Evaluation of fuel diversity in Solid Oxide Fuel Cell system. *Int. J. Hydrogen Energy* **2018**, *43*, 27. [CrossRef]
21. Nitta, N.; Wu, F.; Tae Lee, J.; Yushin, G. Li-ion battery materials: Present and future. *Mater. Today* **2015**, *18*, 252–264. [CrossRef]
22. Zubi, G.; Dufo-Lopez, R.; Carvalho, M.; Pasaoglu, P. The lithium-ion battery: State of the art and future perspective. *Renew. Sustain. Energy Rev.* **2018**, *89*, 292–308. [CrossRef]
23. Simakin, A.V.; Voronov, V.V.; Shafeev, G.A.; Brayner, R.; Bozon-Verduraz, F. Nanodisks of Au and Ag produced by laser ablation in liquid environment. *Chem. Phys. Lett.* **2001**, *348*, 182–186. [CrossRef]
24. Chen, Y.H.; Yeh, C.S. Laser ablation method: Use of surfactants to form the dispersed Ag nanoparticles. *Colloids Surf. A* **2002**, *197*, 133–139. [CrossRef]
25. Sharma, P.; Lotey, G.S.; Singh, S. Solution-combustion: The versatile route to synthesize silver nanoparticles. *J. Nanopart. Res.* **2011**, *13*, 2553–2561. [CrossRef]

26. Yin, B.; Ma, H.; Wang, S.; Chen, S. Electrochemical synthesis of silver nanoparticles under protection of poly(N-vinylpyrrolidone). *J. Phys. Chem. B* **2003**, *107*, 8898–8904. [CrossRef]
27. Ishizaki, T.; Watanabe, R. A New One-Pot Method for the Synthesis of Cu Nanoparticles for Low Temperature Bonding. *J. Mater. Chem.* **2012**, *22*, 25198–25206. [CrossRef]
28. Ouhaibi, A.; Ghamnia, M.; Dahamni, A.; Heresanu, V.; Fauquet, C.; Tonneau, D. The effect of strontium doping on structural and morphological properties of ZnO nanofilms synthesized by ultrasonic spray pyrolysis method. *J. Sci. Adv. Mater. Devices* **2018**, *3*, 29–36. [CrossRef]
29. Qiu, S.; Dong, J.; Chen, G. Preparation of Cu nanoparticle from water-in-oil microemulsions. *J. Colloid Interface Sci.* **1999**, *216*, 230. [CrossRef]
30. Wu, S.H.; Chen, D.H. Synthesis of high-concentration Cu nanoparticles in aqueous CTAB solutions. *J. Colloid Interface Sci.* **2004**, *273*, 165. [CrossRef]
31. Zhu, H.T.; Zhang, C.Y.; Yin, Y.S. Rapid synthesis of copper nanoparticles by sodium hypophosphite reduction in ethylene glycol under microwave irradation. *J. Cryst. Growth* **2004**, *270*, 722. [CrossRef]
32. Stopic, S.; Schroeder, M.; Weirich, T.; Friedrich, B. Synthesis of TiO_2 Core/RuO_2 Shell Particles using Multistep Ultrasonic Spray Pyrolysis. *Mater. Res. Bull.* **2013**, *48*, 3633–3635. [CrossRef]
33. Jokanovic, V.; Spasic, A.M.; Uskokovic, D. Designing of nanostructured hollow TiO2 spheres obtained by ultrasonic spray pyrolysis. *J. Colloid Interface Sci.* **2004**, *278*, 342–352. [CrossRef]
34. Emil, E.; Alkan, G.; Gurmen, S.; Rudolf, R.; Jenko, D.; Friedrich, B. Tuning the Morphology of ZnO Nanostructures with the Ultrasonic Spray Pyrolysis Process. *Metals* **2018**, *8*, 569. [CrossRef]
35. Çakmak, T.; Kaya, E.E.; Küçük, D.; Ebin, B.; Balci, O.; Gürmen, S. Novel Strategy for One-Step Production of Attenuated Ag-Containing AgCu/ZnO Antibacterial-Antifungal Nanocomposite Particles. *Powder Metall. Met. Ceram.* **2020**, *59*, 261–270. [CrossRef]
36. Stopic, S.; Dvorak, P.; Friedrich, B. Synthesis of nanopowder of copper by ultrasonic spray pyrolysis method. *World Metall. Erzmetall* **2005**, *58*, 191–197.
37. Stopic, S.; Friedrich, B.; Dvorak, P. Synthesis of nanosized spherical silver powder by ultrasonic spray pyrolysis. *Metall* **2006**, *60*, 377–382.
38. Jankovic, B.; Stopic, S.; Bogovic, J.; Friedrich, B. Kinetic and thermodynamic investigations of non-isothermal decomposition process of a commercial silver nitrate in argon atmosphere used as the precursors for ultrasonic spray pyrolysis USP. *Chem. Eng. Process.* **2014**, *82*, 71–87. [CrossRef]
39. Bogovic, J.; Schwinger, A.; Stopic, S.; Schroeder, J.; Gaukel, V.; Schuhmann, P.; Friedrich, B. Controlled droplet size distribution in ultrasonic spray pyrolysis. *Metallurgica* **2011**, *10*, 455–459.
40. Emil Kaya, E.; Kaya, O.; Alkan, G.; Gürmen, S.; Stopic, S.; Friedrich, B. New Proposal for Size and Size-Distribution Evaluation of Nanoparticles Synthesized via Ultrasonic Spray Pyrolysis Using Search Algorithm Based on Image-Processing Technique. *Materials* **2020**, *13*, 38. [CrossRef] [PubMed]

Article

Characterization of Defined Pt Particles Prepared by Ultrasonic Spray Pyrolysis for One-Step Synthesis of Supported ORR Composite Catalysts

Gözde Alkan [1], Milica Košević [2,*], Marija Mihailović [2], Srecko Stopic [3], Bernd Friedrich [3], Jasmina Stevanović [2,4] and Vladimir Panić [2,4,5]

1. Deutsches Zentrum für Luft- und Raumfahrt (DLR), Linder Höhe 1, 51147 Cologne, Germany; goezde.alkan@dlr.de
2. Institute of Chemistry, Technology and Metallurgy, University of Belgrade, Njegoševa 12, 11000 Belgrade, Serbia; marija.mihailovic@ihtm.bg.ac.rs (M.M.); j.stevanovic@ihtm.bg.ac.rs (J.S.); panic@ihtm.bg.ac.rs (V.P.)
3. IME Process Metallurgy and Metal Recycling, RWTH Aachen University, 52056 Aachen, Germany; sstopic@ime-aachen.de (S.S.); bfriedrich@ime-aachen.de (B.F.)
4. Centre of Excellence in Environmental Chemistry and Engineering—ICTM, University of Belgrade, 11000 Belgrade, Serbia
5. Chemical-Technological Department, State University of Novi Pazar, 36900 Novi Pazar, Serbia
* Correspondence: milica.kosevic@ihtm.bg.ac.rs; Tel.: +3811-136-40231

Citation: Alkan, G.; Košević, M.; Mihailović, M.; Stopic, S.; Friedrich, B.; Stevanović, J.; Panić, V. Characterization of Defined Pt Particles Prepared by Ultrasonic Spray Pyrolysis for One-Step Synthesis of Supported ORR Composite Catalysts. *Metals* 2022, *12*, 290. https://doi.org/10.3390/met12020290

Academic Editors: Laichang Zhang and Thomas Gries

Received: 26 December 2021
Accepted: 2 February 2022
Published: 8 February 2022

Publisher's Note: MDPI stays neutral with regard to jurisdictional claims in published maps and institutional affiliations.

Copyright: © 2022 by the authors. Licensee MDPI, Basel, Switzerland. This article is an open access article distributed under the terms and conditions of the Creative Commons Attribution (CC BY) license (https:// creativecommons.org/licenses/by/ 4.0/).

Abstract: Polygonal Pt nanoparticles were synthesized using ultrasonic spray pyrolysis (USP) at different precursor concentrations. Physicochemical analysis of the synthesized Pt particles involved thermogravimetric, microscopic, electron diffractive, and light absorptive/refractive characteristics. Electrochemical properties and activity in the oxygen reduction reaction (ORR) of the prepared material were compared to commercial Pt black. Registered electrochemical behavior is correlated to the structural properties of synthesized powders by impedance characteristics in ORR. The reported results confirmed that Pt nanoparticles of a characteristic and uniform size and shape, suitable for incorporation on the surfaces of interactive hosts as catalyst supports, were synthesized. It is found that USP-synthesized Pt involves larger particles than Pt black, with the size being slightly dependent on precursor concentration. Among ORR-active planes, the least active (111) structurally defined the synthesized particles. These two morphological and structural characteristics caused the USP-Pt to be made of lower Pt-intrinsic capacitive and redox currents, as well as of lower ORR activity. Although being of lower activity, USP-Pt is less sensitive to the rate of ORR current perturbations at higher overpotentials. This issue is assigned to less-compact catalyst layers and uniform particle size distribution, and consequently, of activity throughout the catalyst layer with respect to Pt black. These features are considered to positively affect catalyst stability and thus promote USP synthesis for improved properties of host-supported Pt catalysts.

Keywords: electrocatalysis; nanocatalyst; noble metal nanoparticles

1. Introduction

There are numerous contemporary studies dealing with the catalytic improvements of energetics-important electrochemical processes such as hydrogen evolution (HER) and the reduction and oxidation of oxygen or small organic molecules [1]. Particularly, investigations into oxygen reduction reaction (ORR) kinetics are of the highest interest due to the ORR rate-determining characteristics for energy conversion in fuel cells (FCs) and metal–air batteries (M–O_2) [2–5]. The meeting point of FCs and M–O_2 (involving alkaline metal–air batteries, such as Li–O_2 and Na–O_2) clearly exists on the cathode side – in both FC and M–O_2, the electrons for anodic oxidations are provided by ORR at the cathode [6]. Green technology progress not just in energy storage but in the water-splitting domain urges for controllable catalysts for both HER and OER [7,8]. New modifications of morphology and/or electronic structures of transition metal-based catalysts have offered multifunctional solutions in energy storage and conversion as well as in demanded water electrolysis applications [9]. Among various electrode materials and catalysts, such as nanostructured metals, metal oxides, hydroxides, phosphides, and chalcogenides, Pt-based catalysts are still at the forefront owing to their high activity with acceptable stability and slow chemical degradation [10,11].

Besides material selection, surface properties, such as the structure, size, and shape of the nanoparticles, also have a great impact, especially on rather sluggish ORR activity, due to the intrinsic sensitivity of this complex heterogeneous reaction regarding the catalyst surface. Therefore, the synthesis method considerably affects the catalyst ORR activity. The examination of various synthesis methods for Pt nanoparticles (Pt NPs) themselves as well as various Pt-based composites has also been the focus of much research. Chemical precipitation, ion implantation, laser ablation, and chemical reduction have been the most investigated methods so far [12]. In most of the studies, additives were used to control the shape of the fine Pt NPs, such as etchants, adsorbates, surfactants, polymers, or foreign metal ions [4,13,14]. However, the simple synthesis of fine spherical Pt NPs without any additive is rare, especially in bottom-up approaches. Recently, the pulsed laser ablation method in liquids (PLAL) has been utilized by Lau et al. [15]. The successful synthesis of ligand-free pure Pt NPs was achieved and it was reported that possible toxic cross-effects and additional nanoparticle purification steps such as filtration, dialysis, and centrifugation were avoided.

Among various bottom-up approaches, ultrasonic spray pyrolysis (USP) was reported as the one that easily enables the generation of ultrafine, uniform or complex structures with controlled stoichiometry as well as chemical and phase content [16–19], which could be very promising for the synthesis of Pt nanocatalyst with good ORR activity [20,21]. There are a few studies dealing with Pt-based catalyst synthesis by USP, especially in hybrid form with metal oxides such as SiO_2, CeO_2, Al_2O_3, and $FeAl_2O_4$ [22,23], and also with Pt/TiO_2, as carried out by Košević et al. [24], which is of great concern as an interactive Pt support. In these studies, the catalysts synthesized through USP were reported to exhibit superior catalytic activity with respect to wet chemical, colloidal, and dry impregnation methods [25–28]. However, there is a lack of systematic studies dealing with the synthesis of pure nanoparticles through controlled USP.

Therefore, we aimed to synthesize Pt NPs by a precursor solution concentration-dependent USP process and analyze its influence on morphological, structural, and electrochemical properties of Pt NPs. The ORR activity of Pt NPs synthesized exclusively by USP was elucidated and compared to commercial Pt powder to reveal the potential of the further use of USP in the synthesis of complex Pt-based electrocatalysts. In this way, USP was introduced as a novel approach for the simple synthesis of Pt NPs, whose structural and morphological characteristics can be finely tuned via easily-controllable pyrolytic parameters such as precursor concentration and temperature.

2. Materials and Methods

2.1. Material Synthesis

H_2PtCl_6 H_2O (Sigma Aldrich) was used as Pt precursor. In total, 2 g or 4 g of H_2PtCl_6 $6H_2O$ was dissolved in 1 L of de-ionized water to obtain the USP-feeding solutions. Two-zones ultrasonic spray pyrolysis was utilized for the formation of Pt nanoparticles, whose details can be found elsewhere [10,17]. In the first heating zone, droplets experience evaporation, while in the second heating zone and with the addition of hydrogen, the material reduction into Pt NPs takes place. The precursor solution was atomized by an ultrasonic generator (1.7 MHz) with fine droplets subsequently transported into the heating zones to experience evaporation, thermal reduction, and precipitation into Pt metallic nanoparticles. As carrier and reaction gases, 1.5 L/min N_2 in the first heating zone and

1.5 L/min N_2 + 0.5 L/min H_2 in the second heating zone were utilized. The residence time was estimated as $t_\text{residence} = \frac{V_r \cdot T_\text{room}}{r_F \cdot T_r}$, which provides the value of 3.01 s.

2.2. Material Characterization

2.2.1. Composition, Morphology, and Structural Characterization

To reveal the effect of precursor solution concentration on morphology, size, and purity of Pt nanoparticles, two different concentrations were examined. UV-Vis (Agilent, Santa Clara, CA 95051, USA) and DLS measurements (Malvern Panalytical Ltd, Kassel, Germany), scanning transmission electron microscopy (STEM), and thermal gravimetric analysis (TGA, NETZSCH, Selb, Germany) analyses were performed with typical instrumentations and conditions. The morphology and elemental compositions of the synthesized powders were analyzed by STEM Tecnai F20 (FEI Company, Eindhoven, The Netherlands), and a system (EDAX Inc., Mahwah, NJ, USA) equipped with energy dispersive spectroscopy (EDX) operated at 200KV for the analysis of characteristic X-ray emissions.

2.2.2. Electrochemical Characterization

Electrochemical characterization of the prepared Pt NPs was performed by linear sweep polarization measurements (LSV) and galvanostatic electrochemical impedance spectroscopy (GEIS) during the oxygen reduction reaction (ORR) as well as by cyclic voltammetry (CV). LSV and CV were conducted at sweep rates of 1 and 50 mV s^{-1}, respectively; LSV and GEIS measurements were performed at a working electrode (WE) rotation speed of 1500 rpm. In total, 0.5 M H_2SO_4 purged with N_2 (CV) or O_2 (ORR and GEIS) was used as an electrolyte.

It is known that HSO_4^- and SO_4^{2-} are interfering anions in ORR due to their competing adsorption onto Pt active sites that are required to be occupied by oxygen adatoms [29]. If one needs to step deep into an analysis of ORR kinetics, synthetic solutions of hardly adsorbing anions, e.g., $HClO_4$, are to be applied. However, the working environment of FC Pt-based catalysts is prone to anion adsorption from cheap and abundant solutions. Therefore, some studies deal with ORR in H_2SO_4 solution [30–32] as well. Our goal was to examine USP-Pt in such "competing adsorption" conditions and to compare the registered behavior to Pt black in the same environment.

All electrochemical measurements were performed in a three-electrode cell with an SCE reference electrode (all potentials in the paper are provided on an SCE scale) and a platinum plate as a counter electrode on potentiostat/galvanostat Bio-Logic SP200 (Bio-Logic SAS, Grenoble, France). WE was prepared from a powdered sample as follows: 3 mg of the synthesized USP powder was dispersed in 1 mL of distilled water and ultrasonically homogenized for 1 h (40 kHz, 70 W). The obtained suspension was pipetted onto WE to form a 0.31 mg cm^{-2} Pt layer onto a glassy carbon disk electrode (0.196 cm^2) that served as a current connector and was room-dried. Bearing in mind that the literature data [2] showed that Pt oxide formation could influence the structure of the Pt and hence its activity, CV measurements prior to LSV and GEIS measurements were conducted in the two different potential regions. Particularly, the cathodic potential was fixed at −0.2 V while the anodic limit was set to 0.55 or 1.25 V. Upon recording the stable CV curve in both applied potential regions, i.e., −0.2–0.55 without Pt oxide formation and −0.2–1.25 V with Pt oxide formation, LSV measurement was performed that started from 0.55 V and from open circuit potential, E_oc, respectively. GEIS was recorded down the polarization curve with a sinusoidal current of 30 μA amplitude, in a single sine mode, within a frequency range of 300 kHz–20 mHz, and with 20 points per decade.

Results of the electrochemical characterization were compared to those of commercial Pt black (Alfa Aesar, 25.0–29.8 m^2/g, d: 4.68–5.58 nm).

3. Results

3.1. DLS, STEM, and TGA Characterization of Pt Samples

In order to reveal the thermal breakdown behavior of the Pt precursor, H_2PtCl_6 $6H_2O$, for determining the USP reaction temperature, TGA analysis was performed in an inert atmosphere. The results are presented in Figure 1.

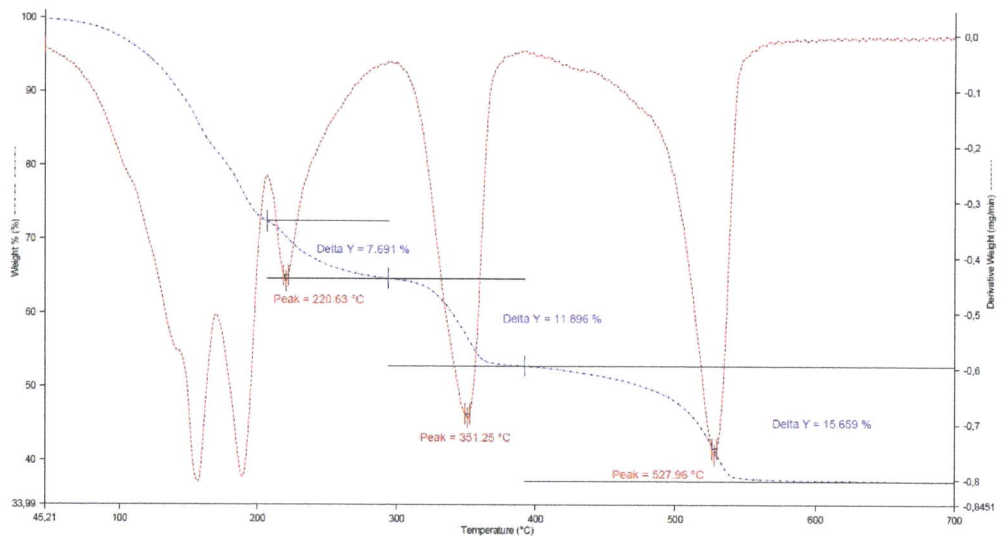

Figure 1. TGA curve of H_2PtCl_6 $6H_2O$ precursor in N_2.

It was previously reported by Rowston and Ottaway [16] and Schweizer and Kerr [18] that the thermal decomposition of H_2PtCl_6 takes place through stepwise reactions, as follows:

$$H_2PtCl_6 \rightarrow PtCl_4 + 2HCl \tag{1}$$

$$PtCl_4 \rightarrow PtCl_2 + Cl_2 \tag{2}$$

$$PtCl_2 \rightarrow Pt + Cl_2 \tag{3}$$

As revealed by Figure 1, mass loss began at early temperatures due to the loss of chemically bound water (the two peaks at around 170 and 190 °C). The mass loss related to the peak at 170 °C was below 20 %, which corresponds to the stoichiometric loss of five water molecules (17.4%). The two peaks at 190 and 221 °C can be associated with the joint loss of the remaining 6th crystalline water molecule and two HCl molecules, according to Reaction (1), as the corresponding sum of mass losses of 17% was quite close to the stoichiometric 17.6%. It follows that the last water molecule was lost at the temperature of 190 °C by overlapping with the start of the precursor decomposition to $PtCl_4$, which ends up at 300 °C. The loss of the remaining chlorine and the generation of metallic Pt through Reactions (2) and (3) is represented by the two well-separated peaks at 351 and 523 °C with an overall mass loss of 27.5%, which is negligibly different from the stoichiometric 27.4%. Hence, complete transformation into Pt took place at around 550 °C.

In order to facilitate the reduction for shorter residence times at a slightly lower temperature of 500 °C with respect to TGA while simultaneously having a defined structure and size of Pt particles, H_2 was utilized in USP synthesis.

TEM micrographs along with corresponding selected area (electron) diffraction (SAED) analysis of USP-synthesized Pt nanoparticles with different USP precursor concentrations are represented in Figure 2.

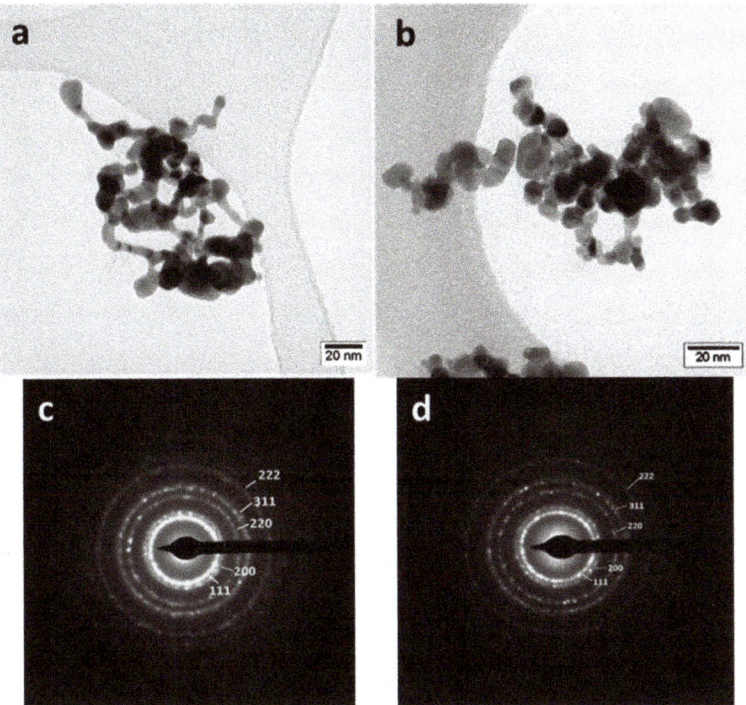

Figure 2. TEM micrographs of Pt nanoparticles synthesized by precursor concentrations of: (**a**) 2; (**b**) 4 g /L; and corresponding selected area (electron) diffraction analysis (**c**) and (**d**), respectively.

Both samples consisted of fine crystals (5–15 nm) with polygonal soft-edge morphology. Slightly larger and more agglomerated particles were obtained with a higher precursor concentration (Figure 2a,b). A string-like 1D agglomeration of the smallest particles appeared a more pronounced at a lower concentration. The additional effect of precursor concentration can be observed when Figure 2a,b is analyzed. A lower concentration (2 g/L), which can be considered as a lower driving force for particle growth, resulted in higher nucleation rates and growth of the crystals to a smaller extent, as we previously reported [17]. Consequently, the particles from the lower concentration appeared smaller and less defined and hence tended to form a string-like 1D agglomeration.

The SAED images in Figure 2c,d represent the characteristic diffraction of a ring pattern with some brighter and more distinct spots in the rings, which indicates the presence of some larger crystallites. However, the rings were still relatively continuous, which means that the crystallites were small, in the nm range, and in a random orientation. The electron diffraction spots could be described by a cubic crystal structure of Pt FCC, space group Fm $3\bar{\ }$ m, with indices as shown in the pattern.

The optical properties of nanoparticles were also examined in a comparative manner by UV-Vis spectroscopy, as provided in Figure 3.

As shown in Figure 3, both observed maximum absorbance peaks appeared at 272 nm, which is slightly higher than the absorbance values reported in the literature (~262 nm) where Pt nanoparticles exhibited a prevailing size of 5–6 nm [20,21]. A slight red shift could be due to an aggregation effect, as revealed in the TEM micrographs shown in Figure 2. When the surface plasmon-induced absorbances of two samples are compared, it can be seen that peak position did not change and the higher concentration sample resulted in slightly higher absorbance. A slight increase in absorbance may be due to more pronounced roundness of the particles.

Although consisting of slightly larger and more agglomerated particles, 4 g/L was determined as a more suitable precursor solution due to the well-defined particles, and this sample was analyzed in terms of electrochemical properties and compared with the commercial Pt powder.

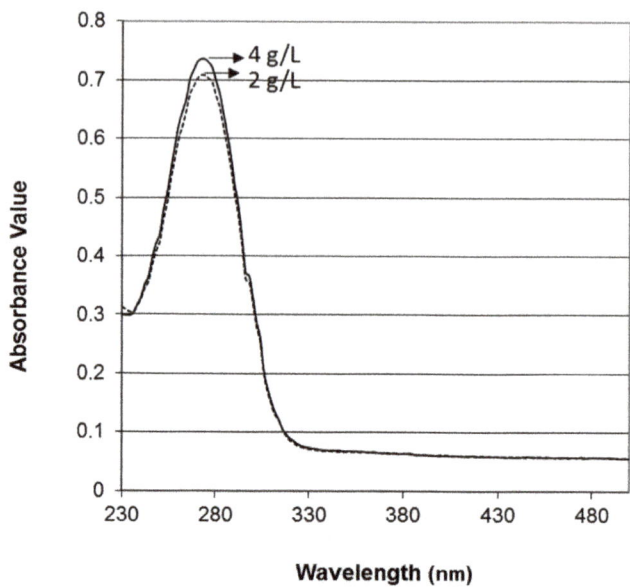

Figure 3. Absorbance vs. wavelength spectra of the synthesized samples.

3.2. Electrochemical Properties of Obtained Pt Particles

3.2.1. Cyclic Voltammetry

Stable cyclic voltammograms of USP-synthesized Pt (USP-Pt) and Pt black are shown in Figure 4. Both curves had a shape that was characteristic of platinum. Well-resolved Pt oxide formation/reduction peaks were observed for both samples in a wider potential range, with higher CV currents for Pt black. This also holds for hydrogen adsorption/desorption peaks. However, in the double-layer region (around 0.15 V), the currents for USP-Pt were smaller, which indicates the formation of slightly smaller Pt particles. Similar findings are valid if CVs in the narrower potential region are considered. With respect to the wider potential range, hydrogen adsorption/desorption peaks were less pronounced because the surface had not been continuously renewed and reconstructed by reversible oxide formation/reduction. The ratio between CV currents of USP-Pt and Pt black appeared to not be affected by cycling limits. These basic electrochemical properties show that clean Pt particles of typical characteristics can be synthesized by a simple USP synthesis approach.

3.2.2. Linear Sweep Voltammetry

LSV curves for the ORR of USP-Pt and Pt black obtained after CV measurements in shorter and wider potential ranges (Figure 5) represent typical polarization curves for ORR on Pt. The formation of Pt oxide (the case of CV in wider potential range) had a beneficial influence on the ORR activity of both USP-Pt and Pt black samples. Namely, the samples showed better ORR activity after CV measurements in a wider potential range as ORR takes place at more anodic potentials, except in the region of a limiting diffusion current (potentials negative to 0.3 V). The reversible oxide formation can cause the growth of Pt particles and hence reduce the real surface area, which consequently decreases the apparent limiting diffusion current. This effect is more pronounced for USP-Pt due to initially larger Pt particles.

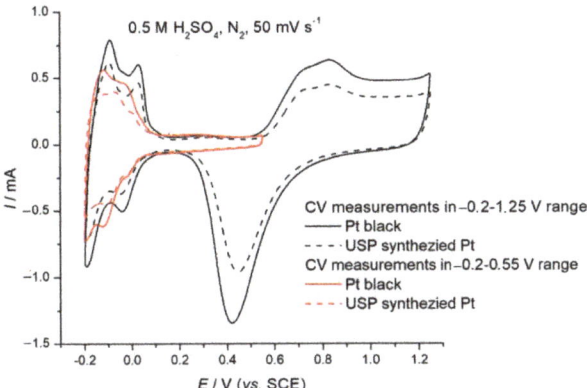

Figure 4. Cyclic voltammograms of USP-Pt and Pt black in shorter (−0.2–0.55) and wider (−0.2–1.25) potential ranges. Electrolyte: de-aerated 0.5 M H_2SO_4, sweep rate 50 mV s^{-1}.

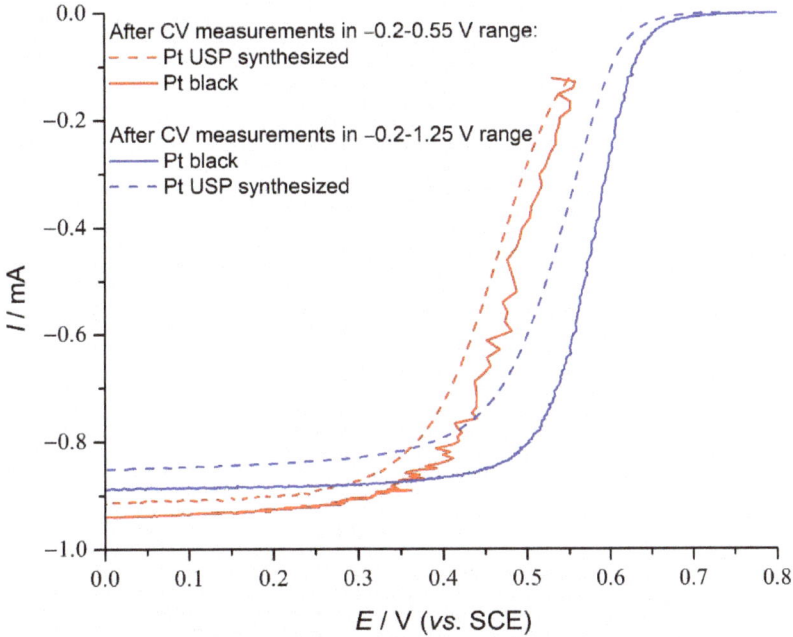

Figure 5. Quasi-steady-state polarization curves of USP-Pt and Pt black obtained after CV measurements in shorter and wider potential ranges. Electrolyte: 0.5 M H_2SO_4 purged with O_2, room temperature, 1500 rpm, sweep rate: 1 mV/s.

ORR required the application of more negative potential for USP-Pt, i.e., apparent currents were higher for Pt black within all the applied potential range. This indicates that Pt black was mainly more active for ORR due to geometric issues, i.e., the Pt black layer on the GC working electrode had a slightly larger real surface area. In addition, the sole low-index Pt (111) plane was found by electron dispersion (Figure 2) for USP-Pt, which is the least active plane in comparison to the Pt (100) and Pt (101) facets [32]. This finding can additionally affect the lower ORR activity of USP-Pt, and can particularly cause the registered higher ORR overpotentials of USP-Pt in comparison to Pt black. The limiting current of around 4.6 mA cm^{-2} that was found for USP-Pt is in accordance with the limiting currents ranging 4–6 mA cm^{-2} reported for various materials [4,5,8]. Namely, platinum-based hybrid materials Pt-Er@PC-900 and Pt-ErPCN-900, i.e., Pt/Er nanoparticles decorated on Cd-MOF derived hierarchical carbon, showed excellent ORR activity with a limiting current of 5.5 mA cm^{-2} while the limiting current of commercial Pt/C was 4.7 mA cm^{-2} [8]. A new class of non-platinum electrocatalysts with high stability and activity in ORR was developed by Ibraheem et al. [4,5]. The ORR activity of these new hybrid non-platinum materials was comparable to the activity of commercial Pt/C catalysts and even better when in comparison to IrO$_2$. Namely, strongly coupled Fe2NiSe4@Fe-NC hybrid material showed excellent stability in ORR with a limiting current of around 5.5 mA cm^{-2} [4]. High activity and stability in ORR were preserved when this hybrid material was comprised of P instead of Se, i.e., a limiting current of NiFeP supported on three-dimensional, interconnected Fe,N-decorated carbon (NiFeP@3D-FeNC) was around 5.5 mA cm^{-2} [5]. Even unsupported NiFeP material showed good ORR activity (limiting current of 4.2 mA cm^{-2}), while unsupported Fe$_2$NiSe$_4$ exhibited a lower limiting current of 1.5 mA cm^{-2}.

3.2.3. Galvanostatic Electrochemical Impedance Spectroscopy (GEIS)

The GEIS measurements of the samples were conducted at different steady-state currents depending on whether they were performed after a narrower or wider range of potential had been applied in preceding CV measurements. From Figure 5, the currents analogue to the potential of Pt oxide formation (around 0.6 V from CV in Figure 4) were −250 µA for Pt black and −150 µA for USP-Pt. Therefore, to avoid the formation of Pt oxide, applied currents in GEIS performed after CV in a shorter potential range were in a range from the diffusion-limited current to −250, i.e., −150 µA. Currents applied in GEIS performed after CV in a wider potential range (Figure 3) included current values within the ORR region from Figure 5. Given that GEIS results did not differ by much whether they were obtained after CV in narrower or wider potential regions, only the GEIS results gained after CV in the wider region are shown in this paper (Figures 6 and 7). This is in accordance with the CV findings that reversible oxide formation does not really affect the activity of investigated Pt samples.

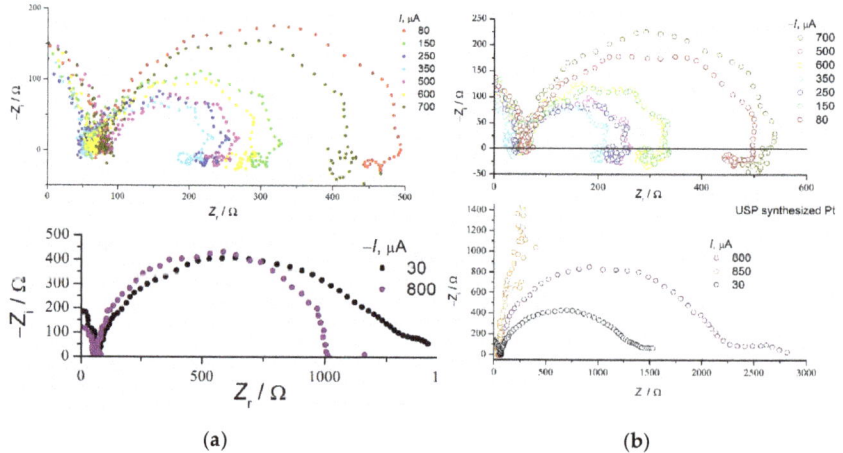

Figure 6. Nyquist presentation of GEIS results (Pt black (**a**) and USP-Pt (**b**)) after CV in a wider (−0.2–1.25 V) range. Electrolyte: O$_2$ purged 0.5 M H$_2$SO$_4$, 1500 rpm.

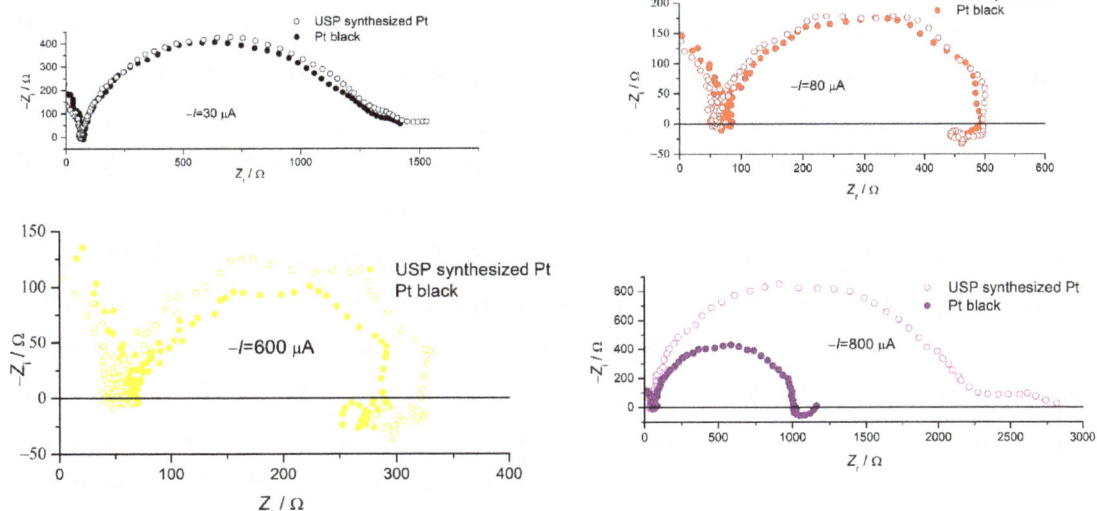

Figure 7. Detailed Nyquist comparison of GEIS results of Pt black and USP-Pt at specific currents. Electrolyte: O_2 purged 0.5 M H_2SO_4, 1500 rpm.

A similar trend in ORR activity was observed for both samples (Figure 3). A charge transfer loop is registered as the main GEIS feature in the investigated frequency range at all applied steady-state currents. The increase in currents in the mixed activation/diffusion region (below −350 µA) induces the decrease in a loop diameter due to the decrease in charge transfer resistance. For the cathodic currents higher than −350 µA, the loop diameter increased with the current due to the intensification of the diffusion limitation of ORR. All loops were of similar shape, except for the loop that registered at the lowest current of −30 µA, which was clearly followed in the low-frequency region by an additional small loop. For the USP-Pt sample, this small loop appeared better developed and at somewhat higher frequencies with respect to the data for Pt black. This indicates that the associated ORR kinetics issues are sensitive to the structure of the catalyst layer, which appears to be more compact in the case of Pt black due to smaller particles and larger real surface area. This seems to also affect the difference in loop features at the highest applied current of −800 µA. The low-frequency loop was uniquely preserved under pronounced diffusion control for USP-Pt. This apparently caused the loop diameter for USP-Pt at −800 µA to be almost three times larger than that for Pt black. However, the loops for USP-Pt and Pt black at other applied currents appeared quite similar in diameter. This difference with respect to polarization measurements (Figure 5), which indicated the higher activity of Pt black, deserves further analysis, according to Figure 7.

Figure 7 presents a comparison of some GEIS results at specific currents, taken from Figure 6 (the values of the currents are shown within).

Although the LSV measurement (Figure 2) showed better ORR activity of the Pt black within the whole current range, GEIS results indicated almost the same activity at lower currents, i.e., up to −500 µA (the loops were of quite similar diameters). However, with the increase in the applied current in GEIS, starting from I = −600 µA (Figure 4), better activity of the Pt black appeared. Finally, at the highest applied current (−800 µA), the Pt black activity was doubled in comparison to the USP-Pt activity. It then follows that the differences in activity are strictly connected to the onset of pure diffusion limitations; therefore, they are not connected to the chemical structure of the investigated samples but to the morphology of the electrode layers, as already discussed.

In order to comment on the differences between polarization and GEIS data, a comparison of ORR activity registered in GEIS and LSV measurements for Pt black and USP-Pt is presented in Figure 8. For this comparison, the square root of the potentials was collected from GEIS data at chosen frequencies (indicated in the figure) from low- and high-frequency regions and plotted against the steady-state GEIS current. Both USP-Pt and Pt black are more active if the sinusoidal perturbation of the current is of higher frequency, and they are even of higher activity, especially in a mixed reaction control, with respect to the quasi-steady-state data. This indicates the distribution of the activity throughout the catalyst layer, with more active sites situated in the outer regions because these sites respond to fast current perturbations. The main difference between the two samples is that this distribution of the active sites is more pronounced for Pt black. USP-Pt does not lack activity at low-frequency perturbations with respect to standard polarization. This means that all active sites would be available during the ORR operation, which is to be expected as being more stable with respect to Pt black. Apparently, these valuable features of USP-Pt in comparison to Pt black are due to a less-compact layer structure caused by larger Pt particles.

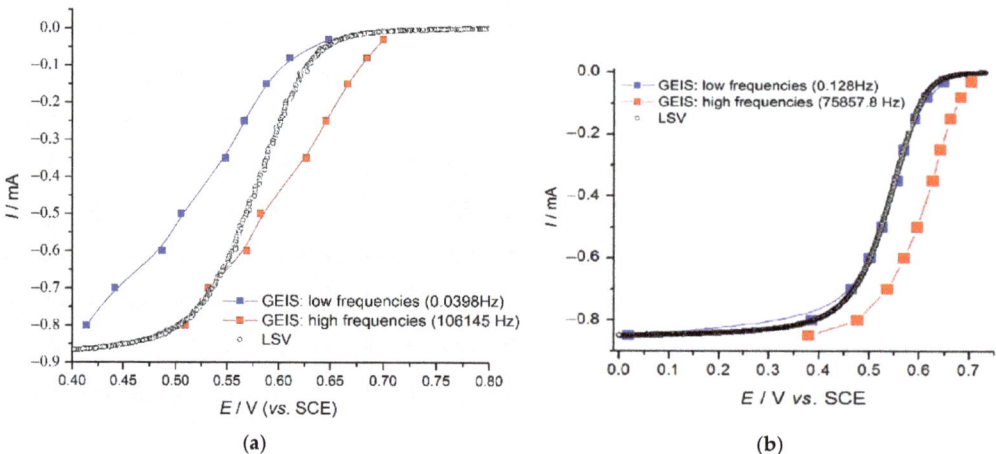

Figure 8. Comparison of Pt black (**a**) and USP-Pt (**b**) ORR activity collected from GEIS and LSV measurements.

4. Conclusions

Ultrasonic spray pyrolysis (USP) was used for the synthesis of Pt particles, which started from using $H_2PtCl_6 \times 6H_2O$ water solution as a precursor without using any synthesis additives to tune structural and morphological properties. The USP synthesis temperature for low residence time was chosen according to thermogravimetric measurements. Results showed that polygonal round-edge and phase-pure Pt nanoparticles with face cubic centered structures were successfully synthesized. The particle size, defined shape, and agglomeration were found to be sensitive to precursor concentration, and higher concentrations were found to be beneficial due to the synthesis of more defined particles.

Electrochemical characterization revealed good electrocatalytic activity of the synthesized material, which was comparable to commercial Pt black. It is found that USP-Pt is of a lower real surface area due to the larger particle size with respect to Pt black. As a consequence, USP-Pt appears less active in quasi-steady-state polarizations in the oxygen reduction reaction (ORR). However, upon dynamic perturbations performed by galvanostatic impedance measurements (GEIS), the difference in ORR activities between the two investigated powdered Pt was found negligible. Only the diffusion-limited currents were found to be higher for Pt black due to the larger real surface area, i.e., the somewhat smaller particles.

The beneficial features of USP-synthesized Pt were found upon the comparison of steady-state and dynamic (GEIS) electrocatalytic data in ORR. USP-Pt is less sensitive to the rate of current perturbations due to larger particles and, consequently, a less-compact catalyst layer. A more defined response of USP-Pt in ORR polarization could be considered as an indication of its higher stability with respect to Pt black.

Author Contributions: Conceptualization, J.S. and B.F.; methodology, V.P. and M.M.; formal analysis, G.A. and M.K.; investigation, G.A. and M.K.; data curation, G.A. and M.K.; writing—original draft preparation, G.A. and M.K.; writing—review and editing, V.P.; supervision, V.P., B.F. and J.S.; project administration, M.M. and S.S. All authors have read and agreed to the published version of the manuscript.

Funding: This work was funded by the Serbian Ministry of Education, Science and Technological Development Grant No. 451-03-9/2021-14/200026.

Conflicts of Interest: The authors declare no conflict of interest.

References

1. Pei, Y.; Hu, M.; Xia, Y.; Huang, W.; Li, Z.; Chen, S. Electrochemical preparation of Pt nanoparticles modified nanoporous gold electrode with highly rough surface for efficient determination of hydrazine. *Sens. Actuators B Chem.* **2020**, *304*, 127416. [CrossRef]
2. Mello, R.L.S. Preparation and electrochemical characterization of Pt nanoparticles dispersed on niobium oxide. *Eclet. Quím.* **2003**, *28*, 69–76. [CrossRef]
3. Yasin, G.; Ibrahim, S.; Ibraheem, S.; Ali, S.; Iqbal, R.; Kumar, A.; Tabish, M.; Slimani, Y.; Nguyen, T.A.; Xu, H.; et al. Defective/graphitic synergy in a heteroatom-interlinked-triggered metal-free electrocatalyst for high-performance rechargeable zinc–air batteries. *J. Mater. Chem. A* **2021**, *9*, 18222. [CrossRef]
4. Ibraheem, S.; Chen, S.; Peng, L.; Li, J.; Li, L.; Liao, Q.; Shao, M.; Wei, Z. Strongly coupled iron selenides-nitrogen-bond as an electronic transport bridge for enhanced synergistic oxygen electrocatalysis in rechargeable zinc-O2 batteries. *Appl. Catal. B Environ.* **2020**, *265*, 118569. [CrossRef]
5. Ibraheem, S.; Chen, S.; Li, J.; Li, W.; Gao, X.; Wang, Q.; Wei, Z. Three-dimensional Fe,N-decorated carbon-supported NiFeP nanoparticles as an efficient bifunctional catalyst for rechargeable zinc−O2 batteries. *ACS Appl. Mater. Inter.* **2019**, *11*, 699–705. [CrossRef]
6. Olabi, A.G.; Sayed, E.T.; Wilberforce, T.; Jamal, A.; Alami, A.H.; Elsaid, K.; Rahman, S.M.A.; Shah, S.K.; Abdelkareem, M.A. Metal-Air Batteries—A Review. *Energies* **2021**, *14*, 7373. [CrossRef]
7. Ibraheem, S.; Yasin, G.; Kumar, A.; Mushtaq, M.A.; Ibrahim, S.; Iqbal, R.; Tabish, M.; Ali, S.; Saad, A. Iron-cation-coordinated cobalt-bridged-selenides nanorods for highly efficient photo/electrochemical water splitting. *Appl. Catal. B Environ.* **2022**, *304*, 120987. [CrossRef]
8. Yasin, G.; Ibraheem, S.; Ali, S.; Arif, M.; Ibrahim, S.; Iqbal, R.; Kumar, A.; Tabish, M.; Mushtaq, M.A.; Saad, A.; et al. Defects-engineered tailoring of tri-doped interlinked metal-free bifunctional catalyst with lower gibbs free energy of OER/HER intermediates for overall water splitting. *Mater. Today Chem.* **2022**, *23*, 100634. [CrossRef]
9. Nadeem, M.; Yasin, G.; Arif, M.; Tabassum, H.; Bhatti, M.H.; Mehmood, M.; Yunus, U.; Iqbal, R.; Nguyen, T.A.; Slimani, Y.; et al. Highly active sites of Pt/Er dispersed N-doped hierarchical porous carbon for trifunctional electrocatalyst. *Chem. Eng. J.* **2021**, *409*, 128205. [CrossRef]
10. Motsoeneng, R.G.; Modibedi, R.M.; Mathe, M.K.; Khotseng, L.E.; Ozoemena, K.I. The synthesis of PdPt/carbon paper via surface limited redox replacement reactions for oxygen reduction reaction. *Int. J. Hydrog.* **2015**, *40*, 16734–16744. [CrossRef]
11. Vidal-Iglesias, F.J.; Ara, R.M.; Solla-Gullo, J.; Herrero, E.; Feliu, J.M. Electrochemical Characterization of Shape-Controlled Pt Nanoparticles in Different Supporting Electrolytes. *ACS Catal.* **2012**, *2*, 901–910. [CrossRef]
12. Kim, J.W.; Lim, B.; Jang, H.-S.; Hwang, S.J.; Yoo, S.J.; Ha, J.S.; Cho, E.A.; Lim, T.-H.; Nam, S.W.; Kim, S.-K. Size-controlled synthesis of Pt nanoparticles and their electrochemical activities toward oxygen reduction. *Int. J. Hydrog. Energy* **2010**, *36*, 706–712. [CrossRef]
13. Stepanov, A.L.; Golubev, A.N.; Nikitin, S.I.; Osin, Y.N. A review on the fabrication and properties of platinum nanoparticles. *Rev. Adv. Mater. Sci.* **2014**, *38*, 160–175.
14. Tao, A.R.; Habas, S.; Yang, P. Shape control of colloidal metal nanocrystals. *Small* **2008**, *4*, 310–325. [CrossRef]
15. Lau, M.; Gökce, B.; Marzun, G.; Rehbock, C.; Barcikowski, S. Rapid nanointegration with laser-generated nanoparticles. *Lasers Manuf. Conf.* **2015**, *109*, 1–3.
16. Alkan, G.; Diaz, F.; Matula, G.; Stopic, S.; Friedrich, B. Scaling up of nanopowder collection in the process of ultrasonic spray pyrolysis. *World Metall Erzmetall* **2017**, *70*, 97–101.

17. Alkan, G.; Rudolf, R.; Bogovic, J.; Jenko, D.; Friedrich, B. Structure and Formation Model of Ag/TiO2 and Au/TiO2 Nanoparticles Synthesized through Ultrasonic Spray Pyrolysis. *Metals* **2017**, *7*, 389. [CrossRef]
18. Messing, G.L.; Zhang, S.C.; Jayanthi, G.V. Ceramic Powder Synthesis by Spray Pyrolysis. *J. Am. Ceram. Soc.* **1993**, *76*, 2707–2726. [CrossRef]
19. Gurav, A.; Kodas, T.; Pluym, T.; Xiong, Y. Aerosol processing of materials. *Aerosol Sci. Technol.* **1993**, *19*, 411–452. [CrossRef]
20. Jung, C.H.; Yun, J.; Qadir, K.; Naik, B.; Yun, J.Y.; Park, J.Y. Catalytic activity of Pt/SiO2 nanocatalysts synthesized via ultrasonic spray pyrolysis process under CO oxidation. *Appl. Catal. B Environ.* **2014**, *154–155*, 171–176. [CrossRef]
21. Jung, C.H.; Yun, J.; Qadir, K.; Park, D.; Yun, J.Y.; Park, J.Y. Pt/oxide nanocatalysts synthesized via the ultrasonic spray pyrolysis process: Engineering metal-oxide interfaces for enhanced catalytic activity. *Res. Chem. Intermed.* **2016**, *42*, 211–222. [CrossRef]
22. Muñoz-Fernandez, L.; Alkan, G.; Milošević, O.; Rabanal, M.E.; Friedrich, B. Synthesis and characterisation of spherical core-shell Ag/ZnO nanocomposites using single and two–steps ultrasonic spray pyrolysis (USP). *Catal. Today* **2019**, *321–322*, 26–33. [CrossRef]
23. Rowston, W.B.; Ottaway, J.M. Determination of noble metals by carbon furnace atomic-absorption spectrometry. Part 1. Atom formation processes. *Analyst* **1979**, *104*, 645–659. [CrossRef]
24. Košević, M.; Zarić, M.; Stopić, S.; Stevanovic, J.; Weirich, T.; Friedrich, B.; Panic, V. Structural and Electrochemical Properties of Nesting and Core/Shell Pt/TiO2 Spherical Particles Synthesized by Ultrasonic Spray Pyrolysis. *Metals* **2020**, *10*, 11. [CrossRef]
25. Schweizer, A.E.; Kerr, G.T. Thermal Decomposition of Hexachloroplatinic Acid. *Inorg. Chem.* **1978**, *17*, 2326–2327. [CrossRef]
26. Alkan, G.; Rudolf, R.; Emil, E.; Jenko, D.; Friedrich, B.; Gurmen, S. Tuning the Morphology of ZnO Nanostructures with the Ultrasonic Spray Pyrolysis Process. *Metals* **2018**, *8*, 569.
27. Gharibshahi, E.; Saion, E. Influence of dose on particle size and optical properties of colloidal platinum nanoparticles. *Int. J. Mol. Sci.* **2012**, *13*, 14723–14741. [CrossRef]
28. Nguyen, T.B.; Nguyen, T.D.; Nguyen, Q.D.; Nguyen, T.T. Preparation of platinum nanoparticles in liquids by laser ablation method. *Adv. Nat. Sci. Nanosci. Nanotechnol.* **2014**, *5*, 035011. [CrossRef]
29. Du, C.; Sun, Y.; Shen, T.; Yin, G.; Zhang, J. 7 Applications of RDE and RRDE Methods in Oxygen Reduction Reaction. In *Rotating Electrode Methods and Oxygen Reduction Electrocatalysts*, 1st ed.; Xing, W., Yin, G., Zhang, J., Eds.; Elsevier: Amsterdam, The Netherlands, 2014; pp. 231–277.
30. Zorko, M.; Martins, P.F.B.D.; Connell, J.G.; Lopes, P.P.; Markovic, N.M.; Stamenkovic, V.R.; Strmcnik, D. Improved Rate for the Oxygen Reduction Reaction in a Sulfuric Acid Electrolyte using a Pt(111) Surface Modified with Melamine. *ACS Appl. Mater. Interfaces* **2021**, *13*, 3369–3376. [CrossRef]
31. Devivaraprasad, R.; Ramesh, R.; Naresh, N.; Kar, T.; Singh, R.K.; Neergat, M. Oxygen reduction reaction and peroxide generation on shape-controlled and polycrystalline platinum nanoparticles in acidic and alkaline electrolytes. *Langmuir* **2014**, *30*, 8995–9006. [CrossRef]
32. Grgur, B.; Marković, N.M.; Ross, P.N. Temperature dependent oxygen electrochemistry on platinum low index single crystal surfaces in acid solutions. *Can. J. Chem.* **1997**, *75*, 1465–1471. [CrossRef]

Article

Synthesis of Silica Particles Using Ultrasonic Spray Pyrolysis Method

Srecko Stopic [1,*], Felix Wenz [1], Tatjana-Volkov Husovic [2] and Bernd Friedrich [1]

1. IME Process Metallurgy and Metal Recycling, RWTH Aachen University, 52056 Aachen, Germany; felix.wenz@rwth-aachen.de (F.W.); bfriedrich@ime-aachen.de (B.F.)
2. Metallurgical Engineering Department, Faculty of Technology and Metallurgy, Karnegijeva 4, 11120 Belgrade, Serbia; tatjana@tmf.bg.ac.rs
* Correspondence: sstopic@ime-aachen.de; Tel.: +49-176-7826-1674

Abstract: Silica has sparked strong interest in hydrometallurgy, catalysis, the cement industry, and paper coating. The synthesis of silica particles was performed at 900 °C using the ultrasonic spray pyrolysis (USP) method. Ideally, spherical particles are obtained in one horizontal reactor from an aerosol. The controlled synthesis of submicron particles of silica was reached by changing the concentration of precursor solution. The experimentally obtained particles were compared with theoretically calculated values of silica particles. The characterization was performed using a scanning electron microscope (SEM) and energy-dispersive X-ray spectroscopy (EDS). X-ray diffraction, frequently abbreviated as XRD, was used to analyze the structure of obtained materials. The obtained silica by ultrasonic spray pyrolysis had an amorphous structure. In comparison to other methods such as sol–gel, acidic treatment, thermal decomposition, stirred bead milling, and high-pressure carbonation, the advantage of the ultrasonic spray method for preparation of nanosized silica controlled morphology is the simplicity of setting up individual process segments and changing their configuration, one-step continuous synthesis, and the possibility of synthesizing nanoparticles from various precursors.

Keywords: silica; ultrasonic spray pyrolysis; synthesis

Citation: Stopic, S.; Wenz, F.; Husovic, T.-V.; Friedrich, B. Synthesis of Silica Particles Using Ultrasonic Spray Pyrolysis Method. *Metals* **2021**, *11*, 463. https://doi.org/10.3390/met11030463

Academic Editor: Petros E. Tsakiridis

Received: 4 February 2021
Accepted: 8 March 2021
Published: 11 March 2021

Publisher's Note: MDPI stays neutral with regard to jurisdictional claims in published maps and institutional affiliations.

Copyright: © 2021 by the authors. Licensee MDPI, Basel, Switzerland. This article is an open access article distributed under the terms and conditions of the Creative Commons Attribution (CC BY) license (https://creativecommons.org/licenses/by/4.0/).

1. Introduction

The formation of silica from olivine in different metallurgical processes was studied very frequently in the last 50 years. Stopic [1] presented different ways for the deposition of silica in hydrometallurgical processes. As mentioned, the production of silica by the olivine route is a cheaper method than the commercial methods such as neutralization of sodium silicate solutions and flame hydrolysis because of the low cost of raw materials and the low energy requirements [2]. The produced silica has a specific surface area between 100 and 400 m^2/g, primary particles between 10 and 25 nm (agglomerated in clusters), and an SiO_2 content above 95%. Due to the high pozzolanic properties and the dispersion state, this silica powder can be applied successfully in concrete.

Mohanray et al. [3] prepared silica from corncob ash by the precipitation method. First, received corncob ash was calcined at 550 °C, 650 °C, and 750 °C for 2 h to remove the volatiles in the sample and determine the amorphous structure of silica. The thermally treated corncob ash was mixed with various concentrations of sodium hydroxide to extract pure silica using 1% of polyvinyl alcohol (PVA) as the dispersing agent. In the last step, spherical nano silica with a particle size of 25 nm was prepared from pure silica by the precipitation method. Unfortunately, starting from corn cob ash calcination and mixing with NaOH [3], precipitation with 1% polyvinyl alcohol (PVA) cannot ensure a controlled silica particle size and their purity. The impurities originate from the calcination process.

In the study of Huan et al. [4], nanosilica was synthesized by the sol–gel method from tetraethoxysilane (TEOS) with base catalysts and volumetric ratio TEOS/C2H5OH/

H2O/NH4OH: 5/30/1/1. The results showed that the prepared nanosilica were in an amorphous phase with an average size of about 60–100 nm and could be used for lead removal from waste water. Generally, the sol–gel method enables a high-purity amorphous silica powder; however, the process yields a low percentage. This method cannot ensure the formation of nonagglomerated spherical silica particles.

Powder precursors for sol–gel synthesis are very expensive, but Oi et al. [5] studied a new precursor for low-cost alternatives. A high surface area was used to form an anion surfactant sodium dodecyl sulfate, which regulates the molar concentration. The particles' size variability was changed by the precursor molar ratio of the sodium silicate solution with hydrochloric acid. The nanostructured silica particles were obtained over a range of particle sizes from 0.5 to 1 μm, with a high specific surface area and cubic and spherical shapes by changing the surfactant pH values and drying methods with the sol–gel process. An increase in pH value from 1 to 5 decreases the surface area from 858 to 630 m^2/g. The study of Nandanwar et al. [6] deals with the sol–gel synthesis of nanosilica, providing a basic understanding of the effect of calcination temperature on the growth of SiO_2 by the hydrolysis of TEOS with ethanol, deionized water, and catalyst mixture.

High-purity nanosilica was synthesized by Kim et al. [7] using acid treatment and surface modification from blast-furnace slag generated in the steel industry. Blast-furnace slag was treated with nitric acid to extract high-purity insoluble silica. Silica particles were produced using filtration and a surface modified by cation surfactant cetyltrimethyl ammonium bromide (CTAB). The size of silica particles was smallest when the modification temperature was 60 °C. The average size of silica particles modified with 3 wt.% CTAB was 107.89 nm, while the average size of unmodified silica was 240.38 nm. An acidic treatment can lead to the formation of silica gel and blockage of the whole process. The filtration is a required operation in this process. This treatment contains many operations in order to obtain silica. Due to these characteristics, we need to find other simple methods for the synthesis of very fine silica without an acidic treatment.

Synthesis of silicon dioxide nanoparticles in low-temperature atmospheric-pressure plasma was performed by Kretushev et al. [8]. Results of studies confirm that spherical nanoparticles within the range of 20–60 nm can be successfully prepared in low-temperature atmospheric-pressure plasma created with a high-frequency discharge maintained in the α mode between two plane-parallel grid electrodes. The degree of tetraethoxysilane decomposition is 80–95% and only slightly depends on the reaction parameters. Nanoparticles with a predominantly spherical shape are synthesized in the region of the high-frequency discharge.

A novel synthesis route is proposed by Stopic et al. [9] based on CO_2 absorption/sequestration in an autoclave by forsterite (Mg_2SiO_4), which is part of the mineral group of olivines. Therefore, it is a feasible and safe method to bind carbon dioxide in carbonate compounds such as magnesite forming the spherical nanosilica at the same time, between 250 and 500 nm, as shown in Figure 1.

In contrast to sol–gel, acidic treatment, and hydrothermal synthesis using some acid and alkaline solutions, this synthesis method in an autoclave takes place in water solution at 175 °C and above 100 bar. This method is an environmentally friendly process related to the capture of carbon dioxide and the preparation of silica.

As the conventional methods for the synthesis of nanosilica from rice husk ash are energy- and time-consuming, Phoohinkong et al. [10] studied the synthesis of nanosilica from real available materials such as rice husk ash via sodium silicate solution. Nanosilica particles were obtained via alkaline extraction and a fast acid precipitation method at room temperature by adding inorganic salts and without surfactant. The flow synthesis was investigated at ambient temperature, varying the concentration of hydrochloric acid and sodium chloride, and the flow-rate while fixing the concentration of sodium silicate. The results revealed that the sodium chloride is significantly inorganic salt for the prepared nanosilica, with uniform spherical morphology (80–150 nm). In this synthesis, the silica nanoparticles, with a diameter around 10 nm and aggregate particles of around 50 to 200 nm, were prepared.

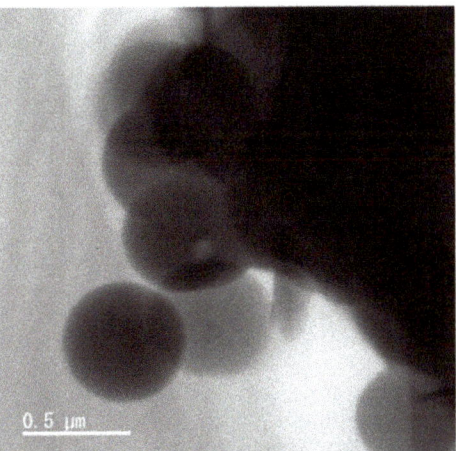

Figure 1. Nanosilica-obtained carbonation of olivine in an autoclave under high pressure.

Production of metal carbonate and nanosilica below 100 nm was enabled in stirred bead milling, as reported by Wang and Forssberg [11]. It is shown that the stirred bead mill with very small beads can be used as efficient equipment for the production of the colloidal particles in the nanoscale from the feed materials of several microns in size at high energy consumptions. Generally, it is concluded that an intense comminution of carbonate minerals in the stirred bead mills leads to a progressive loss in crystallinity of the basal planes of the crystal structure. An intensive mechanical treatment of silica gives the structural changes and the amorphization.

Akhayere et al. [12] reported the synthesis of nanosilica from barley grass waste—an environmental burden—using varying temperatures during preparation. The temperatures used during the investigation were 400, 500, 600, and 700 °C, studying its effects on the mechanical properties of the nanosilica for use in environmentally friendly applications. Using the Brunauer–Emmett–Teller (BET) methodology, the surface area corresponds to 150 m^2/g. The results of this study showed improved and stable mechanical properties with the increase in temperature during synthesis.

A novel low-temperature vapor-phase hydrolysis method for the production of nanosilica using silicon tetrachloride was reported by Chen et al. [13]. Silica nanoparticles were obtained by the hydrolysis of silicon tetrachloride vapor with water vapor at a low temperature range (150–250 °C). Silica nanoparticles with a specific surface area of 418 m^2/g and an average size of 141.7 nm were prepared at a temperature of 150 °C and with a residence time of 5 s. It was an amorphous mesoporous material, with an approximately spherical shape and a mass friction demission of 2.29.

Silica nanoparticles were prepared by ultrasonic spray pyrolysis (USP) between 300 and 600 °C using tetraethylorthosilicate (TEOS) as a precursor, as mentioned by Ratanathavorn et al. [14]. The particle size decreased from 347 to 106 nm when the synthesis temperature increased from 300 °C to 500 °C. Comparing two types of cream perfume, with and without silica, by applying cream perfume on a glass slide at 37 °C for 5 h, it was found that the odor of cream perfume with silica lasted longer than cream perfume without silica. The particles agglomerated and had irregular forms.

Citakovic [15] mentioned that the physical properties of nanomaterials have high significance for their application. Especially, a significant difference in the values of some physical parameters include the melting point, change in the unit-cell parameters, change in the magnetic and optical characteristics, conductivity of the material, etc. The surface-to-volume ratio is an important parameter that has an impact on new characteristics in comparison to those of bulk materials. Silica is present in many different crystalline forms that vary in levels of fibrogenicity according to the degree of crystallization [16].

At present, nanosilica materials are prepared using several methods, including precipitation, sol–gel, acidic treatment, alkaline extraction, flow synthesis, stirred bead milling, the thermal decomposition technique, high-pressure carbonation, and low-temperature atmospheric pressure. However, with their high cost of preparation, many operations and morphological characteristics of particles have limited their wide application. By contrast, ultrasonic spray pyrolysis as a very simple method offers many advantages for the synthesis of oxidic particles as mentioned by Stopic et al. [17,18].

Generally, our aim was to reach the synthesis of nanosilica using the ultrasonic spray pyrolysis method. In contrast to previously mentioned work under high-pressure conditions in an autoclave [14], our aim was to obtain ideally spherical silica particles in a short residence time in dynamic conditions. In order to reach these aims, the concentration of precursor solution was adjusted for this purpose. Generally, an understanding of silica formation was studied via different calculations of the residence time and particle size. Regarding the previous literature analysis, our main aim is the testing of ultrasonic spray pyrolysis as a simple method for the synthesis of spherical silica particles suitable for lead treatment as mentioned by Huan [4].

2. Experimental Part
2.1. Material

The LUDOX 30 wt.% colloidal silica (SiO_2), SAFA 420811, VWR International GmbH, Darmstadt, Germany was used for preparation of precursor solution. The chosen volume of this concentrated solution was diluted in 900 mL of distilled water in order to prepare a suitable precursor for the synthesis of silica particles. The chemical analysis of solution was performed using ICP-OES analysis (SPECTRO ARCOS, SPECTRO Analytical Instruments GmbH, Kleve, Germany). The prepared precursor solution of different concentrations and the obtained suspension after the ultrasonic spray pyrolysis method was prepared for SEM and EDS analysis. First, the sample was shaken, a few drops were injected onto the aluminum slide with a pipette, allowed to dry, and finally evaporated with carbon together. A typical picture of this precursor material is shown in Figure 2:

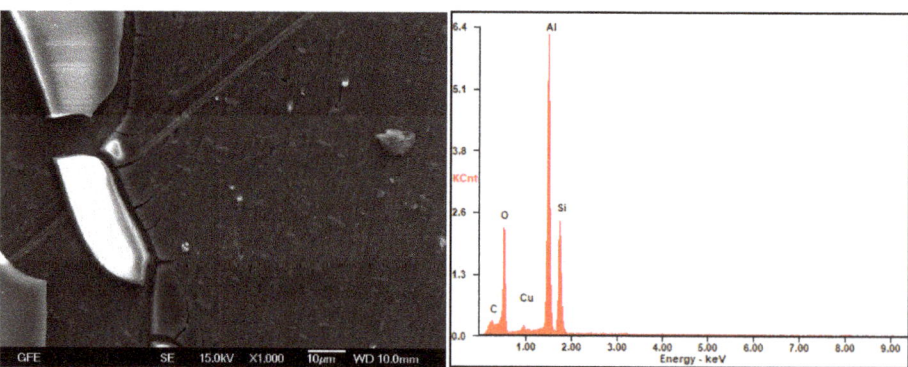

Figure 2. SEM and EDS analysis of diluted colloidal solution after evaporation.

The SEM analysis was performed on the JSM 7000F by JEOL (construction year 2006, JEOL Ltd., Tokyo, Japan) and EDX analysis using the Octane Plus-A by Ametek-EDAX (construction year, 2015, AMETEK Inc., Berwyn, PA, USA), with software Genesis V 6.53 by Ametek-EDAX, revealing an irregular structure of silica precursor, as shown in Figure 2. XRD analysis of silica powders was performed using a Bruker D8 Advance with a LynxEye detector (Bruker AXS, Karlsruhe, Germany). X-ray powder diffraction patterns were collected on a Bruker-AXS D4 Endeavor diffractometer in Bragg–Brentano geometry, equipped with a copper tube and a primary nickel filter providing Cu Kα1,2 radiation (λ = 1.54187 Å).

2.2. Procedure

Synthesis of silica was performed by the transformation of water solution of the chosen precursor to an aerosol in a strong ultrasonic field with an additional thermal decomposition of droplets at elevated temperatures in an inert atmosphere, as shown in Figure 3. The formed droplets of aerosol were transported with carrier gas to the laboratory tubular furnace (Ströhlein, Selm, Germany) in order to be transformed into nanosized particles. Due to the thermal stability of a quartz tube in a furnace, the maximal reaction temperature amounts to 1000 °C. The heating rate was 30 °C/min. Thermal decomposition of the precursor was performed at 900 °C. According to our previous work [17,18], an increase in gas temperature and aerosol velocity decreases the residence time of droplets in the reactor. A decrease in droplet size and an increase in gas temperature lead to a decreased particle size. The collection of powder was performed in two bottles filled with alcohol or distilled water. The obtained suspension was sent to SEM and EDS analysis. The scanning electron microscope was used to examine morphological characteristics of powders such particle sizes and shape. EDS analysis was performed for elemental analysis of the obtained silica powders.

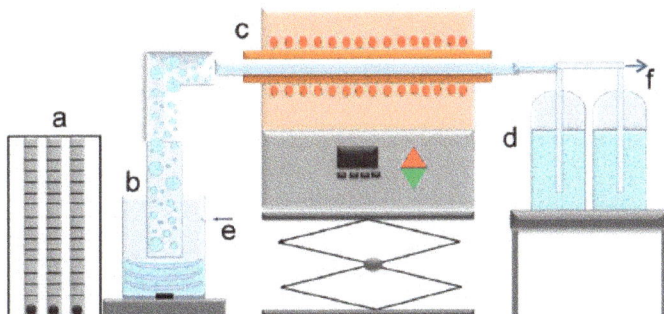

Figure 3. One-step ultrasonic spray pyrolysis lab-scale horizontal equipment: (**a**) Gas flow regulation; (**b**) ultrasonic aerosol generator; (**c**) furnace with the wall-heated reactor; (**d**) collection bottles; (**e**) gas inlet, (**f**) gas outlet.

Very fine aerosol droplets of precursor solution based on colloidal silica were obtained with an ultrasonic atomizer (PRIZNano, Kragujevac, Serbia), using three transducers with a frequency of 1.75 MHz in an ultrasonic field. The aerosol was carried with a nitrogen flow rate between 0.5 and 1.5 L/min into a quartz tube (1.0 m length and 0.021 m diameter) at 900 °C and placed in a previously mentioned Ströhlein furnace. The flow rate was measured using a special flowmeter gas unit (YOKOGAWA Deutschland GmbH, Ratingen).

2.3. Prediction of Particle Size

The formation of SiO_2 will be first defined via the diameter of an aerosol droplet (d_d), as shown with Equation (1) [19,20]:

$$d_d = 0.34 \left(\frac{8 \pi \sigma}{\rho_L f^2} \right)^{\frac{1}{3}}. \qquad (1)$$

where: d_d—diameter of aerosol droplet, f—ultrasound frequency; ρ_L—density of water solution; σ—surface tension of water solution.

Using the following values: f—1.75 MHz; ρ_L—1.02 g/cm³; σ—0.07 J/m², the calculated aerosol droplet amounts to 2.86 μm, as shown in Figure 4 [21]. As shown with

Equation (1) and Figure 4, the aerosol droplet can be decreased by increasing the ultrasonic frequency of an ultrasonic transducer.

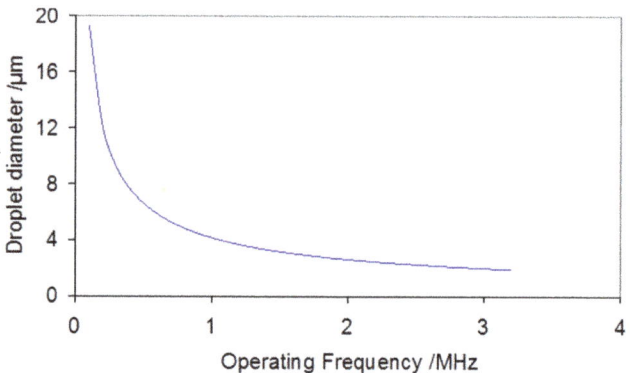

Figure 4. Dependence of aerosol droplet size of the operating frequency.

Assuming that the velocities of the droplet and carrier gas are equal, the droplet velocity was calculated from the ratio of the carrier gas flow (q) to the reaction zone area (A), as shown via Equation (2) [21].

$$v = \left(\frac{q}{A}\right) \quad (2)$$

Using q—0.5 dm^3/min and A—0.38 × 10^{-3} m^2, the droplet velocity amounts to 0.021 m/s.

According to our previous work, the residence time of droplets in the reaction tube can be calculated using Equation (3) [21].

$$t = \left(\frac{V \cdot \text{To}}{q \cdot \text{Tr}}\right) \quad (3)$$

where: V—volume of heating zone in tube (m^3); q—flow rate (L/min); Tr—reaction temperature (K); and To—room temperature. Using the following values: v—0.21 × 10^{-3} m^3, q—0.5–1.5 L/min; Tr—1173 K; and To—298 K, the residence times were calculated at room temperature and 1173 K, as shown in Figure 5. This residence time depends on the reaction temperature and flow rate of carrier gas. An increase in flow rate from 0.5 to 2.5 L/min leads to a residence time of a few seconds in the tubular reactor. An increase in temperature from 25 °C to 900 °C decreases the residence time from 14 to 4 s using a flowrate of 0.5 L/min.

The particle size (d_p) depends on the droplet size and concentration of solution (c). This correlation between the concentration and other precursor characteristics and the final particle size, under the assumption that no precursor is lost in the process, can be described with the following Equation (4) derived from one basic equation reported by Messing et al. [22]:

$$d_p = d_d \left(\frac{M_p}{M_{SiO_2}} * \frac{c}{\rho}\right)^{0.33} \quad (4)$$

where d_p is the diameter of the particle, d_d is the diameter of the aerosol droplet, M_p is the molar mass of the precursor (g/mol), ρ is the density of silica particles, and c is the concentration of the precursor solution.

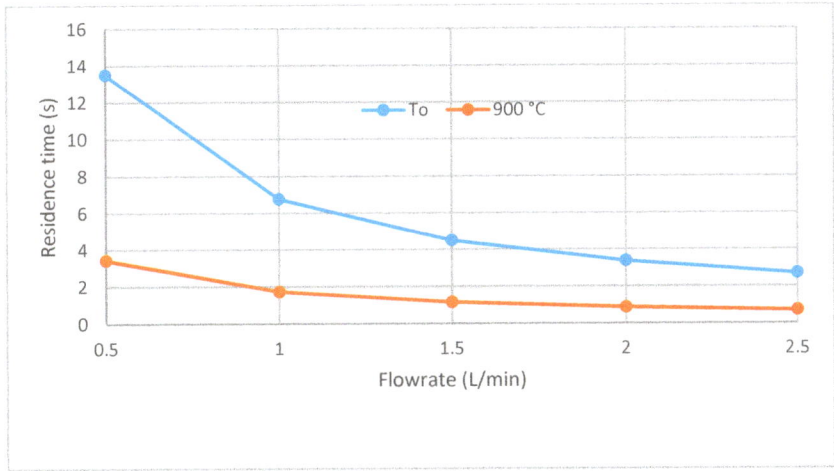

Figure 5. Calculated values of the residence time at different temperatures.

Using the following values: M_{SiO2} = 60.09 g/mol; ρ_{SiO2} = 2.65 × 10³ kg/m³; and concentrations of solution (g/cm³): 60, 30, 15, 7.5, 1.5, 0.15, and 0.1, the obtained values for particles sizes are presented in Table 1:

Table 1. Calculated particle size depending on concentration of precursor solution.

Concentration (mol/L)	1	0.5	0.25	0.125	0.025	0.0025	0.0017
Concentration (g/cm³)	60	30	15	7.5	1.5	0.15	0.10
Particle size (nm)	810	643	508	412	283	110	96

The obtained values of particle sizes are presented in Figure 6.

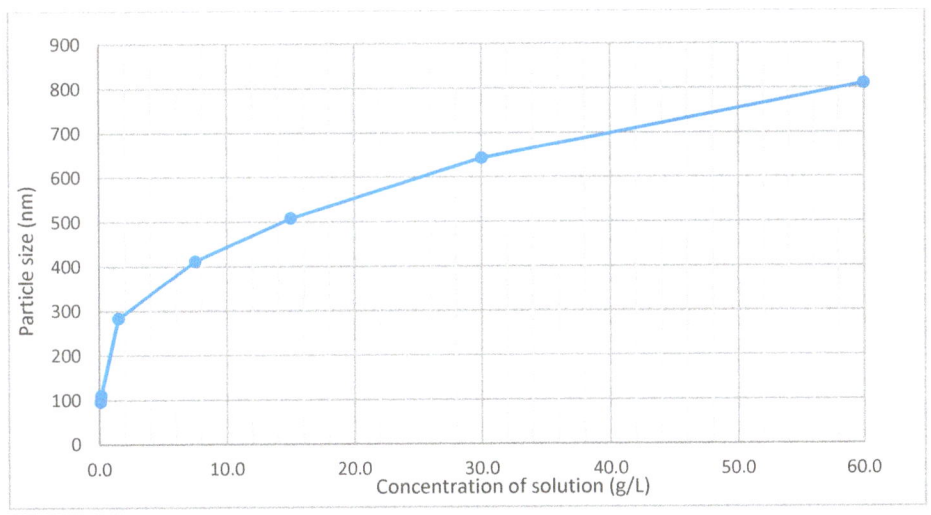

Figure 6. Relationship between particle size and concentration of precursor solution.

As expected by Equation (4), a decrease in the precursor concentration of solution decreases the particle size. A comparison of the calculated data with those published in the literature by Kim [7] confirmed that ultrasonic spray pyrolysis can also prepare particle sizes of 100 nm using a small concentration of solution of 0.1 mg/L (0.002 mol/L). In order to validate the prediction of particle size from Table 1, the following experimental concentrations (0.5 and 0.125 mol/L) were tested in our experimental work.

3. Results and Discussion

SEM and EDS analysis of the obtained particles at 900 °C using precursor solution concentrations of 0.50 and 0.125 mol/L found very fine spherical particles after ultrasonic spray pyrolysis, as shown in Figures 7a and 8a. Using small concentrations of solution such as 0.125 mol/L, the silica particle is ideally spherical, and single without agglomeration, which is a typical case for higher concentrations. The presence of large particles of about 1 µm is confirmation of a collision of droplets during the transport of an aerosol. Some satellite spherical particles are observed at primary large particles. The measurement of particle size was performed using Software Image Pro Plus, Media Cybernetics, USA [23].

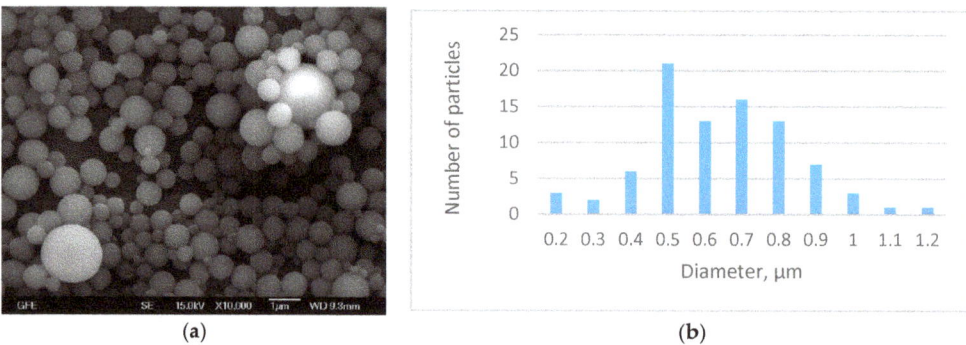

Figure 7. (a) SEM analysis of particles obtained at 900 °C using 0.50 mol/L precursor solution. (b) Particle size distribution.

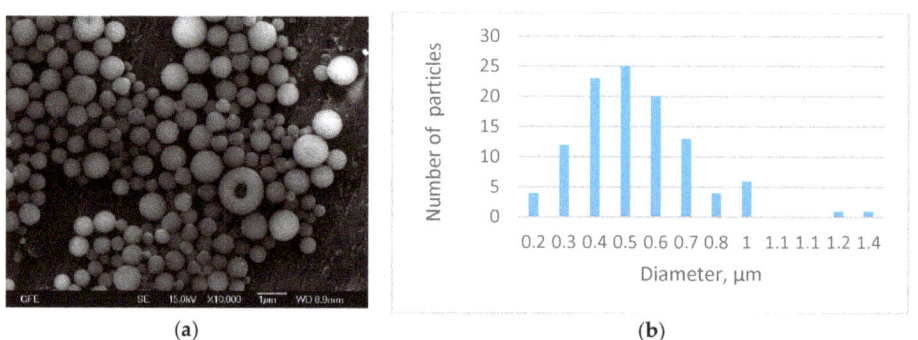

Figure 8. (a) SEM analysis of particles obtained at 900 °C using 0.125 mol/L precursor solution. (b) Particle size distribution.

A decrease in solution concentration from 0.5 to 0.125 mol/L leads to smaller particle size, as shown in Figures 7b and 8b, respectively. EDS analysis has confirmed the presence of silicon and oxygen together with elements such as Al, Cu, and C, which are used for the preparation of samples for characterization, as shown in Figure 9.

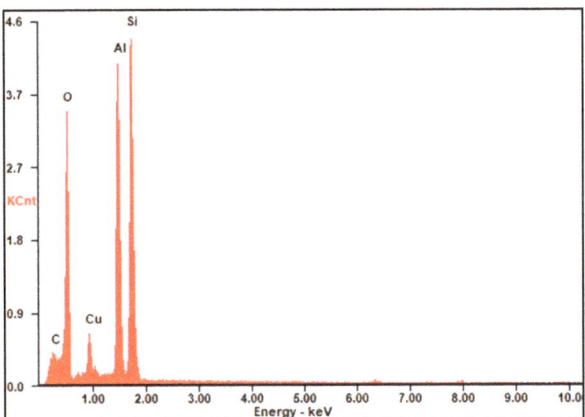

Figure 9. Typical EDS analysis for both obtained powders.

The presence of elements such as Cu, C, and Al are not connected with our ultrasonic spray pyrolysis synthesis. Additionally, ICP-OES analysis of Si in solution before and after USP synthesis was included in our consideration. The concentration of Si decreased from 13,600 to 74.6 mg/L for the 0.5 M precursor solution. A similar behavior was revealed for the 0.125 M solution, where the concentration of Si decreased from 3750 to 1 mg/L, which confirms the full transformation of the used precursor to SiO_2.

For the sample with a concentration of 0.5 mol/L, most particles are around 0.6 μm in diameter, with an average diameter value of 0.68 μm, which suggests similar particle diameters. The sample with a concentration of 0.125 mol/L could be characterized with most particles around 0.50 μm and with an average diameter 0.60 μm, which suggests the existence of larger particles but with a lower quantity, and the lower diameter was compared with particles from the precursor solution of 0.50 mol/L. The calculated particle sizes are situated between the maximal and minimal values of the measured diameters, as shown in Table 2.

Table 2. The values of theoretical and measured silica particles.

Concentration of Solution, mol/L	Calculated Value of Diameter, μm	Measured Maximal Diameter, μm	Measured Average Diameter, μm	Measured Minimal Diameter, μm
0.50	0.64	1.20	0.69	0.16
0.125	0.41	1.35	0.61	0.24

According to our previous laser diffraction measurement of produced aerosol from an ultrasonic generator between 0.8 and 2.5 MHz reported by Bogovic et al. [24], the obtained values of droplet size are, in all cases, higher than theoretically predicted, as shown with Equation (1), due to the immediate coagulation that occurs in the aerosol production chamber. As mentioned previously by Tsi et al. [25], only 5–10% of the particle sizes obtained in spray pyrolysis of 6–9 μm precursor droplets were of the sizes predicted by the one-particle-per-droplet mechanism. Differences between calculated and experimentally obtained values of particle sizes may be partially due to the approximate values used for surface tension and the density of aqueous solution, micro-porosity of particles, and mostly due to the coalescence/agglomeration of aerosol droplets at a high flow rate for the carrier gas (turbulence effects). In Equation (4) [22,26], also based on the assumption of one particle per droplet, the influence of temperature on the mean particle size was not taken into account.

XRD analysis of powder obtained at 900 °C has shown an amorphous structure of the prepared silica powder, as shown in Figure 10. A Hill-like peak in the range of [2Θ] = 21–24 indicates the absence of any ordered crystalline structure and a highly disordered structure of silica. The same X-ray diffraction patterns of nanosilica were reported by Huan et al. [4]. The extracted nanosilica from tetraethoxysilane (TEOS) has a high lead treatment efficiency from waste water. We hope that spherical amorphous silica particles prepared by ultrasonic spray pyrolysis have the same properties. Chen et al. [13] reported from a comparison with crystalline silica that the amorphous structure has more advantages such as nontoxicity, better interaction, and pollution adsorbent. In comparison to particles obtained by Huan [4], we produced ideally spherical nonagglomerated particles.

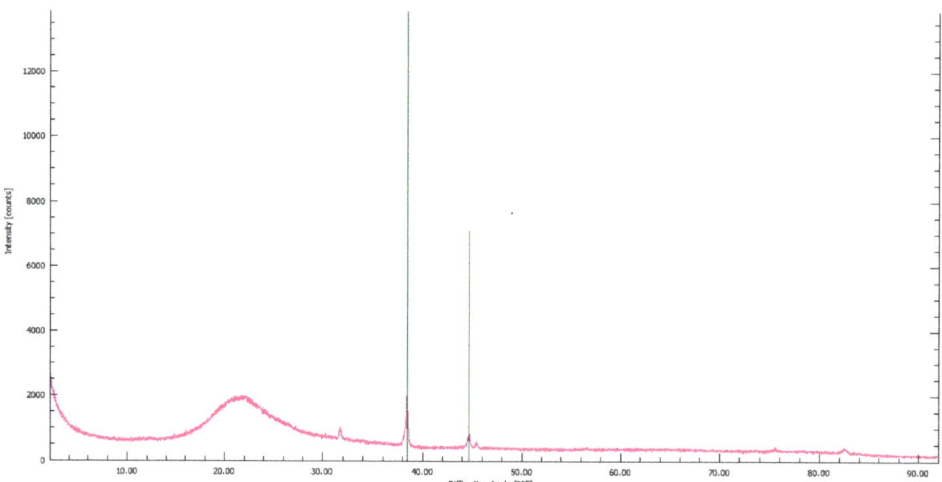

Figure 10. XRD analysis of silica powder obtained by ultrasonic spray pyrolysis.

The crystalline peaks in Figure 10 belong to aluminum (sample holder). In order to obtain a fully crystalline structure, ultrasonic spray pyrolysis shall be performed at higher temperatures and longer residence times. According to Figures 5 and 6, we found that the chosen residence time in the furnace is sufficient for the complete transformation of precursor to the aimed SiO_2. Regarding the synthesis of silica using the previously mentioned sol–gel, high-pressure carbonation, and other methods, ultrasonic spray pyrolysis enables a controlled particle size and morphology using different concentrations of solution.

4. Conclusions

Synthesis of silica powder was performed from a high concentrated colloidal solution (30%) at 900 °C using the ultrasonic spray pyrolysis method. This method enables the production of very fine spherical silica particles from an irregular structure in one horizontal reactor. The controlled synthesis of particles was reached by changing the concentration of precursor solution from 0.5 to 0.125 mol/L. A decrease in concentration from 0.5 to 0.125 mol/L leads to a decrease in measured average diameter from 690 to 610 nm. A Hill-like peak in the range of [2Θ] = 21–24 obtained by XRD analysis indicates the absence of any ordered crystalline structure and a highly disordered structure of silica with high purity. A comparison of theoretically and measured diameter values of prepared silica has shown relatively good agreement, where a deviation amounts to 17% for average diameter. A disadvantage of this method is the collision of droplets during their transport using carrier gas, and especially low efficiency, due to losses in the dissolved precursor on the construction elements of the reactor.

Author Contributions: Conceptualization, F.W. and S.S.; funding acquisition, B.F.; investigation, F.W.; methodology, S.S. and F.W.; supervision, S.S. and B.F.; writing—original draft, F.W, T.-V.H. and S.S. All authors have read and agreed to the published version of the manuscript.

Funding: This research received no external funding.

Institutional Review Board Statement: Not applicable.

Informed Consent Statement: Not applicable.

Conflicts of Interest: The authors declare no conflict of interest.

References

1. Stopic, S.; Friedrich, B. Deposition of silica in hydrometallurgical processes. *Vojnoteh. Glas. Mil. Tech. Cour.* **2020**, *68*, 65–78. [CrossRef]
2. Lazaro, A.; Brouwers, H.; Quercia, G.; Geus, J. The properties of amorphous nano-silica synthesized by the dissolution of olivine. *Chem. Eng. J.* **2012**, *211–212*, 112–121. [CrossRef]
3. Mohanraj, K.; Kannan, S.; Barathan, S.; Sivakumar, G. Preparation and characterization of nano SiO_2 from corn cob ash by precipitation method. *Optoelectron. Adv. Mater. Rapid Commun.* **2012**, *6*, 394–397.
4. Huan, N.X.; Anh, T.N.; Hang, N.T.; Nhung, D.T.; Thanh, N.V. Nanosilica synthesis and application for lead treatment in water. *J. Viet. Environ.* **2018**, *9*, 255–263.
5. Ui, S.; Choi, I.; Choi, S. Synthesis of high surface area mesoporous silica powder using anionic surfactant. *ISRN Mater. Sci.* **2014**, *2019*, 1–6. [CrossRef]
6. Nandanwar, N.; Singh, P.; Fozia, Z.; Haque, F. Synthesis and characterization of SiO_2 nanoparticles by sol-gel process and its degradation of methylene blue. *Am. Chem. Sci. J.* **2015**, *5*, 1–10. [CrossRef]
7. Kim, S.; Seo, S.; Jung, S. Preparation of high purity nano silica particles from blast-furnace slag. *Korean J. Chem. Eng.* **2010**, *27*, 1901–1905. [CrossRef]
8. Kretushev, I.; Mishin, M.; Aleksandrov, S. Synthesis of silicon dioxide nanoparticles in low temperature atmospheric pressure plasma. *Russ. J. Appl. Chem.* **2014**, *87*, 1581–1586. [CrossRef]
9. Stopic, S.; Dertmann, C.; Koiwa, I.; Kremer, D.; Wotruba, H.; Etzold, S.; Telle, R.; Knops, P.; Friedrich, B. Synthesis of nanosilica via olivine mineral carbonation under high pressure in an autoclave. *Metals* **2019**, *9*, 708. [CrossRef]
10. Phoohinkong, W.; Kitthawee, U. Low-cost and fast production of nano-silica from rice husk ash. *Adv. Mater. Res.* **2014**, *979*, 216–219. [CrossRef]
11. Wang, Y.; Forssberg, E. Production of carbonate and silica nano-particles in stirred bead milling. *Int. J. Miner. Process.* **2006**, *81*, 1–14. [CrossRef]
12. Akhayere, E.; Kavaz, D.; Vaseashta, A. Synthesizing nano silica nanoparticles from Barley Grain waste: Effect of temperature on mechanical properties. *Pol. J. Environ. Stud.* **2019**, *28*, 2513–2521. [CrossRef]
13. Chen, X.; Jiang, J.; Yan, F.; Tian, S.; Li, K. A novel low temperature vapor phase hydrolysis method for the production of nano-structured silica materials using silicon tetrachloride. *RSC Adv.* **2014**, *4*, 8703. [CrossRef]
14. Ratanathavorn, W.; Bouhod, N.; Modsuwan, J. Synthesis of silica nanoparticles by ultrasonic spray pyrolysis technique for cream perfume formulation. *J. Food Health Bioenviron. Sci.* **2018**, *11*, 1–5.
15. Citakovic, N. Physical properties of nanomaterials. *Vojnoteh. Glas. Mil. Tech. Cour.* **2019**, *67*, 159–171. [CrossRef]
16. Byrne, J.; Baugh, J. The significance of nanoparticles in particles-induced pulmonary fibrosis. *Mc Gill J. Med.* **2008**, *11*, 43–50.
17. Stopic, S.; Schroeder, M.; Weirich, T.; Friedrich, B. Synthesis of TiO_2 Core/RuO_2 Shell Particles using Multistep Ultrasonic Spray Pyrolysis. *Mater. Res. Bull.* **2013**, *48*, 3633–3635. [CrossRef]
18. Košević, M.; Stopic, S.; Bulan, A.; Kintrup, J.; Weber, R.; Stevanović, J.; Panic, V.; Friedrich, B. A continuous process for the ultrasonic spray pyrolysis synthesis of RuO_2-TiO_2 particles and their application as an active coating of activated titanium anode. *Adv. Powder Technol.* **2017**, *28*, 43–49. [CrossRef]
19. Lang, R.J. Ultrasonic atomization of liquids. *J. Acoust. Soc. Am.* **1962**, *34*, 6–8. [CrossRef]
20. Peskin, R.L.; Raco, R.J. Ultrasonic atomization of liquids. *J. Acoust. Soc. Am.* **1963**, *35*, 1378–1381. [CrossRef]
21. Stopic, S. *Synthesis of Metallic Nanosized Particles by Ultrasonic Spray Pyrolysis*; Shaker GmbH: Kohlsheid, Germany, 2015; p. 117.
22. Messing, G.; Zhang, S.; Jayanthi, G. Ceramic powder synthesis by spray pyrolysis. *J. Am. Ceram. Soc.* **1993**, *76*, 2707–2726. [CrossRef]
23. Francisco, J.S.; Pinto de Moraes, H.; Dias, E. Evaluation of the image-pro plus 4.5 software for automatic counting of labeled nuclei by PCNA immunohistochemistry. *Braz. Oral Res.* **2004**, *18*, 100–104. [CrossRef]
24. Bogovic, J.; Schwinger, A.; Stopic, S.; Schroeder, J.; Gaukel, V.; Schuhmann, P.; Friedrich, B. Controlled droplet size distribution in ultrasonic spray pyrolysis. *Metall* **2011**, *10*, 455–459.
25. Tsai, S.C.; Song, Y.L.; Tsai, C.S.; Yang, C.; Chiu, W.; Lin, H. Ultrasonic spray pyrolysis for nanoparticles synthesis. *J. Mater. Sci.* **2004**, *39*, 3647–3657. [CrossRef]
26. Stopić, S.; Friedrich, B.; Dvorak, P. Synthesis of nanosized spherical silver powder by ultrasonic spray pyrolysis. *Metall* **2006**, *60*, 377–382.

Article

Atomic Layer Deposition of aTiO$_2$ Layer on Nitinol and Its Corrosion Resistance in a Simulated Body Fluid

Rebeka Rudolf [1,*], Aleš Stamboli´c [2] and Aleksandra Kocijan [2]

[1] Faculty of Mechanical Engineering, University of Maribor, Smetanova ulica 17, 2000 Maribor, Slovenia
[2] Institute of Metals and Technology, Lepi pot 11, 1000 Ljubljana, Slovenia; ales.stambolic@gmail.com (A.S.); aleksandra.kocijan@imt.si (A.K.)
* Correspondence: rebeka.rudolf@um.si

Citation: Rudolf, R.; Stamboli´c, A.; Kocijan, A. Atomic Layer Deposition of aTiO$_2$ Layer on Nitinol and Its Corrosion Resistance in a Simulated Body Fluid. *Metals* **2021**, *11*, 659. https://doi.org/10.3390/met11040659

Academic Editor: Kewei Gao

Received: 23 March 2021
Accepted: 15 April 2021
Published: 18 April 2021

Publisher's Note: MDPI stays neutral with regard to jurisdictional claims in published maps and institutional affiliations.

Copyright: © 2021 by the authors. Licensee MDPI, Basel, Switzerland. This article is an open access article distributed under the terms and conditions of the Creative Commons Attribution (CC BY) license (https://creativecommons.org/licenses/by/4.0/).

Abstract: Nitinol is a group of nearly equiatomic alloys composed of nickel and titanium, which was developed in the 1970s. Its properties, such as superelasticity and Shape Memory Effect, have enabled its use, especially for biomedical purposes. Due to the fact that Nitinol exhibits good corrosion resistance in a chloride environment, an unusual combination of strength and ductility, a high tendency for self-passivation, high fatigue strength, low Young's modulus and excellent biocompatibility, its use is still increasing. In this research, Atomic Layer Deposition (ALD) experiments were performed on a continuous vertical cast (CVC) NiTi rod (made in-house) and on commercial Nitinol as the control material, which was already in the rolled state. The ALD deposition of the TiO$_2$ layer was accomplished in a Beneq TFS 200 system at 250 °C. The pulsing times for TiCl$_4$ and H$_2$O were 250 ms and 180 ms, followed by appropriate purge cycles with nitrogen (3 s after the TiCl$_4$ and 2 s after the H$_2$O pulses). After 1100 repeated cycles of ALD depositing, the average thickness of the TiO$_2$ layer for the CVC NiTi rod was 52.2 nm and for the commercial Nitinol, it was 51.7 nm, which was confirmed by X-ray Photoelectron Spectroscopy (XPS) and Scanning Electron Microscope (SEM) using Energy-dispersive X-ray (EDX) spectroscopy. The behaviour of the CVC NiTi and commercial Nitinol with and without the TiO$_2$ layer was investigated in a simulated body fluid at body temperature (37 °C) to explain their corrosion resistance. Potentiodynamic polarisation measurements showed that the lowest corrosion current density (0.16 µA/cm^2) and the wider passive region were achieved by the commercial NiTi with TiO$_2$. Electrochemical Impedance Spectroscopy measurements revealed that the CVC NiTi rod and the commercial Nitinol have, for the first 48 h of immersion, only resistance through the oxide layer, as a consequence of the thin and compact layer. On the other hand, the TiO$_2$/CVC NiTi rod and TiO$_2$/commercial Nitinol had resistances through the oxide and porous layers the entire immersion time since the TiO$_2$ layer was formatted on the surfaces.

Keywords: nitinol; continuous vertical cast (CVC), NiTi rod; atomic layer deposition; corrosion properties; potentiodynamic test; electrochemical impedance spectroscopy

1. Introduction

Nitinol is a group of alloys that are in the equiatomic composition range of nickel and titanium. It shows unique properties, such as superelasticity and a Shape-Memory Effect. It also exhibits good corrosion resistance in a chloride environment, an unusual combination of strength and ductility, a high tendency for self-passivation, high fatigue strength, low Young's modulus and excellent biocompatibility. Its properties have enabled its use, especially for biomedical purposes (orthodontic treatments, cardiovascular surgery for stents and guide wires, orthopaedic surgery for various staples and rods, maxillofacial and reconstructive surgery). In addition, Nitinol has been used in the Aerospace, Automotive, Marine and Chemical industries and Civil and Structural Engineering [1–7].

Nitinol has a tendency for self-passivation in a physiological saline solution. The passive films formed on the surface consist mainly of amorphous titanium dioxide [6]. However, the naturally formed oxides on the metal surfaces are thin and do not prevent the corrosion process or Ni leaching. The main problem of Nitinol is its high Ni content. Ni releasing can induce toxic, allergic and hypersensitive reactions or tissue necrosis after long-term implantation [8–10]. Leaching of Ni can arise when a strongly acidic fluid attacks the surface of the alloy. This corrosion is accompanied by nickel release from an implant into the surrounding body fluid and tissue, which can enhance an allergic reaction in a sensitive organism. Another path leading to the accumulation of Ni in the surface layers can be the type of surface treatment itself. Low-temperature (60–160 °C) pre-treatment protocols or high-temperature annealing in the air used for deposition of a thick TiO_2 layer onto a Nitinol surface results in Ni accumulation in the surface depth [11–13]. This hidden Ni can easily be released through the defective surfaces, exceeding the Ni release from non-treated material by two to three orders of magnitude. A high concentration of Ni close to the metal–oxide interface will yield larger particles. The larger particles will induce severe local strain in the lower oxide layer, leading to local rupture and cracking. Such cracks can then extend towards the surface and act as channels, explaining the much greater Ni release. To prevent corrosion and, consequently, Ni release, a coating of appropriate thickness must be formed on the NiTi surface. Titanium oxide coatings are useful enough to suppress nickel ions' out-leaching [14–18].

Titanium dioxide is formed when titanium is subjected to oxidising conditions. TiO_2 is an electron excess conductor with oxygen deficiency. Due to the electroneutrality condition oxygen vacancies must be compensated by a corresponding number of negative charges. This can be achieved by O^{2-} ions that are incorporated into the lattice, resulting in electroneutrality of the complete crystal. The disorder of the oxide determines which species are mobile in the course of oxidation. The driving force for diffusion is the concentration gradient of the vacancies in the oxide, such as, in TiO_2 oxygen ions diffuse inward since the oxygen vacancy concentration is highest at the oxide/metal interface. The disorder in oxides not only determines the location for scale growth, but also influences the ability of the oxide scale to close the cracks by oxide regrowth during high-temperature exposure [19]. Oxidation of Nitinol occurs as follows:

$$NiTi + O_2 \rightarrow Ni_3Ti + TiO_2 \rightarrow Ni_4Ti + TiO_2 \rightarrow Ni + TiO_2 \qquad (1)$$

Due to the four-times lower Gibbs free energy of formation of titanium oxide than that of nickel oxide, titanium oxide growth is preferred on the Nitinol surface. Observations are consistent with a model of oxygen absorption on the NiTi surface that reacts with outward diffusing Ti to form TiO_2. During the early stages of oxidation, the growth of the TiO_2 layer is the only contributor to thickness, and, therefore, oxide formation is relatively rapid. However, the preferential oxidation of Ti creates a Ti-depleted (Ni-rich) zone at the $NiTi/TiO_2$ interface. The formation of the Ni-rich layer increases the effective diffusion distance with an associated decrease in overall oxidation kinetics. Therefore, continued oxide growth involves the simultaneous nucleation and growth of titanium oxides and Ni-rich phases. Ultimately, these processes lead to the formation of a protective oxide scale, which prevents further oxidation of the base material [20,21]. Naturally grown titanium oxide on a Nitinol surface is approximately 5 nm thick, and thicker titanium oxide should be formed to achieve better properties [22].

The most common methods employed for layer formation are anodisation, plasma spraying, Atomic Layer Deposition (ALD), etc. ALD is an interesting technique for producing TiO_2 thin films due to its simplicity, reproducibility, high conformity of thin films and excellent control of the layer thickness at the angstrom level. The thin film is formed as a result of repeated deposition cycles. At least two subsequent self-limited surface reactions are used to form a new layer [23–26]. In the literature, there are results for a commercial flat-annealed NiTi foil and NiTi wires (ϕ = 0.2 mm) [27], and for Al_2O_3 and Pt, ALD coatings on NiTi thin films [28]. None of the studies, however, dealt with the study of

ALD deposition on the surface of a continuous vertical cast (CVC) NiTi rod (ϕ = 11 mm), which was vacuum remelted before casting. Namely, the production of Nitinol is still very complex, and, therefore, new faster processes resulting in the shape of a rod with the smallest diameter possible are being sought.

The ALD technique for producing TiO_2 on a CVC NiTi rod made in-house [29] was tested in this study. For comparison, a commercial Nitinol was used to evaluate how the ALD coating works. CVC NiTi rod-testing production is explained in our previous works [30,31], while the commercial Nitinol was already in the rolled state. With the ALD technique, the thin layer of titanium oxide was deposited on both types of specimens. The formation of TiO_2 was investigated by X-ray Photoelectron Spectroscopy (XPS) and Scanning Electron Microscopy (SEM) using Energy-dispersive X-ray (EDX) spectroscopy. TiO_2-protected specimens were tested on corrosion behaviour in simulated body fluid to clarify their corrosion resistance and confirm the positive effect of the formatted TiO_2 layer in order to reduce the corrosion rate.

2. Materials and Methods

2.1. Atomic Layer Deposition of a TiO_2 Layer

The process of depositing the layer onto the surface with ALD is cyclic. Each cycle is set up from 4 steps (see Figure 1). In the first step, the first precursor ($TiCl_4$) is pulsed into the reactor. $TiCl_4$ is adsorbed to the specimen's surface and reacts there with the reactive sites (initially, these are organic impurities). In the second step, nitrogen is pulsed into the reactor to purge the excess of the reactant and by-products. In the third step, the second precursor (H_2O) is pulsed into the reactor. H_2O is adsorbed to the specimen's surface and reacts there with the reactive sites. In the last step, nitrogen is again pulsed into the reactor to purge the excess of the reactant and by-products. 1100 cycles were made in this study. The simplified equation for this reaction is:

$$TiCl_4 + 2\ H_2O \rightarrow TiO_2 + 4\ HCl \tag{2}$$

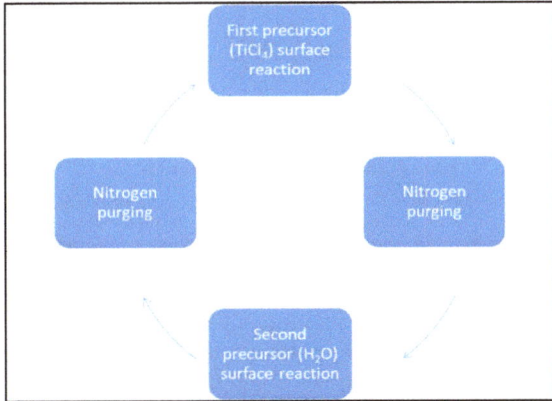

Figure 1. Steps of Atomic Layer Deposition (ALD) deposition.

The deposition of a TiO_2 layer on the surface of a continuous vertical cast (CVC) NiTi rod and commercial Nitinol was accomplished in a Beneq TFS 200 system at 250 °C. The pulsing times for $TiCl_4$ and H_2O were 250 ms and 180 ms, followed by appropriate purge cycles with nitrogen (3 s after the $TiCl_4$ and 2 s after the H_2O pulses). The chemical compositions of specimens were as follows: Commercial Nitinol (45 wt.% Ti, 55 wt.% Ni) and the CVC NiTi rod made in-house (composition determined by XRF analysis in-house: Ti 38.9 wt.%, 59.8 wt.% Ni). The detailed microstructure investigation revealed that the CVC

NiTi rod containing over 50 at. % Ni, consisted of Ti$_2$Ni and cubic NiTi, with corresponding EDX spectra in the field of investigation—see Figure 2. The chemical composition of the CVC NiTi rod varied through the cross and longitudinal sections because the drawing process was not optimal; the manufacturing problem is described in more detail in our previous study [29].

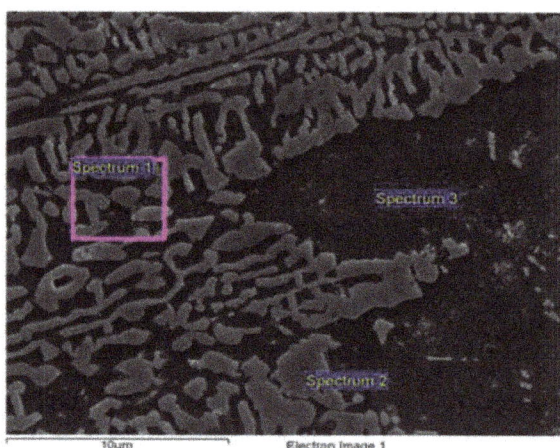

EDX· Spectrum	Ti·(in·at.%)	Ni·(in·at.%)
1	38.38	61.62
2	31.68	68.32
3	40.14	59.86
mean	36.73	63.26
max	40.14	68.32
min	31.68	59.86

Figure 2. Microstructure of the continuous vertical cast (CVC) NiTi rod with corresponding Energy-dispersive X-ray (EDX) spectrum.

The X-ray Photoelectron Spectroscopy (XPS) analyses were performed in order to identify the oxidation states of the elements on the surface of the Ti-oxide films and calculate their surface composition. XPS analyses were carried out on the PHI-TFA XPS spectrometer (Physical Electronics Inc., MN, USA), equipped with a monochromatic Al source. The analysed area was 0.4 mm in diameter, and the analysed depth was about 3–5 nm. High energy resolution XPS spectra were taken with a pass energy of 29 eV, energy resolution of 0.6 eV and energy step of 0.1 eV. Quantification of the surface composition was performed from the XPS peak intensities, taking into account the relative sensitivity factors provided by the instrument manufacturer [32]. Two places on every specimen were analysed, and the average composition was calculated. Chemical bonding of the elements was deduced from the high-energy resolution XPS spectra using reference XPS1. The Multipak software package (version 9.9, ULVAC-PHI Inc., Japan) was used for XPS spectra processing. The error in the binding energy of the measured spectra is ±0.3 eV.

Additionally, a Jeol JSM-7800F Field Emission SEM, equipped with EDX spectroscopy X-Max^N 80, Oxford Instruments (Oxford Instruments, Abingdon, UK), was used for identification of the EDX line chemical composition of the formatted TiO$_2$ analysis on the surface of the CVC NiTi rod and commercial Nitinol. For this purpose, the CVC NiTi rod was cut into a cylinder shape with ϕ = 11 mm and h = 1 cm, using an Accutom 50 (IMT, Ljubljana, Slovenia) electronic saw for precision cutting. The commercial Nitinol, which was in the rolled state in the form of sheets with width of 1.5 mm, was cut to circles with ϕ = 10 mm and a width of 1.5 mm using a water jet cutter (Faculty of Mechanical Engineering, Ljubljana, Slovenia) (Omax, Kent, WA, USA). For easier polishing, the specimens were then hot-pressed into Bakelite. Mechanical polishing was performed on Struers Abramin apparatus (IMT, Ljubljana, Slovenia) (Struers, Copenhagen, Denmark). The grinding was performed with 320-grit SiC abrasive paper, mechanical polishing with MD-Largo (Struers, Cleveland, OH 44145, USA) discs with 9 µm diamond suspension and with peroxide grains in a chemically aggressive suspension—OP-S (colloidal silica). At the end, specimens were cleaned with detergent, washed well with water and put in an ultrasound bath in alcohol. Surface images were recorded at 20,000× magnification, where the SEM working parameters were 0.7 kV voltage and 2 mm working distance. Cross-section images were recorded at 200,000× magnification and SEM working parameters of 10 kV voltage and 4 mm working distance.

2.2. Corrosion Tests

The corrosion tests were performed on the CVC NiTi rod with and without a formatted TiO_2 layer and on commercial Nitinol with and without a formatted TiO_2 layer, respectively. All the measurements were held at body temperature (37 °C). Potentiodynamic polarisation measurements and Electrochemical Impedance Spectrometry (EIS) (Biologic, Seyssinet-Pariset, France) were used to study the electrochemical behaviour of the specimens. All the measurements were recorded by a BioLogic Modular Research Grade Potentiostat/Galvanostat/FRA Model SP-300 (Seyssinet-Pariset, France) with an EC-Lab Software (V 11.27, Biologic, Seyssinet-Pariset, France) and three-electrode cell. In this cell, the specimen, with an exposed area of 1 cm^2, was a working electrode, and a saturated calomel electrode (SCE, 0.242 V vs. SHE) was used as a reference electrode, and the Counter Electrode (CE) was a platinum net. The experiment was held in simulated physiological Hank's solution, containing 8 g/L NaCl, 0.40 g/L KCl, 0.35 g/L $NaHCO_3$, 0.25 g/L $NaH_2PO_4 \times 2H_2O$, 0.06 g/L $Na_2HPO_4 \times 2H_2O$, 0.19 g/L $CaCl_2 \times 2H_2O$, 0.41 g/L $MgCl_2 \times 6H_2O$, 0.06 g/L $MgSO_4 \times 7H_2O$ and 1 g/L glucose, at pH = 7.8. All the chemicals were from Merck, Darmstadt, Germany. The potentiodynamic curves were recorded after 1 h specimen stabilisation at the Open-Circuit Potential (OCP), starting the measurement at 250 mV vs. SCE more negative than the OCP. The potential was then increased, using a scan rate of 1 mV s^{-1} until the transpassive region was reached. Long-term open-circuit potentiostatic electrochemical impedance spectra were obtained for the investigated specimens. The impedance was measured at the OCP, with sinus amplitude of 5 mV peak-to-peak and a frequency range of 65 kHz to 1 mHz, in the sequence of 0 h, 1 h, 2 h, 6 h, 12 h, 24 h, 48 h, 72 h, 96 h, 120 h, 144 h, 168 h and 192 h. The impedance data are presented in terms of Nyquist plots. Zview v3.4d Scribner Associates software (V 3.4d, Southern Pines, NC, USA) was used for the fitting process. All the experiments were repeated 3 times.

3. Results and Discussion

3.1. ALD TiO_2 Thin Layer Deposition

SEM investigation revealed that after 1100 repeated ALD cycles, the average thickness of the TiO_2 layer for the CVC NiTi rod was 52.2 nm, and for the commercial Nitinol 51.7 nm, as visible in Figure 3.

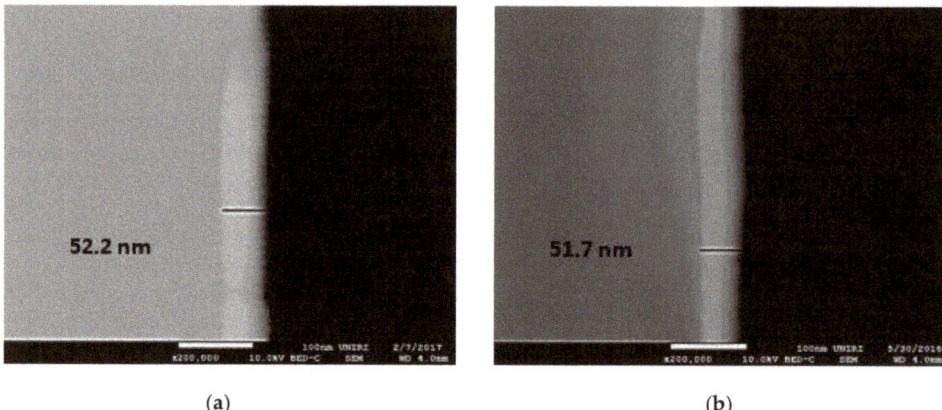

Figure 3. Scanning electron microscopy (SEM) image of the TiO$_2$ layer at 200,000× magnification of the cross-section for the (**a**) CVC NiTi rod and (**b**) Commercial Nitinol.

As seen in Figure 4, the deposited TiO$_2$ layer is quite similar on both surfaces (CVC NiTi rod and commercial Nitinol). The particles are quite different in size (from 50 nm to 1 µm), while their shape is fairly tetrahedral with quite pointed edges, and the grain boundaries are clearly visible.

Figure 4. SEM image at 20,000× magnification of a TiO$_2$ layer on the surface of: (**a**) CVC NiTi rod and (**b**) Commercial Nitinol.

Figure 5a shows an XPS survey spectrum from the TiO$_2$/com.NiTi, and Figure 5b shows a survey spectrum from the TiO$_2$/CVC NiTi rod. Both spectra are similar. They contain the following peaks: Ti 2p$_{3/2}$ at 458.6 eV, C 1s at 284.8 eV, O 1s at 530.0 eV, Ti 3p at 38.2 eV, Ti 3s at 62.5 eV, N 1s at 401.0 eV, O KLL Auger peak at 974 eV and Ti LMM Auger peak at 1106 eV. No traces of a Cl 2p peak were found, which would be expected at 198 eV. From Ti 2p, O 1s, C 1s and N 1s spectra, a surface composition for TiO$_2$/com. NiTi was calculated to be: 45.6 at.% of O, 17.9 at.% of Ti, 35.7 at.% of C and 0.9 at.% of N. Surface composition for the TiO$_2$/CVC NiTi rod was: 42.0 at.% of O, 16.5 at.% of Ti, 40.4 at.% of C and 1.1 at.% of N. The O/Ti ratio for both specimens was the same, i.e., 2.5. The presence of carbon atoms and part of the oxygen atoms are probably related to surface contamination and/or specimen preparation, taking into account that part of the oxygen is related to surface contamination. The Ti/O ratio of 2.5 indicated the TiO$_2$ composition of the deposited films on both specimens. XPS data on the surface composition show that the TiO$_2$/CVC NiTi rod had a higher concentration of carbon, which may be related to the higher degree of surface contamination.

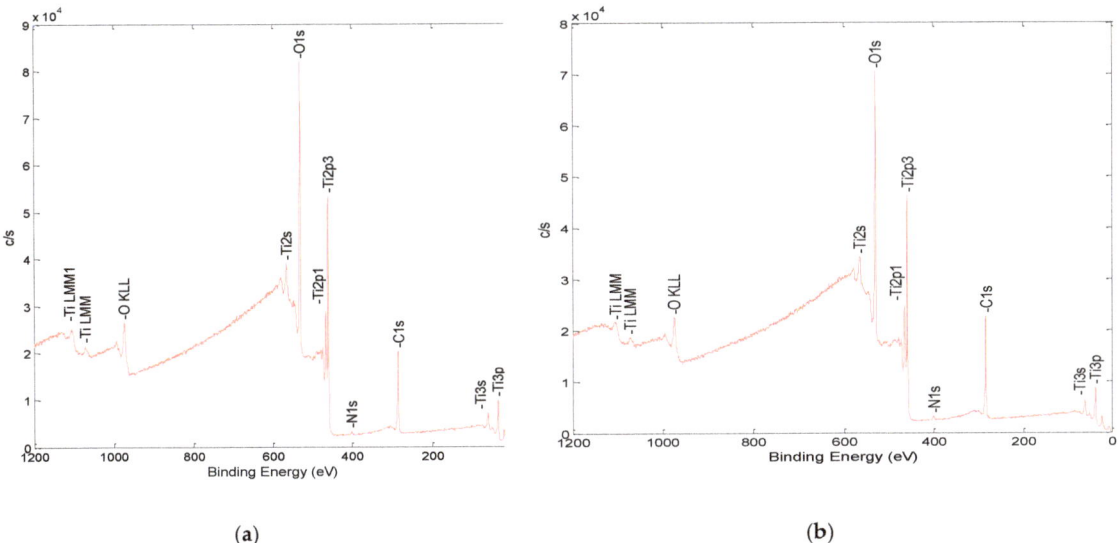

Figure 5. XPS survey spectra from (**a**) the TiO$_2$/com.NiTi and (**b**) the TiO$_2$ CVC NiTi rod.

In order to get an insight into the surface chemistry, high-energy resolution XPS spectra of Ti 2p, O 1s and C 1s were acquired on both specimens. The high-energy resolution XPS spectra Ti 2p, O 1s and C 1s from TiO$_2$/com.NiTi were deconvoluted into different chemical components and are shown in Figure 6. The high-energy resolution XPS spectra Ti 2p, O 1s and C 1s from the TiO$_2$/CVC NiTi rod were deconvoluted into different chemical components and are shown in Figure 7. It is visible that the high-energy resolution XPS spectra from both specimens are very similar. On the surface of both specimens, the Ti 2p$_{3/2}$ peak is at 458.6 eV, and the Ti 2p$_{1/2}$ peak is at 464.5 eV. This binding energy is related to the Ti(4+) oxidation state, which shows the presence of the TiO$_2$ compound (Figures 6a and 7a). No presence of Ti was identified in the lower oxidation states, at least inside the sensitivity of the XPS method. The Ti 2p$_{3/2}$ peak was very narrow, indicating an ordered TiO$_2$ film. The Full Width at Half Maximum (FWHM) of the Ti 2p$_{3/2}$ peak for both specimens was 1.07 eV. The oxygen spectra O 1s were composed of three peaks, the O1 peak at 530.0 eV, the O2 peak at 531.3 eV and a small peak O3 at 532.6 eV. The O1 peak is related to the O^{2-} anions in the TiO$_2$ lattice. The O2 peak at 531.3 eV may be related to the presence of surface OH-groups or O-vacancies in the oxide. The O3 peak may be related to H$_2$O and/or C-O species, mainly due to surface contamination. The carbon C 1s spectra from both specimens contained four peaks, which were related with different chemical bonds of the carbon atoms: the peak at 284.8 eV (C-C/C-H bonds), the peak at 286.2 eV (C-O/C-OH), the peak at 287.1 eV (O-C-O/C = O) and the peak at 289.5 eV (O = C-O/CO3). XPS results show that the surface of the Ti layer was covered by TiO$_2$ with some surface contamination.

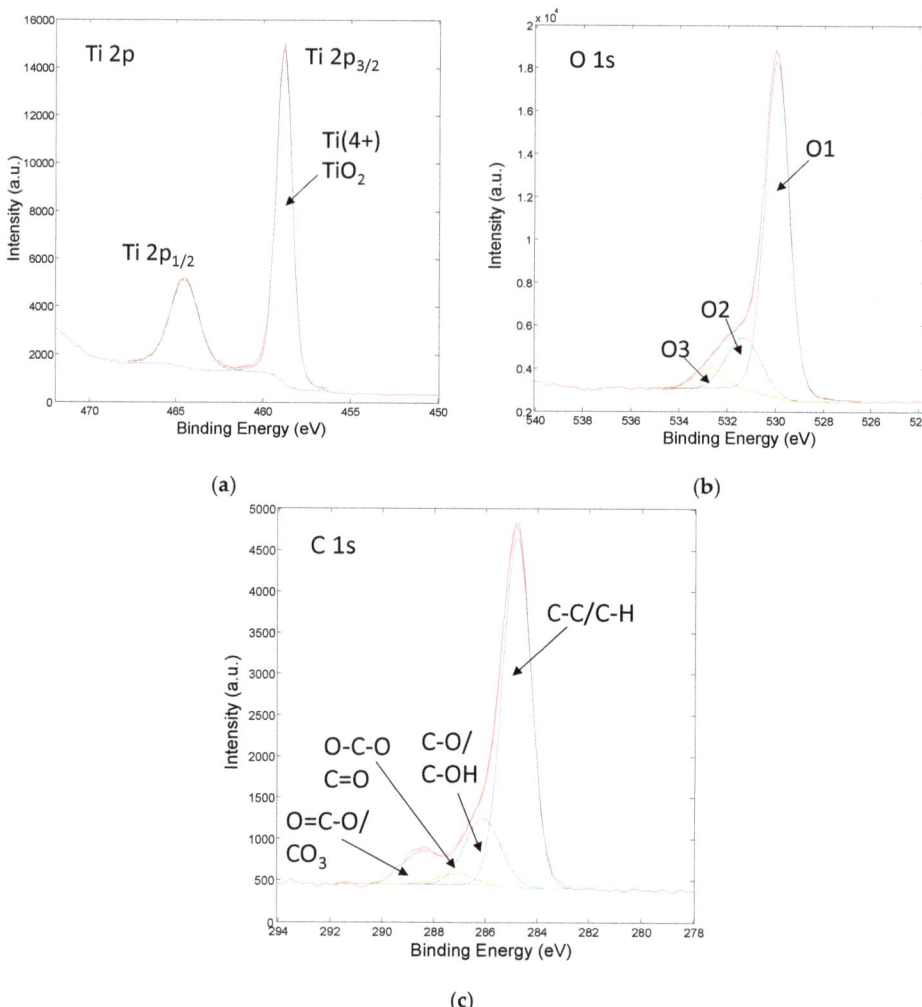

Figure 6. High-energy resolution XPS spectra from the TiO$_2$/com.NiTi surface for: (**a**) Ti 2p, (**b**) O 1s and (**c**) C 1s.

In accordance with the XPS analyses, the results of the performed line EDX analysis on the formatted TiO$_2$ layer confirmed the increased contents of Ti and O (Figure 8) in both specimens, exactly where the TiO$_2$ layer formed. The TiO$_2$ layer is seen as a bright area. The increased concentration profiles for Ti and O are appropriate, as well as the width of the range corresponding to the resulting thickness of the TiO$_2$ layer, as seen with SEM (Figure 3). Due to the extremely thin TiO$_2$ layer, only line EDX analysis could be used in this case, as the usual point/planar EDX analysis would capture a significantly larger volume, and in this way, the signal from the base alloy would be obtained, and the results would be irrelevant.

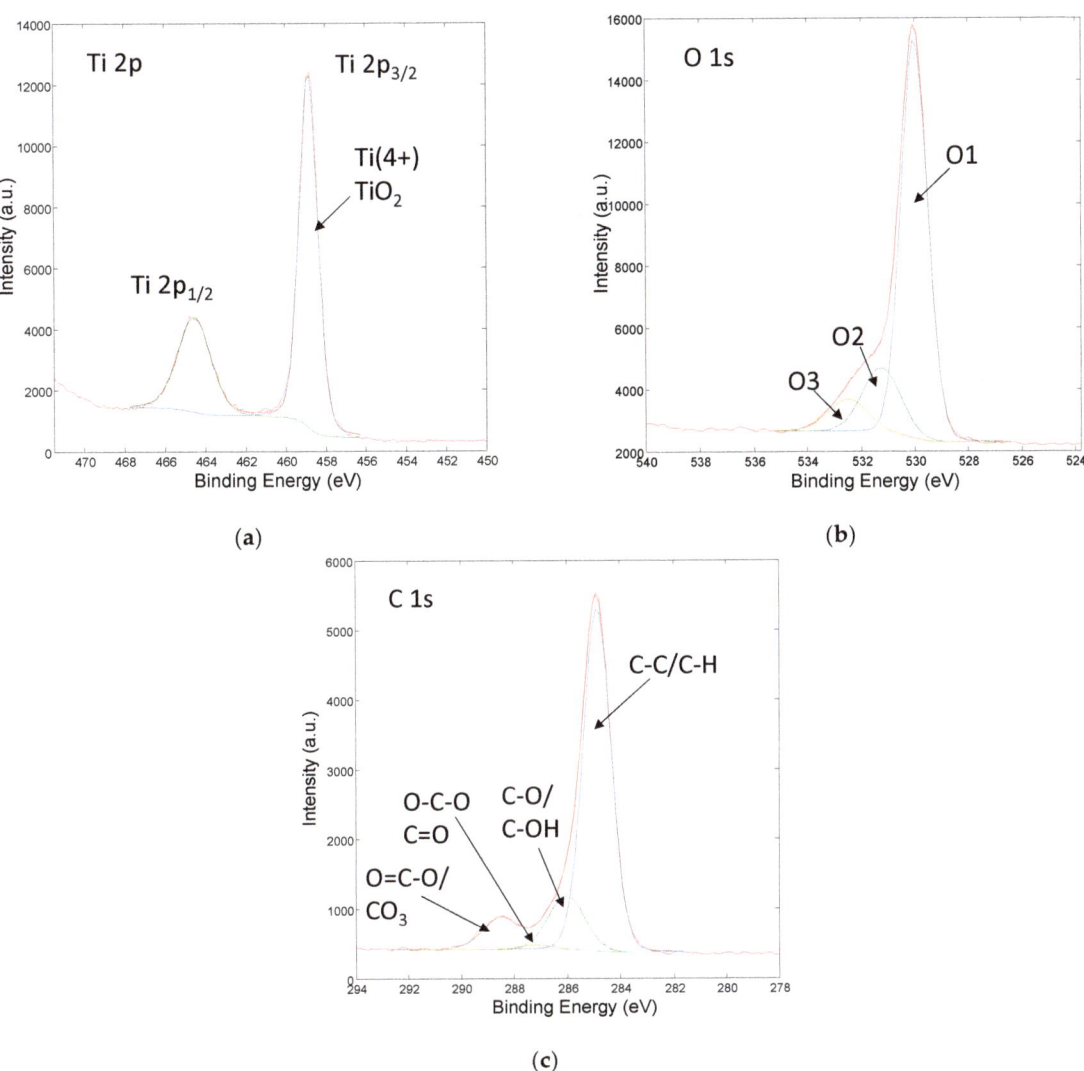

Figure 7. High-energy resolution XPS spectra from the the TiO$_2$ CVC NiTi rod surface for: (**a**) Ti 2p, (**b**) O 1s and (**c**) C 1s.

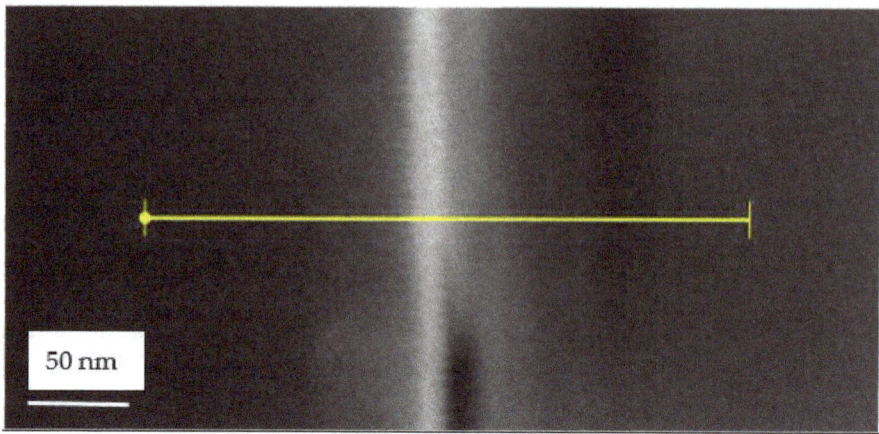
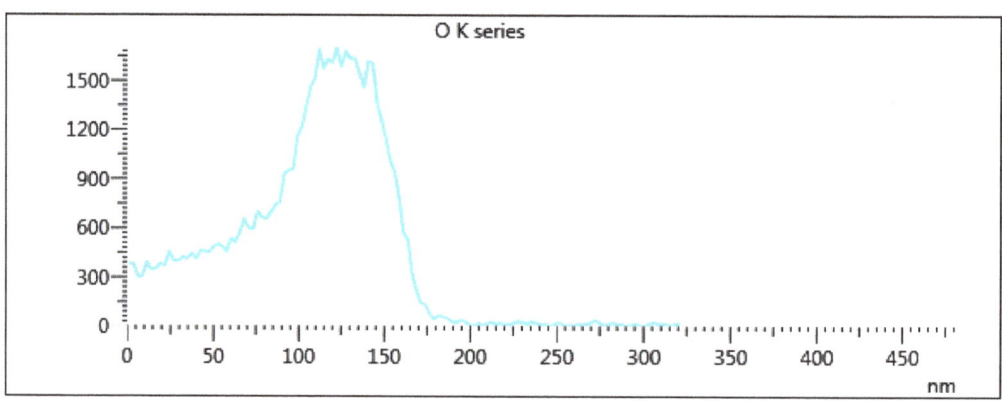
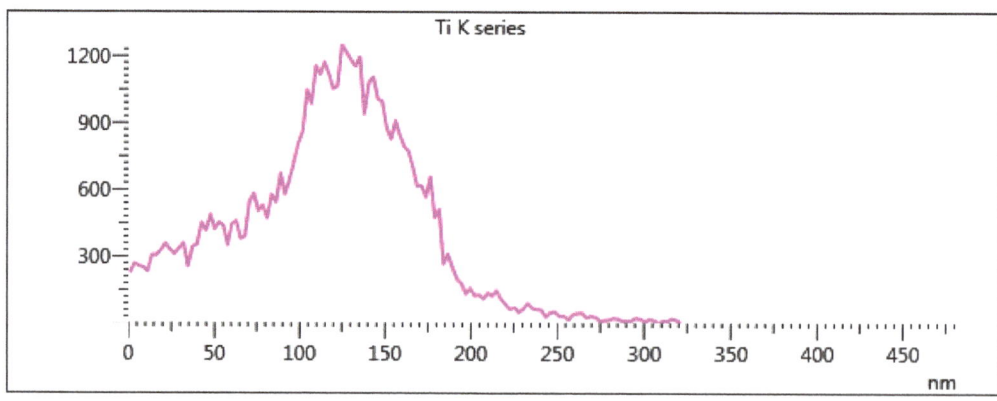

Figure 8. EDX linescan of Ti and O in the area of the formatted TiO_2 (CVC NiTi rod—the environment is on the left side).

3.2. Corrosion Tests

3.2.1. Potentiodynamic Test

Figure 9 shows the potentiodynamic curves for different specimens: CVC NiTi rod, commercial Nitinol (com. NiTi), CVC NiTi rod with a TiO$_2$ layer (TiO$_2$/CVC NiTi rod) and commercial Nitinol with a TiO$_2$ layer (TiO$_2$/com NiTi). All the measurements were carried out in Hank's solution at body temperature (37 °C). The electrochemical parameters of the potentiodynamic test are shown in Table 1. Corrosion potentials (E_{corr}) and corrosion current densities (i_{corr}) were obtained from the Tafel region. Following that region, the specimens exhibited a passive region, which was limited by the breakdown potential (E_{bd}), corresponding to the transpassive oxidation of metal species. The corrosion current density was the lowest for TiO$_2$/com NiTi (0.16 µA/cm^2), which means that the passive layer on these specimens was the most stable and resistant to external influences. This is followed by specimens of com. NiTi (0.30 µA/cm^2), TiO$_2$/CVC NiTi rod (0.34 µA/cm^2) and CVC NiTi rod (0.44 µA/cm^2). The formation of a passive layer is characterised by the width of the passive region. The wider the passive region, the more corrosion-resistant the material is. The CVC NiTi rod had the smallest passive range has a (from −100 mV to 330 mV, which is 430 mV), followed by TiO$_2$/CVC NiTi rod (650 mV), com. NiTi (730 mV) and TiO$_2$/com. NiTi. The CVC NiTi rod had the lowest breakdown potential (329 mV), followed by com. NiTi (634 mV) and TiO$_2$/CVC NiTi rod (643 mV), while TiO$_2$/com. NiTi did not reach the breakdown potential in the measurement range. As the name already suggests, this passive layer breaks down at this potential, which enables the formation of pits on the surface. This is also accompanied by a rapid increase in anode current due to the passivity breakdown. The corrosion rate is, thus, the smallest for TiO$_2$/com. NiTi; the TiO$_2$/CVC NiTi rod and com. NiTi are far away but very close together, while the CVC NiTi rod shows the highest corrosion rate or the lowest corrosion resistance by the potentiodynamic test.

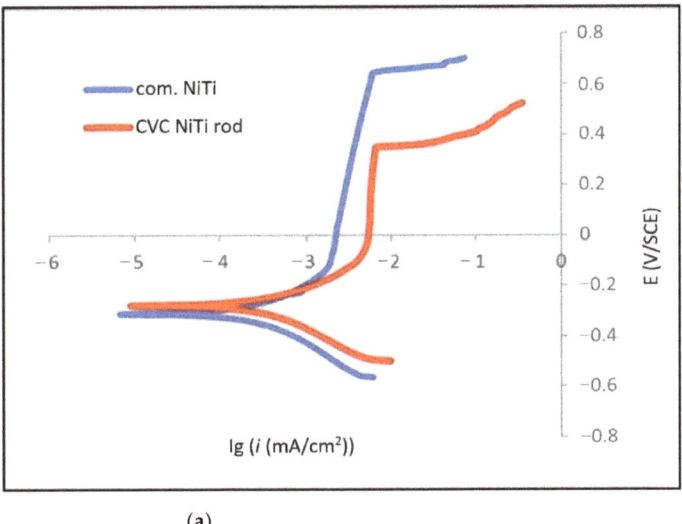

(a)

Figure 9. *Cont.*

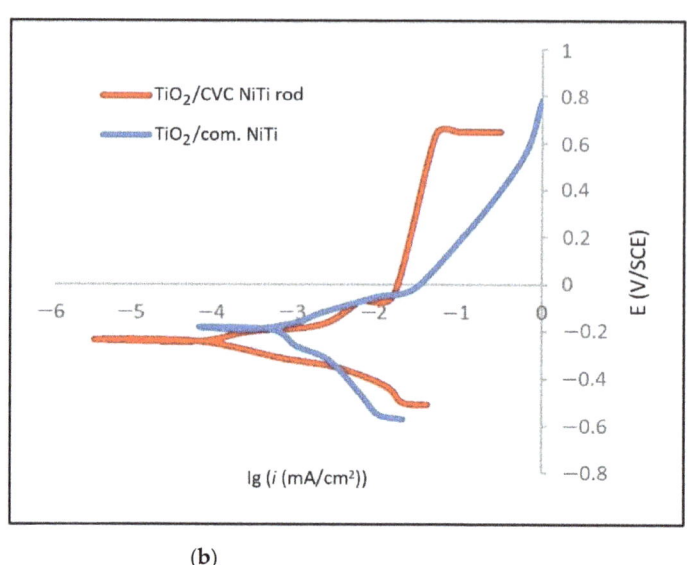

(b)

Figure 9. Potentiodynamic curves for: (**a**) CVC NiTi rod and com. NiTi and (**b**) TiO$_2$/CVC NiTi rod and TiO$_2$/com. NiTi.

Table 1. Electrochemical parameters determined from the potentiodynamic curves.

Sample	E_{corr} (mV)	i_{corr} (µA/cm^2)	E_{bd} (mV)	i_{bd} (µA/cm^2)	Corrosion Rate (mm/Year)	Passive Range (mV)
CVC NiTi rod	−334 ± 4	0.44 ± 0.05	329 ± 4	6.8 ± 0.2	(4.2 ± 0.3) × 10^{-3}	430
com. NiTi	−300 ± 4	0.30 ± 0.03	634 ± 7	6.2 ± 0.2	(2.6 ± 0.2) × 10^{-3}	730
TiO$_2$/CVC NiTi rod	−235 ± 3	0.34 ± 0.03	643 ± 7	6.2 ± 0.2	(3.1 ± 0.2) × 10^{-3}	650
TiO$_2$/com. NiTi	−186 ± 2	0.16 ± 0.02	/	/	(1.1 ± 0.1) × 10^{-4}	/

3.2.2. Electrochemical Impedance Spectroscopy

Electrochemical Impedance Spectroscopy (EIS) measurements were performed at Open-Circuit Potential conditions in simulated body fluid for 8 days. Figure 10 shows (for 4 selected examples: 12 h, 96 h, 168 h and 192 h) the Nyquist impedance diagrams for the CVC NiTi rod, commercial Nitinol, and also for both with a deposited TiO$_2$ layer at different times of immersion. The system response, shown through the Nyqvist plots, shows a typical depressed semicircle shape, and the response was increasing with the immersion time for all specimens.

The analysed data for the Nyquist plots predicted the equivalent circuits shown in Figure 11. For inhomogeneous layers, a similar equivalent circuit was applied by Izquierdo J. et al. [33] and Figueira N. et al. [34]. R_1 represents the resistance through the porous external oxide layer, while R_2 represents the resistance through the inner compact oxide layer. R_S is the resistance of the solution, while CPE$_1$ and CPE$_2$ are constant phase elements corresponding to R_1 and R_2. The use of a Constant Phase Element (CPE) was required to confirm the non-ideal capacitive response observed as a depressed semicircle in the corresponding Nyquist diagrams. The CPE originates from the surface roughness and inhomogeneities present in the TiO$_2$ layers at the microscopic level. The equivalent circuit in Figure 9a has only resistance through the oxide layer and was used only for the specimens of CVC NiTi rod and commercial Nitinol in the first 48 h of immersion. After this exposure time, the equivalent circuit in Figure 9b was applied, and this equivalent circuit was also valid for all the immersion times for the TiO$_2$/CVC NiTi rod and TiO$_2$/commercial

Nitinol. The difference can be explained by not having an oxide layer at the beginning on the surface of the CVC NiTi rod and the commercial Nitinol. Therefore, for the first 48 h of immersion, the resulting layer is still very thin and compact, which can be represented by only 1 resistor. After this time, the oxide layer became thicker and inhomogeneous, so another resistance was added, which represents the resistance through the porous oxide layer. On the surface of the TiO_2/CVC NiTi rod and the TiO_2/commercial Nitinol specimens, a nanosized TiO_2 layer was deposited previously with ALD, so an equivalent circuit with two resistances was used from the beginning of the immersion.

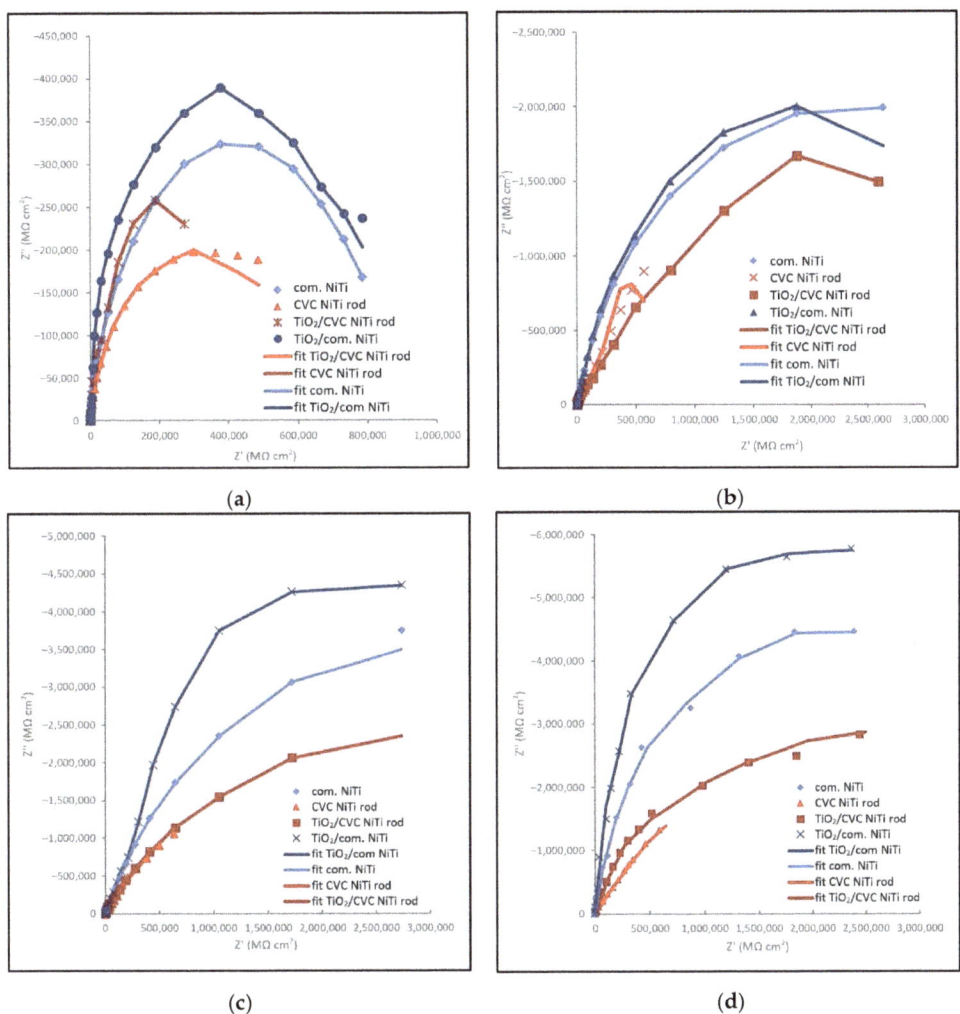

Figure 10. Nyquist diagrams for the CVC NiTi rod, commercial Nitinol, TiO_2/CVC NiTi rod and TiO_2/commercial Nitinol after (**a**) 12 h, (**b**) 96 h, (**c**) 168 h and (**d**) 192 h of immersion.

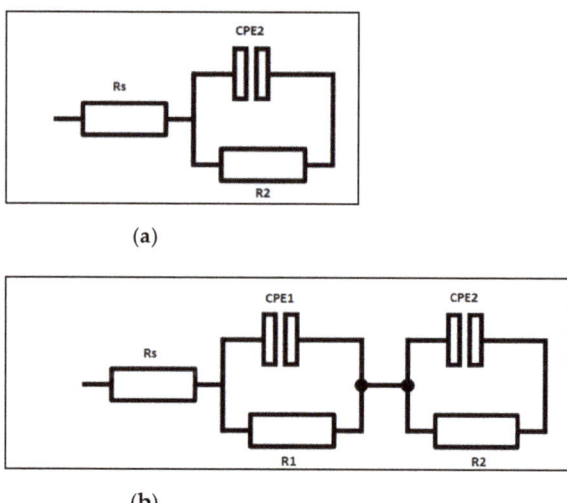

Figure 11. Equivalent circuits for the interpretation of the measured impedance spectra with: (**a**) one resistance and (**b**) two resistances.

Table 2 shows the resistance values of the porous layer (R_1) and the oxide layer (R_2) for all the specimens. The error obtained when fitting the EIS experimental data was below 0.02% for all the specimens. The CVC NiTi rod and the commercial Nitinol, for the first 48 h of immersion, had only resistance through the oxide layer as a consequence of the thin and compact layer (R_2). After this time, resistance through the porous layer was also considered. Both the TiO_2/CVC NiTi rod and TiO_2/commercial Nitinol had resistances through both layers for the total immersion time ($R_1 + R_2$), since the TiO_2 layer was previously deposited on the surface of the specimens. The resistance R_1 was considerably lower for all the specimens, which was not surprising since the solution will penetrate through the porous layer much more easily than through the compact layer. While the resistance through the oxide layer was increasing over the immersion time for all the specimens, the layer was getting thicker, so the resistance through the porous layer was quite constant.

Table 2. Porous corrosion resistance R_1 and oxide corrosion resistance R_2 of different specimens at certain times of immersion.

t (h)	R_{1com} × 10^5 (Ω)	R_{2com} × 10^5 (Ω)	R_{1CVC} × 10^5 (Ω)	R_{2CVC} × 10^5 (Ω)	$R_{1TiO2/CVC}$ × 10^5 (Ω)	$R_{2TiO2/CVC}$ × 10^5 (Ω)	$R_{1TiO2/com}$ × 10^5 (Ω)	$R_{2TiO2/com}$ × 10^5 (Ω)
1	0.00	3.74	0.00	2.08	0.24	2.22	0.82	4.00
6	0.00	6.86	0.00	2.30	0.42	3.29	0.94	8.20
12	0.00	6.49	0.00	2.61	0.79	3.85	1.16	10.20
24	0.00	7.41	0.00	2.71	1.36	4.41	1.34	18.01
48	0.00	10.94	0.00	2.55	2.44	7.58	1.45	25.00
72	4.32	19.65	0.33	7.80	2.59	13.80	1.37	33.81
96	6.92	40.35	0.28	18.00	2.48	22.51	1.56	52.51
120	7.46	56.88	0.34	24.08	2.66	33.00	1.35	68.50
144	8.32	70.29	0.33	28.06	2.68	44.72	1.34	89.00
168	9.01	89.47	0.35	30.25	2.69	56.51	1.26	101.04
192	10.28	81.47	0.31	29.62	2.29	60.31	1.46	113.56

Figure 12 represents the polarisation or the total corrosion resistance R_p as a function of time. R_p can be calculated according to equation:

$$R_p = R_1 + R_2, \qquad (3)$$

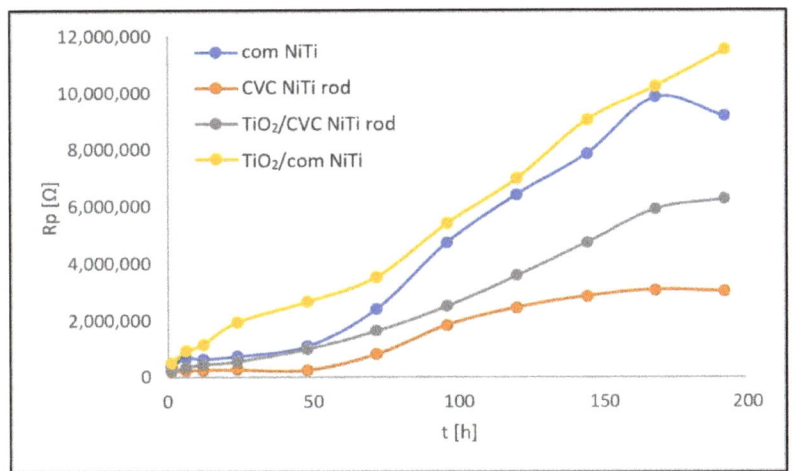

Figure 12. Total corrosion resistance vs time of exposure for all tested specimens.

The diagram of total corrosion resistance (Figure 12) shows clearly that the corrosion resistance increased with time for all 4 specimens. This was a consequence of the formation of a protective oxide layer. In the case when the TiO_2 layer was deposited on the specimen, it is noticeable that the corrosion resistance was increasing in the initial time, while for the CVC NiTi rod and commercial Nitinol, in the initial 48 h, there was no significant change in resistance as a consequence of the thin oxide layer. The highest corrosion resistance was for the TiO_2/commercial Nitinol, followed by the commercial Nitinol and the TiO_2/CVC NiTi rod, and the poorest corrosion resistance was for the CVC NiTi rod. The cause of the poorer corrosion properties of the CVC NiTi rod was inhomogeneity in the chemical composition and higher nickel content than in the commercial NiTi.

4. Conclusions

The following conclusions can be drawn from the current research work:

1. The thickness of the formatted TiO_2 layers on the CVC NiTi rod was 52.2 nm, and on the commercial Nitinol it was 51.7 nm.
2. The formatted TiO_2 layers were confirmed by XPS and SEM/EDX analyses.
3. The high-energy resolution XPS spectra for TiO_2 from both specimens were very similar. The Ti $2p_{3/2}$ peak at 458.6 eV and the Ti $2p_{1/2}$ peak at 464.5 eV were observed on the surface of both specimens. This corresponds to the binding energy, which is related with the Ti(4+) oxidation state. This shows the presence of the TiO_2 compound on the surfaces of both investigated ALD TiO_2-covered specimens.
4. The potentiodynamic test showed that the passive layer on the TiO_2/commercial Nitinol was the most stable and resistant to external corrosion influences. The corrosion stability fell from the commercial Nitinol and the CVC NiTi rod with and without the TiO_2 layer.
5. The corrosion rate was the smallest for TiO_2/commercial Nitinol; the TiO_2/CVC NiTi rod and commercial Nitinol were far away but very close together, while the CVC NiTi rod showed the highest corrosion rate, or the lowest corrosion resistance, by the potentiodynamic test.
6. Electrochemical Impedance Spectroscopy is interpreted using the Nyquist impedance diagrams, where a typical depressed semicircle shape and the response increasing with the immersion time were shown for all specimens.
7. With the help of using the resistance through the porous external oxide layer (R_1) and the resistance through the inner compact oxide layer (R_2), it was determined that

the CVC NiTi rod and the commercial Nitinol had, for the first 48 h of immersion, only resistance through the oxide layer as a consequence of the thin and compact layer (R_2). On the other hand, the TiO_2/CVC NiTi rod and TiO_2/commercial Nitinol had resistances through both layers for the total immersion time ($R_1 + R_2$). The resistance R_1 was considerably lower for all the specimens, which was not surprising since the solution will penetrate through the porous layer much more easily than through the compact layer.
8. It was proven that adding a TiO_2 layer on the Nitinol surface was significant for improving the corrosion resistance, but a decisive role can be still attributed to the chemical composition and microstructure of the substrate that the ALD coating is applied.

Author Contributions: Conceptualization, R.R., A.S. and A.K.; methodology, R.R., A.S. and A.K.; validation, R.R. and A.S.; formal analysis, A.S.; investigation, A.S.; resources, R.R. and A.K.; writing—original draft preparation, R.R. and A.S.; writing—review and editing, R.R.; visualisation, A.S.; supervision, R.R.; funding acquisition, R.R. and A.K. All authors have read and agreed to the published version of the manuscript.

Funding: This research was funded by the Slovenian Research Agency—Applied Project No.: L2-5486 and the young researchers programme—the PhD grant no. 10/2013.

Data Availability Statement: Not Applicable.

Acknowledgments: Special thanks go to Janez Kovač from the Jožef Stefan Institute Ljubljana Slovenia for performing the XPS analysis.

Conflicts of Interest: The authors declare no conflict of interest.

References

1. Halani, R.P.; Kaya, I.; Shin, Y.C.; Karaca, H.E. Phase transformation characteristics and mechanical characterization of nitinol synthesized by laser direct deposition. *Mater. Sci. Eng. A* **2013**, *559*, 836–843. [CrossRef]
2. Frenzel, J.; Zhang, Z.; Neuking, K.; Eggeler, G. High quality vacuum induction melting of small quantities of NiTi shape memory alloys in graphite crucibles. *J. Alloys Compd.* **2004**, *385*, 214–223. [CrossRef]
3. Milošev, I.; Kapun, B. The corrosion resistance of Nitinol alloy in simulated physiological solutions Part 1: The effect of surface preparation. *Mater. Sci. Eng. C* **2012**, *32*, 1087–1096. [CrossRef]
4. Fu, C.H.; Sealy, M.P.; Guo, Y.B.; Wei, X.T. Finite element simulation and experimental validation of pulsed laser cutting of nitinol. *J. Manuf. Process.* **2015**, *19*, 81–86. [CrossRef]
5. Laskovski, A. (Ed.) *Biomedical Engineering, Trends in Materials Science*, 1st ed.; InTech: Rijeka, Croatia, 2011.
6. Simka, W.; Sadkowski, A.; Warczak, M.; Iwaniak, A.; Dercz, G.; Michalska, J.; Maciej, A. Characterization of passive films formed on titanium during anodic oxidation. *Electrochim. Acta* **2011**, *56*, 8962–8968. [CrossRef]
7. Weng, F.; Chen, C.; Yu, H. Research status of laser cladding on titanium and its alloys: A review. *Mater. Des.* **2014**, *58*, 412–425. [CrossRef]
8. Wawrzynski, J.; Gil, J.A.; Goodman, A.D.; Waryasz, G.R. Hypersensitivity to Orthopedic Implants: A Review of the Literature. *Rheumatol. Ther.* **2017**, *4*, 45–56. [CrossRef] [PubMed]
9. Teo, Z.W.W.; Schalock, P.C. Hypersensitivity Reactions to Implanted Metal Devices: Facts and Fictions. *J. Investig. Allergol. Clin. Immunol.* **2016**, *26*, 279–329.
10. Nordberg, G.F.; Gerhardsson, L.; Broberg, K.; Mumtaz, M.; Ruiz, P.; Fowler, B.A. Interactions in Metal Toxicology. In *Handbook on the Toxicology of Metals*, 3rd ed.; Elsevier BV: Amsterdam, The Netherlands, 2007; pp. 117–145.
11. Shabalovskaya, S.; Anderegg, J.; Humbeeck, J.V. Critical overview of Nitinol surfaces and their modifications for medical applications. *Acta Biomater.* **2008**, *4*, 447–467. [CrossRef] [PubMed]
12. Pohl, M.; Glogowski, T.; Kühn, S.; Hessing, C.; Unterumsberger, F. Formation of titanium oxide coatings on NiTi shape memory alloys by selective oxidation. *Mater. Sci. Eng. A* **2008**, *481–482*, 123–126. [CrossRef]
13. Nasakina, E.O.; Sudarchikova, M.A.; Sergienko, K.V.; Konushkin, S.V.; Sevost'yanov, M.A. Ion Release and Surface Characterization of Nanostructured Nitinol during Long-Term Testing. *Nanomaterials* **2019**, *9*, 1569. [CrossRef]
14. Wang, H.R.; Liu, F.; Zhang, Y.P.; Yu, D.Z.; Wang, F.P. Preparation and properties of titanium oxide film on NiTi alloy by micro-arc Oxidation. *Appl. Surf. Sci.* **2011**, *257*, 5576–5580. [CrossRef]
15. Vojtěch, D.; Voděrová, M.; Fojt, J.; Novák, P.; Kubásek, T. Surface structure and corrosion resistance of short-time heat-treated NiTi shape memory alloy. *Appl. Surf. Sci.* **2010**, *257*, 1573–1582. [CrossRef]
16. Hu, T.; Chu, C.; Yin, L.; Pu, Y.; Dong, Y.; Guo, C.; Sheng, X.; Chung, J.; Chu, P. In vitro biocompatibility of titanium-nickel alloy with titanium oxide film by H_2O_2 oxidation. *Trans. Nonferrous Met. Soc. China* **2007**, *17*, 553–557. [CrossRef]

17. Shabalovskaya, S.A.; Tian, H.; Anderegg, J.W.; Schryvers, D.U.; Carroll, W.U.; Van Humbeeck, J. The influence of surface oxides on the distribution and release of nickel from Nitinol wires. *Biomaterials* **2009**, *30*, 468–477. [CrossRef] [PubMed]
18. Tian, H.; Schryvers, T.; Liu, D.; Jiang, Q.; Van Humbeeck, J. Stability of Ni in nitinol oxide surfaces. *Acta Biomater.* **2011**, *7*, 892–899. [CrossRef] [PubMed]
19. Leyens, C.; Peters, M. (Eds.) *Titanium and Titanium Alloys. Fundamentals and Applications*; Wiley: Weinheim, Germany, 2003.
20. Pelton, A.R.; Mehta, A.; Zhu, L.; Trépanier, C.; Imbeni, V.; Robertson, S.; Barney, M. TiNi Oxidation: Kinetics and Phase Transformations. *Solid-to-Solid Transform. Inorg. Mater.* **2005**, *2*, 1029–1034.
21. Zhu, L.; Fino, J.M.; Pelton, A.R. Oxidation of Nitinol. In Proceedings of the SMST-2003, Monterey, CA, USA, 5–8 May 2003.
22. Bauer, S.; Schmuki, P.; von der Mark, K.; Park, J. Engineering biocompatible implant surfaces Part I: Materials and surfaces. *Prog. Mater. Sci.* **2013**, *58*, 263–300. [CrossRef]
23. Saric, I.; Peter, R.; Piltaver, I.K.; Jelovica Badovinac, I.; Salamon, K.; Petravic, M. Residual chlorine in TiO_2 films grown at low temperatures by plasma enhanced atomic layer deposition. *Thin Solid Film.* **2017**, *628*, 142–147. [CrossRef]
24. Piltaver, I.K.; Peter, R.; Šarić, I.; Salamon, K.; Jelovica Badovinac, I.; Koshmak, K.; Nannarone, S.; Delač, M.I.; Petravić, M. Controlling the grain size of polycrystalline TiO_2 films grown by atomic layer deposition. *Appl. Surf. Sci.* **2017**, *419*, 564–572. [CrossRef]
25. Johnson, W.R.; Hultqvist, A.; Bent, F.S. A brief review of atomic layer deposition: From fundamentals to applications. *Mater. Today* **2014**, *17*, 236–246. [CrossRef]
26. Aarik, J.; Aidla, A.; Mändar, H.; Uustare, T. Atomic layer deposition of titanium dioxide from $TiCl_4$ and H_2O: Investigation of growth mechanism. *Appl. Surf. Sci.* **2001**, *172*, 148–158. [CrossRef]
27. Vokoun, D.; Racek, J.; Kaderavek, L.; Kei, C.C.; Yu, Y.S.; Klimša, L.; Šittner, P. Atomic Layer-Deposited TiO2 Coatings on NiTi Surface. *JMEPEG* **2018**, *27*, 572–579. [CrossRef]
28. Vokoun, D.; Klimša, L.; Vetushka, A.; Duchoň, J.; Racek, J.; Drahokoupil, J.; Kopeček, J.; Yu, Y.S.; Koothan, N.; Kei, C.C. Al_2O_3 and Pt Atomic Layer Deposition for Surface Modification of NiTi Shape Memory Films. *Coatings* **2020**, *10*, 746. [CrossRef]
29. Lojen, G.; Stambolić, A.; Šetina, B.; Rudolf, R. Experimental continuous casting of nitinol. *Metals* **2020**, *10*, 505. [CrossRef]
30. Stambolić, A.; Anžel, I.; Lojen, G.; Kocijan, A.; Jenko, M.; Rudolf, R. Continuous vertical casting of a NiTi alloy. *Mater. Tehnol.* **2016**, *50*, 981–988. [CrossRef]
31. Stambolić, A.; Jenko, M.; Kocijan, A.; Žužek, B.; Drobne, D.; Rudolf, R. Determination of mechanical and functional properties by continuous vertical cast NiTi rod. *Mater. Tehnol.* **2018**, *52*, 521–527. [CrossRef]
32. Moulder, J.F.; Stickle, W.F.; Sobol, P.E.; Bomben, K.D. *Handbook of X-Ray Photoelectron Spectroscopy*; Physical Electronics Inc.: Eden Prairie, MN, USA, 1995.
33. Izquierdo, J.; González-Marrero, M.B.; Bozorg, M.; Fernández-Pérez, B.M.; Vasconcelos, H.C.; Santana, J.J.; Souto, R.M. Multiscale electrochemical analysis of the corrosion of titanium and nitinol for implant applications. *Electrochim. Acta* **2016**, *203*, 366–378. [CrossRef]
34. Figueira, N.; Silva, T.M.; Carmezima, M.J.; Fernandes, J.C.S. Corrosion behaviour of NiTi alloy. *Electrochim. Acta* **2009**, *54*, 921–926. [CrossRef]

Article

Spray-Pyrolytic Tunable Structures of Mn Oxides-Based Composites for Electrocatalytic Activity Improvement in Oxygen Reduction

Miroslava Varničić [1], Miroslav M. Pavlović [1,2,*], Sanja Eraković Pantović [1], Marija Mihailović [1], Marijana R. Pantović Pavlović [1,2], Srećko Stopić [3] and Bernd Friedrich [3]

[1] Department of Electrochemistry, Institute of Chemistry, Technology and Metallurgy, National Institute of the Republic of Serbia, University of Belgrade, Njegoševa 12, 11000 Belgrade, Serbia; mima.varnicic@gmail.com or varncic@ihtm.bg.ac.rs (M.V.); sanja@ihtm.bg.ac.rs (S.E.P.); marija.mihailovic@ihtm.bg.ac.rs (M.M.); m.pantovic@ihtm.bg.ac.rs (M.R.P.P.)
[2] Center of Excellence in Environmental Chemistry and Engineering-ICTM, University of Belgrade, Njegoševa 12, 11000 Belgrade, Serbia
[3] Process Metallurgy and Metal Recycling, RWTH Aachen University, Intzestraβe 3, 52072 Aachen, Germany; sstopic@ime-aachen.de (S.S.); bfriedrich@ime-aachen.de (B.F.)
* Correspondence: mpavlovic@tmf.bg.ac.rs

Abstract: Hybrid nanomaterials based on manganese, cobalt, and lanthanum oxides of different morphology and phase compositions were prepared using a facile single-step ultrasonic spray pyrolysis (USP) process and tested as electrocatalysts for oxygen reduction reaction (ORR). The structural and morphological characterizations were completed by XRD and SEM-EDS. Electrochemical performance was characterized by cyclic voltammetry and linear sweep voltammetry in a rotating disk electrode assembly. All synthesized materials were found electrocatalytically active for ORR in alkaline media. Two different manganese oxide states were incorporated into a Co_3O_4 matrix, δ-MnO_2 at 500 and 600 °C and manganese (II,III) oxide-Mn_3O_4 at 800 °C. The difference in crystalline structure revealed flower-like nanosheets for birnessite-MnO_2 and well-defined spherical nanoparticles for material based on Mn_3O_4. Electrochemical responses indicate that the ORR mechanism follows a preceding step of MnO_2 reduction to MnOOH. The calculated number of electrons exchanged for the hybrid materials demonstrate a four-electron oxygen reduction pathway and high electrocatalytic activity towards ORR. The comparison of molar catalytic activities points out the importance of the composition and that the synergy of Co and Mn is superior to Co_3O_4/La_2O_3 and pristine Mn oxide. The results reveal that synthesized hybrid materials are promising electrocatalysts for ORR.

Keywords: MnO_2; cobalt oxide Co_3O_4; perovskite materials; oxygen reduction in alkaline media; electrocatalyst; ultrasonic spray pyrolysis; Pt catalyst

1. Introduction

Economic development and extensive use of fossil fuels has led to fast depletion of energy resources. Hence, the development of clean energy storage and conversion devices, such as metal–air batteries, supercapacitors, fuel cells, and other renewable energy technologies, is in the main focus of numerous researchers and laboratories worldwide [1,2]. The efficiency of these energy devices mainly depends on the electrochemical oxygen reduction reaction (ORR) that occurs at the cathode side, as limiting reaction [3]. High activation barriers, poor rate capability, sluggish kinetics, and serious voltage gap of oxygen electrode reactions limit the performance of energy devices that rely on ORR [4–6].

Until now, the most investigated ORR catalysts have been based on noble metals such as Pt and Pt alloys, to achieve favorable reaction rates [7]. However, the high price and scarcity of the precious metals, inferior stability and sensitivity to CO poisoning, severely limit their widespread applications [2,3,8]. To overcome the above-mentioned issues, lowering the amount of noble metals and exploring new catalytic materials for ORR have triggered extensive research interests.

Transition metals (TMOs) and organic macrocycles represent promising candidates as alternatives to noble metals as catalytic materials for ORR [9–14]. Among them, manganese oxides (Mn_yO_x) are of particular interest because of their prominent advantages of low cost, stability, environmental friendliness, abundance, and considerable catalytic activity toward oxygen reduction reaction [15–20]. Despite insufficient stability in acidic media, Mn-oxides can be applied as promising catalyst in air electrodes for both alkaline fuel cells and metal–air batteries. For example, various oxides have been studied, including perovskite-type, α-Mn_2O_3 (bixbyite), β-MnO_2 (pyrolusite), Mn_3O_4 (spinel), α-MnOOH (manganite), and simple Mn oxides [15–22], and the ORR was found to be highly dependent on the crystal structure of the oxides. Additionally, there are many studies showing manganese-oxide as highly promising material for the metal air batteries. For example, it has been reported the application of MnO_2 at the reduced graphene oxide as hybrid material in Mg-air battery [23], MnO_2 on graphene coated microfibers for Na–air battery [24], and Mn oxide framework for lithium–oxygen batteries [25]. It is possible to increase the activity of manganese oxide by tuning its crystal structure and morphology, doping, compositing, vacancy creation, and hydrogenation. The continuous improvement of the oxygen electrochemical activity of Mn-oxide is still ongoing work.

Cobalt oxide (Co_3O_4) is also one of the well-known materials that has been extensively studied as promising candidate and corrosion resistant ORR catalysts in alkaline media for fuel cells and metal–air batteries. Cobalt oxide belonging to the family of transition metal oxides, is able to display significant morphology modulated catalytic activity for ORR [26,27]. Co_3O_4 has spinal structure, with magnetic Co^{2+} and non-magnetic Co^{3+} at its tetrahedral and octahedral sites [27], that are significant for cobalt oxide catalytic activity. Porous Co_3O_4 nanoplates have been used as ORR catalyst for Zn–air batteries in alkaline medium [28], while flake-particles Co_3O_4 have been employed as catalysts for Li-O_2 batteries [29]. The structure of cobalt oxide has provided the abundant active sites together with ion and electron transport length, which eventually have improved the energy efficiency. Another investigation examines electrocatalyst for the oxygen reduction reaction based on a graphene-supported g-C_3N_4@cobalt oxide core–shell hybrid in alkaline solution with improved stability and activity, approaching to that of 20% Pt–C at the same potential [30]. It has been shown that the sole CoO_x electrode exhibited only the two-electron mechanism with formation of hydrogen peroxide, rather than the four-electron mechanism, while the FCNTs electrode exhibited the two parallel mechanisms favoring four-electron mechanism only at higher overpotential. These results indicate the synergistic effect of the coupling between FCNTs and CoO_x nanoparticles catalyzing the ORR via the direct four-electron mechanism. In another study, cobalt oxide nanocubes incorporated into reduced graphene oxide exhibited better electrocatalytic activity in terms of the current density, overpotential, and stability, compared to commercial Pt/C catalyst for the ORR in an alkaline medium [26]. Unfortunately, their ORR activity alone is generally poor. Thus, for further improvement, other metal atoms or carbon based materials have been incorporated into their catalysts structure [31–35].

La based perovskite materials are considered as a new class of materials in the mixed-oxide family and have attracted increasing attention for potential replacement of the noble metals. They have shown promising catalytic performance for ORR in alkaline media. The activity of these La-based oxides strongly correlated with the covalent bond strength between B-site cation and the oxygenated species. La is located in the middle of the octahedral structure and plays an important stabilizing role [36,37]. Additionally, La_2O_3 contain oxygen vacancies and interstitials with low oxygen vacancy energy, leading to low activation energy. Furthermore, the existing interlayer defect structure of the oxides is also helpful for the active oxygen adsorption, this all has a positive effect in catalyzing the ORR. However, even though lanthanum oxides have outstanding electronic structure, it is not electro-conductive, which limits its electrocatalytic capabilities and brings the necessity to combine it with other oxides and carbon materials [38–40]. Due to its high potential as a stabilizing agent and high activity when mixed with other oxides, in this work it was utilized in the synthesis with Mn- and Co-oxide.

It was shown that combination of oxides, especially TMOs, and their structures, like spinel and perovskite-type oxides, could exhibit excellent ORR activities owing to the combination of metal elements, compared to single-metal oxides [7,41,42]. For example, Co-oxide nanoparticles modified with Mn-oxide nanotube have served as oxygen cathode catalyst for rechargeable zinc–air batteries [43]. La-, Co-, Mn-oxide prepared with carbon nanotubes (CNT) as composite has been successfully used as a bi-functional air electrode in Zn–air batteries [37]. The improved synergy effect has been reported in comparing to single oxide utilization. For example, various structures like honeycomb double-layer MnO_2/Cobalt doped for primary zinc–air batteries [44], 3D hollow sphere Co_3O_4/MnO_2-CNT [45], and core-shell Co_3O_4@MnO_2 [46] have been evaluated as bifunctional catalysts materials and applied for batteries and supercapacitors.

Therefore, we aimed to synthesize and investigate hybrid nanomaterials based on the Mn/Co/La oxides of ordered structure generated by ultrasonic spray pyrolysis (USP) as electrocatalyst for ORR. USP technique was chosen for the synthesis of these materials as it allows a simple single step approach of synthesizing nanomaterials with precisely controllable morphologies and chemical compositions. Different compositions and morphologies were synthesized depending on USP temperature and tested.

One of the issues to be considered is that the future ORR materials should not contain rather electrochemically unstable carbonaceous materials as support. The investigated MnO_x-Co_3O_4 TMO hybrid electrode nanomaterials were carbon free. The influence of manganese oxide type on the ORR as well as difference in composition of Mn and Co was evaluated. The second important issue that we tackled is that the materials should be synthesized using a low-cost, simple technique that can be easily applied also for the large scales such as USP.

2. Experimental

2.1. Chemicals

Lanthanum(III)nitrate hexahydrate $La(NO_3)_3 \times 6H_2O$ (99.9% rare earth oxide), manganese(II)nitrate tetrahydrate $Mn(NO_3)_2 \times 4H_2O$ (99%), and cobalt(II)chloride hexahydrate $CoCl_2 \times 6H_2O$ (99%) were used during the synthesis process and were purchased all from Alfa Aeser, US. For the comparison, commercial manganese (IV) oxide, MnO_2, was used and obtained from Sigma-Aldrich (Saint Louis, MO, USA). Potassium hydroxide and Nafion 117 solution (5 wt.%) were purchased from Sigma-Aldrich. All chemicals were of analytical reagent grade and all solutions were prepared using ultrapure water from Millipore.

2.2. Material Synthesis and Electrode Preparation

2.2.1. Material Synthesis

The synthesis of Co/Mn/La oxide hybrid materials was performed by single-step ultrasonic spray pyrolysis process. The solution for the material synthesis was prepared by mixing starting precursor solutions to give desired stoichiometric mole ratios La:Co:Mn = 3:5:10. The ratio of La:Co:Mn of 3:5:10 was chosen based on previous research [47,48], since we wanted to investigate the influence of Mn in mixture for ORR reaction. Aqueous 0.1 M solutions of $La(NO_3)_3$, $Mn(NO_3)_2$ and $CoCl_2$ were used as precursors. The USP conversion temperature was adjusted and controlled using a thermostated furnace. All powders were synthesized by ultrasonic spray pyrolysis in the equipment with horizontal nebula flow.

Nebula generation from the prepared solutions of precursors took place in an ultrasonic atomizer (Gapusol 9001, RBI/France) with an ultrasonic nebulizer (Prizma Kragujevac, Serbia) to create an nebula-born aerosol [47,48]. The nebula with droplets having a diameter of around 2.3 µm was produced with an ultrasound frequency of 2.5 MHz. The nebulization/aerosol generation was carried out in O_2/N_2 atmosphere as carrier gas, having O_2 to N_2 in volume ratio of 2:1 and continuous flow rate of 3 dm^3 min^{-1}. The synthesis temperatures were set to 500, 600, or 800 °C.

2.2.2. Electrode Preparation

For the electrochemical measurements, the catalyst-modified surface of electrodes was prepared from the glassy carbon disc (Pine Research Instrumentation USA, 5 mm). Prior to the use, glassy carbon disc was polished with alumina slurry kit (Pine Research Instrumentation, Durham, NC, USA) of different grades, and then cleaned ultrasonically in ethanol and water.

The electrodes were prepared in the following way. Firstly, 5 mg mL^{-1} of water suspension of the USP-synthesized powder was agitated in an ultrasonic bath for 30 min in order to form homogeneous ink. Then, 20 μL of the ink were cast by micropipette onto the glassy carbon disc and left to air-dry for 2 h. In the next step, 10 μL of Nafion solution (100:1 diluted commercial Nafion solution) were pipetted onto the catalyst-covered GC disc, as binding agent, and left to dry at room temperature.

2.3. Measurements

2.3.1. Material Characterization

Structural and phase analysis of the synthesized materials was investigated by X-ray diffraction (XRD). The measurements were undertaken on a Philips PW 1050 powder diffractometer with Ni-filtered CuKα radiation at room temperature and scintillation detector within the range 10–82° in steps of 0.05° with the scanning rate of 5 s/step.

Scanning electron microscopy (SEM) with an energy dispersive X-ray spectroscopy (EDS) were employed to analyze morphology and element composition of porous Mn/Co/La oxide hybrid materials. Scanning electron microscope (Zeiss DSM 982 Gemini; Vega TS 5139MM Tescan, Brno, Czech Republic) was employed for the examination of obtained particles on a different magnification level providing different information on the morphology and particle shape and size. The elemental composition was determined by EDS with Si(Bi) X-ray detector connected to SEM and a multi-channel analyzer.

2.3.2. Electrochemical Measurements

Electrochemical measurements were performed using BioLogic potentiostat (BioLogic SAS, SP-240, Grenoble, France). The electrocatalytic properties of the hybrid materials were checked by means of cyclic voltammetry (CV) and linear sweep voltammetry (LSV), using the scan rate of 50 and 2 mV s^{-1}, respectively. In order to check the material activity for oxygen reduction reaction, 3-electrode set-up, using a rotating disk working electrode was employed. The GC modified with synthesized materials as described in Section 2.2.2 were used as working electrode, while saturated calomel (SCE) and Pt electrode were employed as the reference and counter electrode, respectively. All the potentials presented are referred to the SCE. The supporting electrolyte was a 0.1 M aqueous KOH solution. All electrochemical experiments were performed at 25 °C under nitrogen or oxygen atmosphere at 600, 800, 1000, 1500, or 2500 rpm rotation of the working electrode. Prior to every experiment, either in N_2 or O_2 atmosphere, the gas was bubbled through the electrolyte for at least 20 min.

3. Results and Discussion

3.1. XRD Analysis

The hybrid nanomaterials based on rare earth/transition metal oxides were synthesized with facile and cost-effective USP procedure, bearing in mind the methodology as follows. One group of electrocatalyst is based on manganese, cobalt, and lanthanum metals and the effects of three different USP temperatures were investigated (500, 600, and 800 °C). The other group of materials was prepared by the same synthesis procedure, but without manganese component—it was only based on cobalt and lanthanum oxide, in order to investigate the influence of USP-synthesized Mn oxide within hybrid oxide electrocatalysts.

For the sake of comparison, the electrocatalytical performance of commercial manganese oxide was also included in the investigations.

The crystalline structures of the hybrid materials were revealed by XRD analysis. X-ray diffraction patterns of synthesized materials are presented in Figure 1. As can be seen in Figure 1a, the materials synthesized at 500 and 600 °C are composed mainly of manganese (IV) oxide in the form of birnessite, also denoted as δ-MnO_2, as defined by main diffraction peaks at 2θ of 12.4, 25.3, 37, and 66°. The specific XRD peaks correspond to a card no.: JCPDS 00-043-1456 (MnO_2). Birnessite is reported as 2D layered manganese oxide with lamellar structure consisting of edge-sharing MnO_6 octahedra, and is considered to be the most active phase for ORR among other crystalline structures of MnO_2 [49–51]. In addition, the diffraction peaks at 18.9, 31.2, 45, and 59° reveal the presence of Co_3O_4 (JCPDS card no. 01-080-1535). However, La oxide or other compounds which should indicate the La presence are not evidenced. It follows that La is present as poorly crystalline or amorphous lanthanum compound(s). These "La-hided" states of Co-La oxide hybrids corresponds to our recent findings [48].

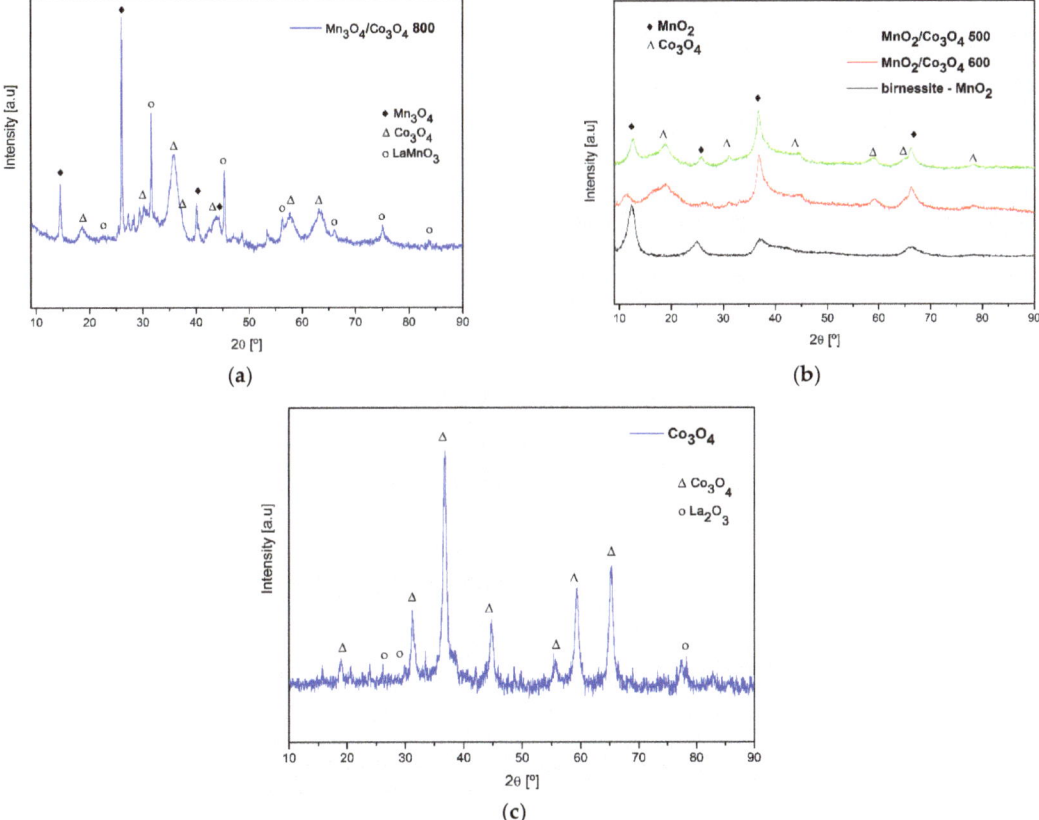

Figure 1. XRD patterns of (**a**) Mn/Co/La oxide hybrid materials synthesized on 500 and 600 °C, and commercial MnO_2; (**b**) Mn/Co/La oxide material synthesized at 800 °C; (**c**) Co/La oxide hybrid material.

XRD pattern of the material sample synthesized at 800 °C is presented in Figure 1b. The presence of Co_3O_4 is clearly confirmed; however, the other type of Mn oxide-Mn_3O_4 is formed. It is evidenced by well-resolved peaks at the positions 14.4, 26.2, 40.1, and 44°

(JCPDS card no. 03-06502776). Additionally, the diffraction peaks related to the JCPDS card no. 01-075-0440 confirms the formation of the perovskite structure-LaMnO$_3$. It can be seen that XRD peaks in Mn-containing compounds synthesized at 800 °C are sharper than those at lower temperatures, which indicate highly crystalline nature of those compounds. The formation of different manganese oxide types at different temperatures is expected. It has been already reported that under these conditions Mn changes its form from Mn(IV) oxide to Mn(II, III) oxide [52]. Synthesized hybrid materials are denoted as MnO$_2$/Co$_3$O$_4$-500, MnO$_2$/Co$_3$O$_4$-600, and Mn$_3$O$_4$/Co$_3$O$_4$-800 in the further text, according to XRD findings.

The structural and phase characteristics of the Mn-free synthesized material, based on cobalt and lanthanum compounds, is presented in Figure 1c. XRD peaks clearly confirm formation of Co$_3$O$_4$. In addition, weak peaks at positions 26.2°, 29.8°, and 78° indicate the formation of La$_2$O$_3$, although it can be assumed that this oxide is present mainly in an amorphous form. Thus, this catalytic material is denoted as Co$_3$O$_4$/La$_2$O$_3$ in further text.

3.2. SEM and EDS Characterization

The morphology of as-synthesized catalytic materials was investigated by SEM. Depending on the preparation temperature the samples morphology appears different, as shown in Figure 2. For the materials prepared at the temperatures 500 and 600 °C, the results indicate formation of spherical grains, Figure 2a,c, with a petal-like structure discovered at the higher-resolution images (Figure 2b,d). It can be seen that the petal-structured grains, having the size of around 2 µm, are built from numerous nanosheets. The nanosheets appear finer and more densely packed at higher USP temperature (Figure 2b,d). The very similar structures were observed by Che et al. reporting the core-shell microspheres composed of Co$_3$O$_4$@MnO$_2$ with flower-like structured Co$_3$O$_4$ as the core onto which MnO$_2$ nanosheets have been subsequently grown [46]. This typical flower-like morphology of birnessite-MnO$_2$ forming micro/nanospheres has been also reported to have high surface area that might exhibit fast electrode kinetics and good stability. MnO$_2$ nanosheets are thus recognized as excellent candidates for electrocatalytic materials for electrochemical oxygen reactions [50,51,53].

On the other hand, the Mn$_3$O$_4$/Co$_3$O$_4$ material synthesized at 800 °C has homogenous dense structure as presented in Figure 3. The enlarged SEM image reveals well-defined submicron particles of the catalytic material (Figure 3b). Similar change in morphology, leading to the formation of defined particles instead of flower-like structure, has been reported and assigned to the presence of Mn$_3$O$_4$ type of Mn oxide, in comparison to the distinguished nanosheets typical for MnO$_2$ [54,55]. The SEM images of Co$_3$O$_4$/La$_2$O$_3$ catalytic material shows highly agglomerated particles with irregular shapes and various sizes ranging from nano- to several µm (Figure 3d). The porous agglomerated particles of Co$_3$O$_4$ material have also been reported, providing large micro- and mesoporous surface [56].

The elemental composition, done by energy dispersive X-ray spectroscopy (EDS), confirms the presence of Mn, Co, La, and O with the atomic ratios presented in Table 1. It can be seen that for MnO$_2$/Co$_3$O$_4$-500, MnO$_2$/Co$_3$O$_4$-600, and Mn$_3$O$_4$/Co$_3$O$_4$-800 materials, the obtained atomic ratio of Mn:Co is close to 2:1, which is in accordance to the projected atomic ratio used for synthesis. On the other hand, the atomic ratio between lanthanum, cobalt, and manganese is smaller than projected ratio. This anticipates that the material is likely structured as separated phases of the oxides of well-resolved crystalline state, MnO$_2$ and Co$_3$O$_4$, covering poorly crystalized La compounds, as found by XRD, which can mask its EDS response. Finally, it was seen in XRD patterns that Co$_3$O$_4$/La$_2$O$_3$ catalytic material exhibits diffraction peaks similar to La$_2$O$_3$ card, which can cause the apparently hidden XRD state of La$_2$O$_3$ by crystalline Co$_3$O$_4$.

3.3. Electrochemical Characterization

To study the electrocatalytic performances of synthesized Mn/Co/La-hybrid materials for oxygen reduction reaction, three-electrode half-cell design was used. Electrodes were prepared with catalytic material as described above in Section 2.2.2 and used as working electrodes in an RDE system to provide constant hydrodynamic conditions.

Figure 2. SEM images of (**a,b**) MnO$_2$/Co$_3$O$_4$-500, (**c,d**) MnO$_2$/Co$_3$O$_4$-600; at lower and higher magnifications.

Cyclic voltammograms were recorded in the potential window between −0.9 and 0.6 V, at the scan rate of 20 mV s^{-1}. Figure 4 presents CV responses of synthesized materials and pristine MnO$_2$ electrode in deaerated 0.1 M KOH. The cyclic voltammograms of MnO$_2$/Co$_3$O$_4$ electrodes exhibit almost featureless shape with capacitive current showing some reversible charge transfer processes at the potentials positive to 0.1 V, with counterparts negative to −0.1 V. This behavior is in accordance with the literature results for other MnO$_2$ nanostructures electrodes reporting the similar behavior in nitrogen atmosphere as well as for pristine MnO$_2$ [44,51,57]. The redox processes appeared suppressed upon increase in synthesis temperature. On the other hand, CV behavior of Co$_3$O$_4$/La$_2$O$_3$ hybrid material shows fully reversible redox transitions of much higher currents at the potentials positive to −0.1 V, which can be assigned to redox transitions of Co. This CV performance has been reported as typical behavior of cobalt oxide nanoparticles [58]. The shape of CV curves of Co$_3$O$_4$/La$_2$O$_3$ is different in comparison to that of MnO$_2$/Co$_3$O$_4$ materials, with considerably higher capacitive currents. It seems that CV fingerprints follow the registered structural organization of the oxides in the Co$_3$O$_4$/La$_2$O$_3$ material. Cobalt oxide particles dictate the CV behavior of Co$_3$O$_4$/La$_2$O$_3$ material in a way to resemble completely the redox processes of pure Co$_3$O$_4$. On the other hand, Mn-based materials are of CV behavior similar to pristine MnO$_2$, since the particle surface composition is of twice as much as nominal loading of manganese with respect to cobalt. CV response of MnO_2/Co_3O_4 synthesized at lower temperature (MnO_2/Co_3O_4-500) strives for the shape more similar to that of Co_3O_4. This could be related to the more spaced petals (Figure 2b) with respect to dense appearance of petals at higher synthesis temperature (MnO_2/Co_3O_4-600, Figure 2d). It follows that Co_3O_4 contributes more to CV response through the more spaced MnO_2-rich petals (Table 1). Additionally, it seems that the absence of petal-like structure and transition from MnO_2 to Mn_2O_3 (Figures 1b and 3d) in the case of Mn_2O_3/Co_3O_4-800 does not affect much the CV response of rather low-current featureless characteristics.

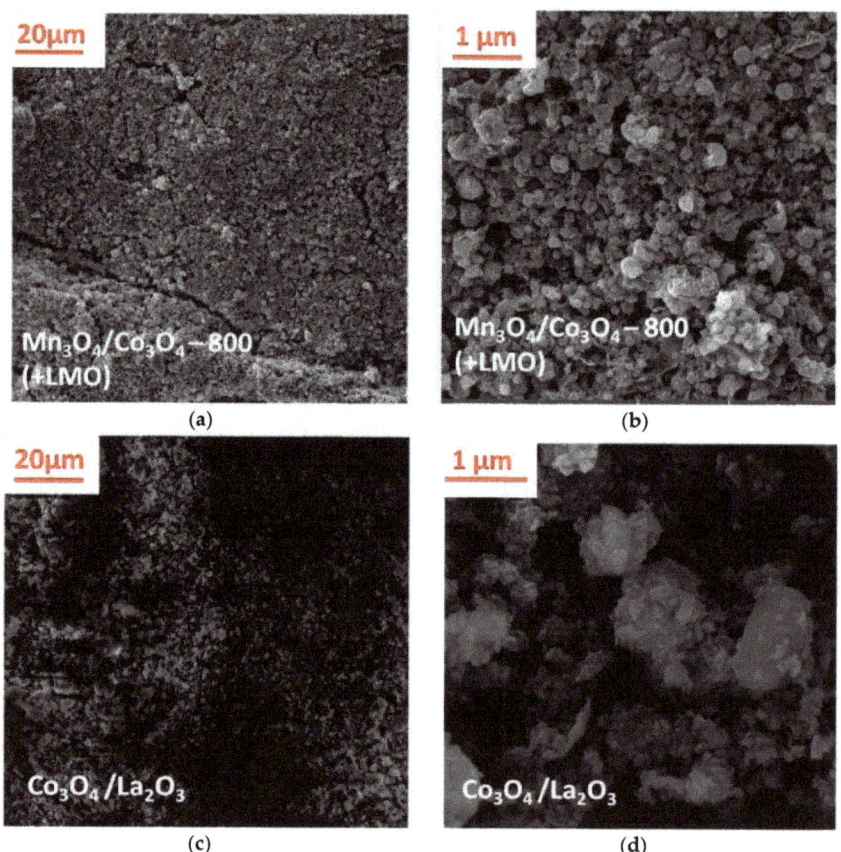

Figure 3. SEM images of (**a**,**b**) Mn_3O_4/Co_3O_4-800 and (**c**,**d**) Co_3O_4/La_2O_3; at lower and higher magnifications.

Table 1. Element analysis of synthesized materials for EDS analysis (at. %).

Element	Sample				
	MnO_2/Co_3O_4-500	MnO_2/Co_3O_4-600	$Mn_3O_4/Co_3O_4/LaMnO_3$-800	Co_3O_4 (+La_2O_3)	MnO_2
O	68.71	62.92	61.78	69.77	60.29
La	2.04	3.78	3.11	2.26	/
Co	9.88	11.24	12.98	27.96	/
Mn	19.37	22.06	22.13	/	32.43

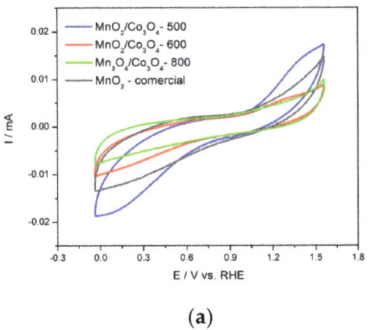
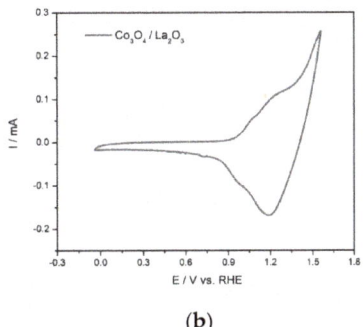

Figure 4. CV performance of synthesized catalytic materials (**a**) comparison between Mn-based materials and commercial MnO_2, (**b**) Co-based electrode; 0.1 M KOH, N_2 atmosphere, 20 mv s^{-1}.

The electrocatalytic activities of the prepared nanocatalysts for ORR were evaluated by means of LSV in O_2-saturated alkaline electrolyte. Figure 5 presents ORR electrochemical performances of Mn/Co/La oxides synthesized at three different temperatures. As can be seen, when the electrolyte was saturated with O_2, remarkable reduction currents are observed, which introduces the synthesized materials as ORR-active. The onset potential of all electrodes was approx. −0.3 V vs. SCE, which is competitive to other TMO based materials [56]. The electrode material at 500° and 600° show similar curve shape and activity, which is expected for similar flower-like birnessite MnO_2, having similar electrochemical activities between 1 and 2.5 mA cm^{-2} (Figure 5a) [51]. The Mn_3O_4/Co_3O_4 material, obtained at 800 °C, exhibits higher ORR activity than MnO_2/Co_3O_4-500 and MnO_2/Co_3O_4-600 materials. This can be ascribed to the presence of Mn_3O_4 as catalyst with mixed oxidation state (+2, +3) in comparison to the MnO_2 with +4 oxidation state of Mn [20]. Additionally, as it is registered by SEM, two different manganese types provide different morphologies. Consequently, the ORR current increases continuously for Mn_3O_4/Co_3O_4-800, whereas the reduction on MnO_2/Co_3O_4-500 and -600 appears stepped, with a transition around −0.7 V. This could be the indication of different ORR mechanisms on MnO_2/Co_3O_4 and Mn_3O_4/Co_3O_4.

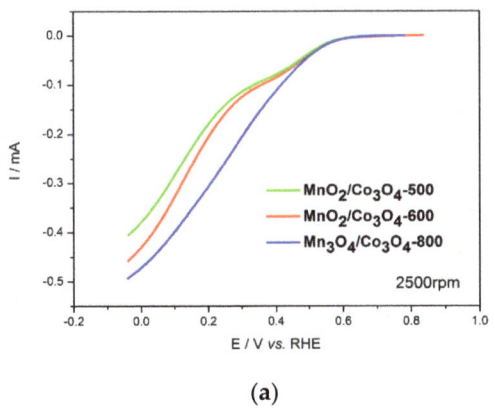
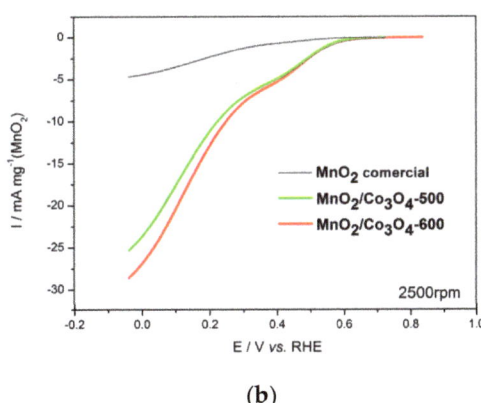

Figure 5. Comparison of ORR activities of Mn-based hybrid materials: (**a**) at different synthesis temperature and (**b**) commercial MnO_2 for ORR per mass of MnO_2.

For the sake of comparison to pristine MnO_2, electrode activities calculated per mass of MnO_2 (mass activity) was performed, as presented in Figure 5b. It can be seen that nanostructured MnO_2/Co_3O_4 materials are of significantly improved ORR activity compared to the pristine manganese oxide. This proves the validity of the hybrid oxides approach of ordered structure for the synthesis of materials for ORR.

In order to study ORR activity further, the series of polarization curves (LSV) in saturated oxygen atmosphere were recorded at different electrode rotating rates between 600 and 2500 rpm. As can be seen in Figure 6a, ORR current is increasing with the increase of the rotation rate at higher overpotentials, due to the improved mass transfer. However, the first reduction step (positive to −0.7 V) for MnO$_2$/Co$_3$O$_4$-500 and -600 appears negligibly dependent on rotation rate. It follows that corresponding process(es) are not directly related to ORR, but to partial reduction of the material induced by the presence of oxygen (please see Equations (3)–(7)).

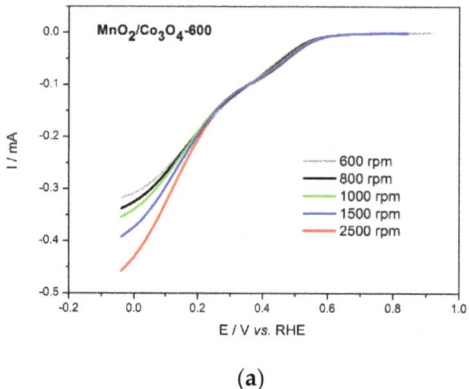
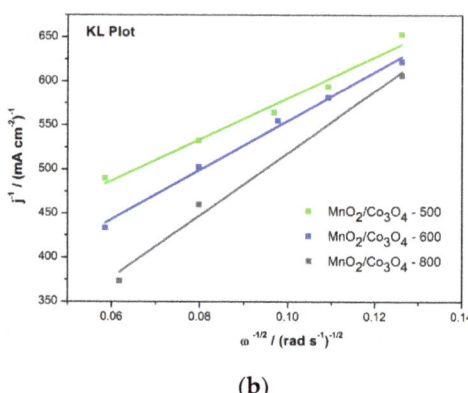

(a) (b)

Figure 6. LSV performance of synthesized materials: MnO$_2$/Co$_3$O$_4$-600 (**a**) at different electrode rotation rates; (**b**) KL-plot of the synthesized materials; 0.1 M KOH, O$_2$ atmosphere.

The number of the electrons that are involved in oxygen reduction reaction is an important parameter for evaluating the catalytic performance of the synthesized materials. Therefore, the ORR was further analyzed using Koutecky–Levich (KL) equation (Equation (1)). The corresponding linear fit that is presented in Figure 6b at E = −1 V, was used to calculate the number of electrons transferred during the oxygen reduction with synthesized materials. In the equation:

$$\frac{1}{j} = \frac{1}{j_L} + \frac{1}{j_k} = \frac{1}{B\omega^{0.5}} + \frac{1}{j_k} \quad (1)$$

$$B = 0.62nFC_0D^{2/3}v^{-1/6} \quad (2)$$

where j corresponds to the measured current density, n is the overall number of electrons exchanged, F stands for Faraday constant (F = 96486 C mol^{-1}), C_0 is the oxygen concentration in 0.1 M KOH (typically C_0 = 1.2 × 10^{-6} mol cm^{-3}), D is oxygen diffusion coefficient (typically D = 1.9 × 10^{-5} cm^2 s^{-1}), v is the kinematic viscosity of the solution (v = 0.01 cm^2 s^{-1}); surface area of the electrode used to calculate current density is A = 0.196 cm^{-2}.

It has been reported that the ORR in alkaline media can proceed either via two-electron pathway, which involves the formation of hydrogen peroxide as an intermediate, or via direct four-electron reduction pathway where oxygen is directly reduced to OH$^-$ [59]. Generally, direct four-electron transfer pathways are more desirable than the partial reduction pathway since it provides a higher rate for ORR. Although the ORR mechanism on Mn-oxide is still not fully understood, the possible pathway suggests the reactions described:

$$MnO_2(s) + H_2O + e^- \leftrightarrow MnOOH(s) + OH^- \quad (3)$$

$$2MnOOH(s) + O_2 \leftrightarrow 2(MnOOH\ldots.O)(s) \quad (4)$$

$$MnOOH(s) + O_2 \leftrightarrow MnOOH\ldots.O_{2,\,ads}\,(s) \qquad (5)$$

$$(MnOOH\ldots.O) + e^- \leftrightarrow MnO_2(s) + OH^- \qquad (6)$$

$$MnOOH\ldots.O_{2,\,ads}\,(s) + e^- \leftrightarrow MnO_2(s) + HO_2^- \qquad (7)$$

The overall reaction of Equations (3), (4), and (6) equals to the four-electron reduction process, whereas the summary of Equations (3), (5), and (7) results in an overall two-electron transfer mechanism oxygen reduction. In the first step, Mn^{4+} is reduced to Mn^{3+}, which is followed by adsorption and reduction of oxygen. Hence, the promotion of Mn^{3+} generation could lead to more effective ORR over stronger oxygen adsorption and accelerated O_2 reduction to OH^-, which results in overall increase in catalytic activity [49]. Furthermore, it has been reported that porous structure can also stabilize Mn^{3+} species at the particle surface [36]. For the synthesized materials, Figure 6b reveals that the number of transferred electrodes, calculated from the slope, for MnO_2/Co_3O_4-500 and -600 catalytic materials are 3.6 and 3.89 which is close to 4 indicating that ORR catalyzed by those materials proceeds via quasi-four-electron ORR mechanism. However, the overall electron transfer number of Mn_3O_4/Co_3O_4-800 material is calculated to be 3, indicating that both of the suggested schemes coexist in the catalyzing process [60]. Even though the higher oxidation states of Mn (Mn^{4+} and Mn^{3+}) are considered as crucial for manifestation of Mn cation defects and oxygen vacancies that are important as catalytically active sites, the morphology of samples at 500, 600, and 800, as well as their mutual interaction with Co-oxide, also plays an important role in catalytic activity.

The Co_3O_4/La_2O_3 material was also checked for the ORR performance at various rotation speeds, as presented in Figure 7. The ORR on this hybrid material starts at approx. -0.35 V vs. SCE, which is in accordance to other Co_3O_4 reported nanomaterials, but still is considerably negative if compared to the commercial 20 wt.% Pt@XC-72 catalyst (-0.16 V) [56]. On the other hand, the current of approx. -3.3 mA cm^{-2} is higher in comparison to the performance of similar nanomaterials, found as -2.5 [56], -1.5 [43], and -1 mA cm^{-2} [58] at similar electrode potentials. ORR on synthesized Co_3O_4/La_2O_3 was studied also using a KL plot shown in Figure 7b. The number of electrons transferred, calculated based on KL equation, is 3.7 suggesting predominantly the pathway of direct four-electron reduction of oxygen. This is fairly comparable to the state-of-the-art electrode based on Pt (20% wt. Pt@XC-72), which is reported to be between 3.8 and 4.03 [56].

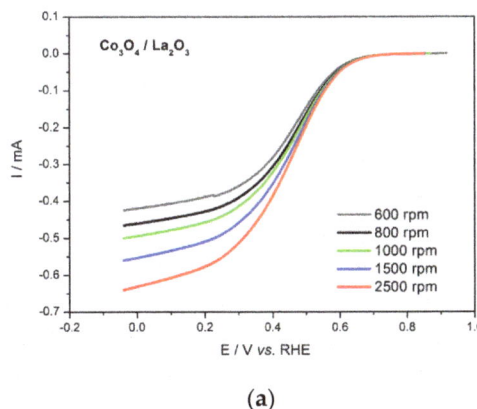
(a)

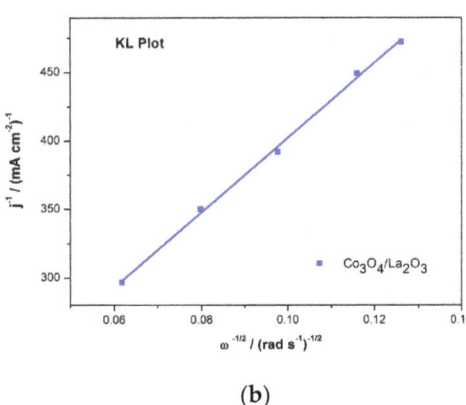
(b)

Figure 7. Reduction of O_2 at the Co-based electrode (a,b) KL plot for Co_3O_4/La_2O_3.

Finally, the comparison of all synthesized oxide combinations as catalytic materials is presented in Figure 8. As can be seen, Co_3O_4/La_2O_3 catalyst outperforms MnO_2/Co_3O_4 and Mn_3O_4/Co_3O_4 materials, in comparison to the onset electrode potential, as well as in electrocatalytic activity toward ORR in the studied potential region. Similar behavior has been reported by Xu et al. showing that Co_3O_4/La_2O_3 supported by carbon nanotubes (CNT) has shown better activity in comparison to the MnO_2/Co_3O_4–CNT. Although CNT have been used to increase catalytically active surface area, the activities of the studied hybrid materials are in the range 1.8–4 mA cm^{-2} that is comparable to our hybrid materials but without addition of a carbon support [37].

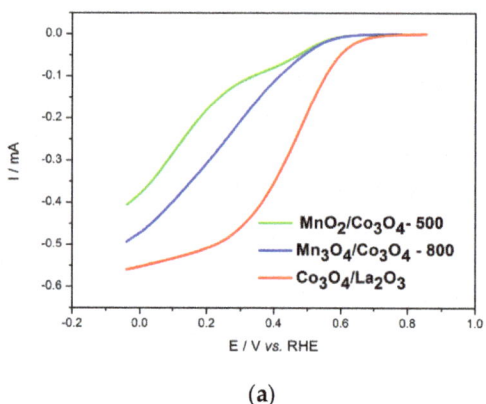
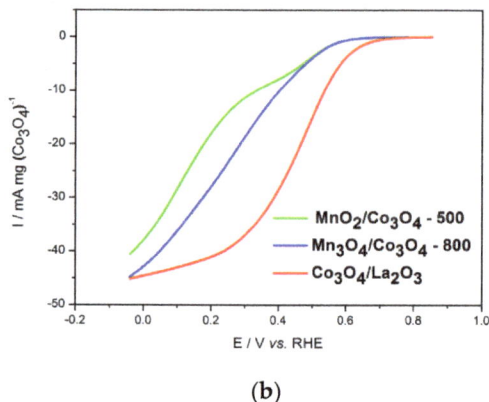

Figure 8. Comparison of (**a**) synthesized materials in mA (**b**) calculated per mass of Co.

As mentioned above, the mechanism on the synthesized electrode most probably proceeds with an additional MnOOH reaction step. It has been reported that this reaction is unfavorable due to the strength of the Mn-O bond which makes the initial reduction more difficult. This leads to lower electrocatalytic activity of the electrodes with high percentage of Mn oxide of birnessite type. It is also speculated that the crystal phase (channel structure) can be another determining criterion for ORR kinetics at manganese oxides.

Additionally, the observed larger inner-spacing between nanoparticles in Co_3O_4/La_2O_3 (Figure 3) in comparison to dense structure of MnO_6 octahedral sheet of birnessite MnO_2, provides more active surface area that is more accessible to the electrolyte (reactant). In addition, it is reported that Co_3O_4 has high affinity toward O_2 molecules, which enables better oxygen transport within its porous structure [61].

This all contributed to the observed behavior that the Co_3O_4/La_2O_3 electrode exhibits better catalytic activity in comparison to the MnO_2/Co_3O_4 electrodes. Additionally, Du et al. stated that Co_3O_4 nanoparticles-modified MnO_2 electrodes have much lower ORR activity in comparison to pure MnO_2 nanomaterial due to partial occupation of active sites on MnO_2 by Co_3O_4 [43]. This likely can be another reason for the lower catalytic activity of our MnO_2/Co_3O_4 oxides catalysts. Although the synergic effect of MnO_2 and Co_3O_4 oxides has been reported in many publications to increase ORR activity [43,45], the investigation of the parameters such as the composition, crystalline structure, and morphology are to be investigated in order to propose the most probable synergy mechanism.

Since the synthesized hybrid materials have different compositions, and consequently structures, the kinetic comparison would be more informative if would be presented as activity (currents) per mol of Co_3O_4 and MnO_2. Figure 9 presents the comparison of molar activities with respect to Co_3O_4 for the samples containing Co oxide, and with respect to MnO_2 for the samples synthesized at 500 and 600 °C, taking into account the compositions found by EDS (Table 1). As can be seen from Figure 9a, the Co_3O_4/La_2O_3 electrode is of higher activity at the beginning (at lower overpotentials) due to a different ORR reaction mechanism occurring in comparison to the Mn-based electrodes (additionally involves the Mn oxide electrochemical transformations). However, in the region of higher overpotentials that are more relevant for the electrochemical devices (fuel cells, batteries), the electrodes based on synergy of Mn/Co oxides outperform the electrode without Mn (Co_3O_4/La_2O_3). This result is somewhat contrary to the presented performance of the current calculated per mass (Figure 8), which indicated the electrode without Mn (Co$_3$O$_4$/La$_2$O$_3$) as the one of best performances. In addition, Figure 9b shows that trends of molar activities of MnO$_2$-containing electrodes are similar to those of mass activity (Figure 5b). This clearly emphasizes the importance of the calculations to take into consideration the mole fractions in hybrid materials of different compositions in order to quantify the synergy effects.

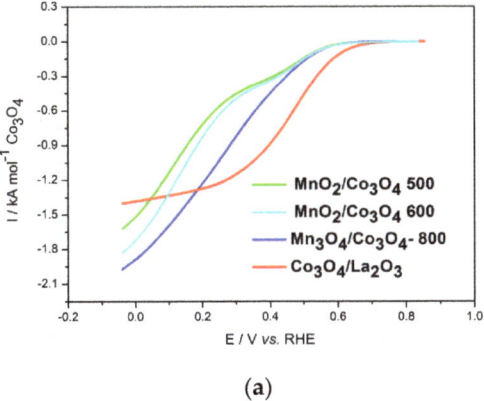
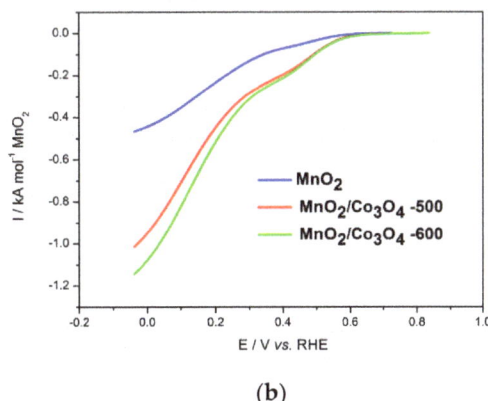

(a)   (b)

Figure 9. Comparison of all materials (**a**) per mol of Co oxide and (**b**) per mol of Mn oxide.

4. Conclusions

In summary, highly active electrocatalysts based on Mn and rare earth oxides with Co$_3$O$_4$ for ORR, have been successfully synthesized using the ultrasonic spray pyrolysis process. For the sake of comparison, the hybrid materials with and without Mn oxides were investigated.

It was shown that different Mn oxides were well incorporated in the Co$_3$O$_4$ matrix. Different synthesis temperatures led to the formation of two different manganese oxides-birnessite type δ-MnO$_2$ at 500 and 600 °C, and manganese (II, III) oxide-Mn$_3$O$_4$ at 800 °C. The catalysts morphology has been also affected by the state of Mn oxide. SEM images reveal flower-like nanosheets for hybrid materials with birnessite-MnO$_2$ and well-defined spherical nanoparticles for material based on Mn$_3$O$_4$ and for the material based only on Co as catalysts (Co$_3$O$_4$/La$_2$O$_3$). The electrochemical performance of MnO$_x$/Co$_3$O$_4$ and Co$_3$O$_4$/La$_2$O$_3$ demonstrate a comparable ORR activity to Pt/C and superior activity to the pristine Mn oxide electrodes. It was shown that the mass activity of synthesized hybrid materials not supported on carbon blacks outperforms the literature values of carbon-based materials. Mass activity performance was compared to the molar activity—calculated per mol of the Mn and Co oxides being in charge for the catalytic performance in oxygen reduction reaction. It was revealed that the synergic coupling of Mn oxides and Co$_3$O$_4$ have better catalytic performance in comparison to the electrodes based on pristine MnO$_2$ and Co$_3$O$_4$/La$_2$O$_3$. It was found that molar and mass activities give different information, since different amounts of active components are affecting the synergistic catalysis.

The crystal structure and morphological characteristics, as well as right amounts of investigated oxides, play the crucial role for high catalytic activity. Taking into account that the investigated materials are very low-cost materials especially compared to the state-of-the-art Pt/C-based electrodes, the demonstrated hybrid materials are promising catalysts for practical application for rechargeable metal–air batteries and fuel cells.

Author Contributions: M.V.: Investigation, methodology, writing—review and editing, writing—original draft. M.M.P.: Validation, visualization, writing—review and editing. S.E.P.: Conceptualization, formal analysis, data curation. M.M.: Conceptualization, formal analysis. M.R.P.P.: Formal analysis, visualization. S.S.: Conceptualization. B.F.: Funding acquisition. All authors have read and agreed to the published version of the manuscript.

Funding: This work was supported by the Ministry of Education, Science and Technological Development of the Republic of Serbia (Grant No. 451-03-9/2021-14/200026). The authors would like to thank the Ministry of Education, Science and Technological Development of the Republic of Serbia and DAAD, Germany, for funding of the Project No.: 57334757.

Data Availability Statement: All available data is contained within the article.

Acknowledgments: The authors would like to thank Tanja Barudžija and Miodrag Mitrić from Vinča Institute for support on the XRD measurements, Đorđe Veljović for SEM-EDS analysis. Special thanks to Vladimir Panić and Jasmina Stevanović for help in results analyses.

Conflicts of Interest: The authors declare that they have no known competing financial interest or personal relationships that could have appeared to influence the work reported in this paper.

References

1. Wang, Y.; Diaz, D.F.R.; Chen, K.S.; Wang, Z.; Adroher, X.C. Materials, technological status, and fundamentals of PEM fuel cells—A review. *Mater. Today* **2020**, *32*, 178–203. [CrossRef]
2. Wang, X.; Li, Z.; Qu, Y.; Yuan, T.; Wang, W.; Wu, Y.; Li, Y. Review of Metal Catalysts for Oxygen Reduction Reaction: From Nanoscale Engineering to Atomic Design. *Chem* **2019**, *5*, 1486–1511. [CrossRef]
3. Ren, X.; Lv, Q.; Liu, L.; Liu, B.; Wang, Y.; Liu, A.; Wu, G. Current progress of Pt and Pt-based electrocatalysts used for fuel cells. *Sustain. Energy Fuels* **2020**, *4*, 15–30. [CrossRef]
4. Xiong, Y.; Xiao, L.; Yang, Y.; DiSalvo, F.J.; Abruña, H.D. High-Loading Intermetallic Pt3Co/C Core–Shell Nanoparticles as Enhanced Activity Electrocatalysts toward the Oxygen Reduction Reaction (ORR). *Chem. Mater.* **2018**, *30*, 1532–1539. [CrossRef]
5. Shi, W.; Wang, Y.-C.; Chen, C.; Yang, X.-D.; Zhou, Z.-Y.; Sun, S.-G. A mesoporous Fe/N/C ORR catalyst for polymer electrolyte membrane fuel cells. *Chin. J. Catal.* **2016**, *37*, 1103–1108. [CrossRef]
6. Gómez-Marín, A.M.; Feliu, J.M. Oxygen Reduction on Platinum Single Crystal Electrodes. In *Encyclopedia of Interfacial Chemistry*; Wandelt, K.B.T.-E., Ed.; Elsevier: Oxford, UK, 2018; pp. 820–830. ISBN 978-0-12-809894-3.
7. Yamada, I.; Takamatsu, A.; Asai, K.; Shirakawa, T.; Ohzuku, H.; Seno, A.; Uchimura, T.; Fujii, H.; Kawaguchi, S.; Wada, K.; et al. Systematic Study of Descriptors for Oxygen Evolution Reaction Catalysis in Perovskite Oxides. *J. Phys. Chem. C* **2018**, *122*, 27885–27892. [CrossRef]
8. Kodama, K.; Nagai, T.; Kuwaki, A.; Jinnouchi, R.; Morimoto, Y. Challenges in applying highly active Pt-based nanostructured catalysts for oxygen reduction reactions to fuel cell vehicles. *Nat. Nanotechnol.* **2021**, *16*, 140–147. [CrossRef]
9. Miura, A.; Rosero-Navarro, C.; Masubuchi, Y.; Higuchi, M.; Kikkawa, S.; Tadanaga, K. Nitrogen-Rich Manganese Oxynitrides with Enhanced Catalytic Activity in the Oxygen Reduction Reaction. *Angew. Chem. Int. Ed.* **2016**, *55*, 7963–7967. [CrossRef]
10. Wang, Y.; Li, Y.; Lu, Z.; Wang, W. Improvement of O2 adsorption for α-MnO2 as an oxygen reduction catalyst by Zr4+ doping. *RSC Adv.* **2018**, *8*, 2963–2970. [CrossRef]
11. Menezes, P.W.; Indra, A.; González-Flores, D.; Sahraie, N.R.; Zaharieva, I.; Schwarze, M.; Strasser, P.; Dau, H.; Driess, M. High-Performance Oxygen Redox Catalysis with Multifunctional Cobalt Oxide Nanochains: Morphology-Dependent Activity. *ACS Catal.* **2015**, *5*, 2017–2027. [CrossRef]
12. Kumar, K.; Canaff, C.; Rousseau, J.; Arrii-Clacens, S.; Napporn, T.W.; Habrioux, A.; Kokoh, K.B. Effect of the Oxide–Carbon Heterointerface on the Activity of Co3O4/NRGO Nanocomposites toward ORR and OER. *J. Phys. Chem. C* **2016**, *120*, 7949–7958. [CrossRef]
13. Wu, Y.; Wang, Y.; Xiao, Z.; Li, M.; Ding, Y.; Qi, M. Electrocatalytic oxygen reduction by a Co/Co3O4@N-doped carbon composite material derived from the pyrolysis of ZIF-67/poplar flowers. *RSC Adv.* **2021**, *11*, 2693–2700. [CrossRef]
14. Wang, Y.; Li, J.; Wei, Z. Transition-metal-oxide-based catalysts for the oxygen reduction reaction. *J. Mater. Chem. A* **2018**, *6*, 8194–8209. [CrossRef]
15. Yin, M.; Miao, H.; Hu, R.; Sun, Z.; Li, H. Manganese dioxides for oxygen electrocatalysis in energy conversion and storage systems over full pH range. *J. Power Sources* **2021**, *494*, 229779. [CrossRef]
16. Nikitina, V.A.; Kurilovich, A.A.; Bonnefont, A.; Ryabova, A.S.; Nazmutdinov, R.R.; Savinova, E.R.; Tsirlina, G.A. ORR on Simple Manganese Oxides: Molecular-Level Factors Determining Reaction Mechanisms and Electrocatalytic Activity. *J. Electrochem. Soc.* **2018**, *165*, J3199–J3208. [CrossRef]
17. Poux, T.; Bonnefont, A.; Kéranguéven, G.; Tsirlina, G.A.; Savinova, E.R. Electrocatalytic Oxygen Reduction Reaction on Perovskite Oxides: Series versus Direct Pathway. *ChemPhysChem* **2014**, *15*, 2108–2120. [CrossRef] [PubMed]
18. Ryabova, A.S.; Napolskiy, F.S.; Poux, T.; Istomin, S.Y.; Bonnefont, A.; Antipin, D.M.; Baranchikov, A.Y.; Levin, E.E.; Abakumov, A.M.; Kéranguéven, G.; et al. Rationalizing the Influence of the Mn(IV)/Mn(III) Red-Ox Transition on the Electrocatalytic Activity of Manganese Oxides in the Oxygen Reduction Reaction. *Electrochim. Acta* **2016**, *187*, 161–172. [CrossRef]
19. Zhong, X.; Oubla, M.; Wang, X.; Huang, Y.; Zeng, H.; Wang, S.; Liu, K.; Zhou, J.; He, L.; Zhong, H.; et al. Boosting oxygen reduction activity and enhancing stability through structural transformation of layered lithium manganese oxide. *Nat. Commun.* **2021**, *12*, 3136. [CrossRef] [PubMed]

20. Dessie, Y.; Tadesse, S.; Eswaramoorthy, R.; Abebe, B. Recent developments in manganese oxide based nanomaterials with oxygen reduction reaction functionalities for energy conversion and storage applications: A review. *J. Sci. Adv. Mater. Devices* **2019**, *4*, 353–369. [CrossRef]
21. Speck, F.D.; Santori, P.G.; Jaouen, F.; Cherevko, S. Mechanisms of Manganese Oxide Electrocatalysts Degradation during Oxygen Reduction and Oxygen Evolution Reactions. *J. Phys. Chem. C* **2019**, *123*, 25267–25277. [CrossRef]
22. Lambert, T.N.; Vigil, J.A.; White, S.E.; Delker, C.J.; Davis, D.J.; Kelly, M.; Brumbach, M.T.; Rodriguez, M.A.; Swartzentruber, B.S. Understanding the Effects of Cationic Dopants on α-MnO2 Oxygen Reduction Reaction Electrocatalysis. *J. Phys. Chem. C* **2017**, *121*, 2789–2797. [CrossRef]
23. Liu, H.; Zhang, J.; Fang, H.; Huang, J.; Wu, X.; He, X.; Song, J.; Li, Z.; Yan, Y.; Xu, W.; et al. Synthesis of δ–MnO2/Reduced Graphene Oxide Hybrid In Situ and Application in Mg–Air Battery. *J. Electrochem. Soc.* **2021**, *168*, 80518. [CrossRef]
24. Khan, Z.; Park, S.; Hwang, S.M.; Yang, J.; Lee, Y.; Song, H.-K.; Kim, Y.; Ko, H. Hierarchical urchin-shaped α-MnO2 on graphene-coated carbon microfibers: A binder-free electrode for rechargeable aqueous Na–air battery. *NPG Asia Mater.* **2016**, *8*, e294. [CrossRef]
25. Bi, R.; Liu, G.; Zeng, C.; Wang, X.; Zhang, L.; Qiao, S.-Z. 3D Hollow α-MnO(2) Framework as an Efficient Electrocatalyst for Lithium-Oxygen Batteries. *Small* **2019**, *15*, e1804958. [CrossRef] [PubMed]
26. Shahid, M.M.; Rameshkumar, P.; Basirun, W.J.; Juan, J.C.; Huang, N.M. Cobalt oxide nanocubes interleaved reduced graphene oxide as an efficient electrocatalyst for oxygen reduction reaction in alkaline medium. *Electrochim. Acta* **2017**, *237*, 61–68. [CrossRef]
27. Shahid, M.M.; Zhan, Y.; Alizadeh, M.; Sagadevan, S.; Paiman, S.; Oh, W.C. A glassy carbon electrode modified with tailored nanostructures of cobalt oxide for oxygen reduction reaction. *Int. J. Hydrogen Energy* **2020**, *45*, 18850–18858. [CrossRef]
28. Tan, P.; Wu, Z.; Chen, B.; Xu, H.; Cai, W.; Ni, M. Exploring oxygen electrocatalytic activity and pseudocapacitive behavior of Co3O4 nanoplates in alkaline solutions. *Electrochim. Acta* **2019**, *310*, 86–95. [CrossRef]
29. Lu, J.; Dey, S.; Temprano, I.; Jin, Y.; Xu, C.; Shao, Y.; Grey, C.P. Co3O4-Catalyzed LiOH Chemistry in Li–O2 Batteries. *ACS Energy Lett.* **2020**, *5*, 3681–3691. [CrossRef]
30. Jin, J.; Fu, X.; Liu, Q.; Zhang, J. A highly active and stable electrocatalyst for the oxygen reduction reaction based on a graphene-supported g-C3N4@cobalt oxide core–shell hybrid in alkaline solution. *J. Mater. Chem. A* **2013**, *1*, 10538–10545. [CrossRef]
31. Al-Hakemy, A.Z.; Nassr, A.B.A.A.; Naggar, A.H.; Elnouby, M.S.; Soliman, H.M.A.E.-F.; Taher, M.A. Electrodeposited cobalt oxide nanoparticles modified carbon nanotubes as a non-precious catalyst electrode for oxygen reduction reaction. *J. Appl. Electrochem.* **2017**, *47*, 183–195. [CrossRef]
32. Yu, J.; Chen, G.; Sunarso, J.; Zhu, Y.; Ran, R.; Zhu, Z.; Zhou, W.; Shao, Z. Cobalt Oxide and Cobalt-Graphitic Carbon Core–Shell Based Catalysts with Remarkably High Oxygen Reduction Reaction Activity. *Adv. Sci.* **2016**, *3*, 1600060. [CrossRef] [PubMed]
33. Liang, Y.; Wang, H.; Diao, P.; Chang, W.; Hong, G.; Li, Y.; Gong, M.; Xie, L.; Zhou, J.; Wang, J.; et al. Oxygen Reduction Electrocatalyst Based on Strongly Coupled Cobalt Oxide Nanocrystals and Carbon Nanotubes. *J. Am. Chem. Soc.* **2012**, *134*, 15849–15857. [CrossRef] [PubMed]
34. Ahmed, J.; Kim, H.J.; Kim, S. Embedded cobalt oxide nano particles on carbon could potentially improve oxygen reduction activity of cobalt phthalocyanine and its application in microbial fuel cells. *RSC Adv.* **2014**, *4*, 44065–44072. [CrossRef]
35. Kostuch, A.; Gryboś, J.; Wierzbicki, S.; Sojka, Z.; Kruczała, K. Selectivity of Mixed Iron-Cobalt Spinels Deposited on a N,S-Doped Mesoporous Carbon Support in the Oxygen Reduction Reaction in Alkaline Media. *Materials* **2021**, *14*, 820. [CrossRef]
36. Zhu, H.; Zhang, P.; Dai, S. Recent Advances of Lanthanum-Based Perovskite Oxides for Catalysis. *ACS Catal.* **2015**, *5*, 6370–6385. [CrossRef]
37. Xu, N.; Qiao, J.; Zhang, X.; Ma, C.; Jian, S.; Liu, Y.; Pei, P. Morphology controlled La2O3/Co3O4/MnO2–CNTs hybrid nanocomposites with durable bi-functional air electrode in high-performance zinc–air energy storage. *Appl. Energy* **2016**, *175*, 536–544. [CrossRef]
38. Wang, N.; Liu, J.; Gu, W.; Song, Y.; Wang, F. Toward Synergy of Carbon and La2O3 in Their Hybrid as Efficient Catalyst for Oxygen Reduction Reaction. *RSC Adv.* **2016**, *6*, 77786–77795. [CrossRef]
39. Liu, K.; Lei, Y.; Wang, G. Correlation between oxygen adsorption energy and electronic structure of transition metal macrocyclic complexes. *J. Chem. Phys.* **2013**, *139*, 204306. [CrossRef] [PubMed]
40. Zhang, X.; Xiao, Q.; Zhang, Y.; Jiang, X.; Yang, Z.; Xue, Y.; Yan, Y.-M.; Sun, K. La2O3 Doped Carbonaceous Microspheres: A Novel Bifunctional Electrocatalyst for Oxygen Reduction and Evolution Reactions with Ultrahigh Mass Activity. *J. Phys. Chem. C* **2014**, *118*, 20229–20237. [CrossRef]
41. Sugawara, Y.; Kobayashi, H.; Honma, I.; Yamaguchi, T. Effect of Metal Coordination Fashion on Oxygen Electrocatalysis of Cobalt–Manganese Oxides. *ACS Omega* **2020**, *5*, 29388–29397. [CrossRef]
42. Li, M.; Xiong, Y.; Liu, X.; Bo, X.; Zhang, Y.; Han, C.; Guo, L. Facile synthesis of electrospun MFe2O4 (M = Co, Ni, Cu, Mn) spinel nanofibers with excellent electrocatalytic properties for oxygen evolution and hydrogen peroxide reduction. *Nanoscale* **2015**, *7*, 8920–8930. [CrossRef] [PubMed]
43. Du, G.; Liu, X.; Zong, Y.; Hor, T.S.A.; Yu, A.; Liu, Z. Co3O4 nanoparticle-modified MnO2 nanotube bifunctional oxygen cathode catalysts for rechargeable zinc–air batteries. *Nanoscale* **2013**, *5*, 4657–4661. [CrossRef]
44. Yang, X.; Peng, W.; Fu, K.; Mao, L.; Jin, J.; Yang, S.; Li, G. Nanocomposites of honeycomb double-layered MnO2 nanosheets/cobalt doped hollow carbon nanofibers for application in supercapacitor and primary zinc-air battery. *Electrochim. Acta* **2020**, *340*, 135989. [CrossRef]

45. Li, X.; Nengneng, X.; Li, H.; Wang, M.H.; Zhang, L.; Qiao, J. 3D hollow sphere Co3O4/MnO2-CNTs: Its high-performance bi-functional cathode catalysis and application in rechargeable zinc-air battery. *Green Energy Environ.* **2017**, *2*, 316–328. [CrossRef]
46. Che, H.; Lv, Y.; Liu, A.; Mu, J.; Zhang, X.; Bai, Y. Facile synthesis of three dimensional flower-like Co3O4@MnO2 core-shell microspheres as high-performance electrode materials for supercapacitors. *Ceram. Int.* **2017**, *43*, 6054–6062. [CrossRef]
47. Eraković, S.; Pavlović, M.M.; Stopić, S.; Stevanović, J.; Mitrić, M.; Friedrich, B.; Panić, V. Interactive promotion of supercapacitance of rare earth/CoO3-based spray pyrolytic perovskite microspheres hosting the hydrothermal ruthenium oxide. *Electrochim. Acta* **2019**, *321*, 134721. [CrossRef]
48. Pavlović, M.M.; Pantović Pavlović, M.R.; Eraković Pantović, S.G.; Stevanović, J.S.; Stopić, S.R.; Friedrich, B.; Panić, V. V The Roles of Constituting Oxides in Rare-Earth Cobaltite-Based Perovskites on their Pseudocapacitive Behavior. *J. Electroanal. Chem.* **2021**, *897*, 115556. [CrossRef]
49. Zhang, T.; Ge, X.; Zhang, Z.; Tham, N.N.; Liu, Z.; Fisher, A.; Lee, J.Y. Improving the Electrochemical Oxygen Reduction Activity of Manganese Oxide Nanosheets with Sulfurization-Induced Nanopores. *ChemCatChem* **2018**, *10*, 422–429. [CrossRef]
50. Chen, B.; Miao, H.; Hu, R.; Yin, M.; Wu, X.; Sun, S.; Wang, Q.; Li, S.; Yuan, J. Efficiently optimizing the oxygen catalytic properties of the birnessite type manganese dioxide for zinc-air batteries. *J. Alloys Compd.* **2021**, *852*, 157012. [CrossRef]
51. Xiao, W.; Wang, D.; Lou, X.W. Shape-Controlled Synthesis of MnO2 Nanostructures with Enhanced Electrocatalytic Activity for Oxygen Reduction. *J. Phys. Chem. C* **2010**, *114*, 1694–1700. [CrossRef]
52. Saputra, E.; Muhammad, S.; Sun, H.; Ang, H.-M.; Tadé, M.O.; Wang, S. Manganese oxides at different oxidation states for heterogeneous activation of peroxymonosulfate for phenol degradation in aqueous solutions. *Appl. Catal. B Environ.* **2013**, *142–143*, 729–735. [CrossRef]
53. Li, Z.; Yang, Y.; Relefors, A.; Kong, X.; Siso, G.M.; Wickman, B.; Kiros, Y.; Soroka, I.L. Tuning morphology, composition and oxygen reduction reaction (ORR) catalytic performance of manganese oxide particles fabricated by γ-radiation induced synthesis. *J. Colloid Interface Sci.* **2021**, *583*, 71–79. [CrossRef] [PubMed]
54. BOSE, V.; BIJU, V. Mixed valence nanostructured Mn3O4 for supercapacitor applications. *Bull. Mater. Sci.* **2015**, *38*, 865–873. [CrossRef]
55. Sankar, V.; Kalpana, D.; Kalai Selvan, R. Electrochemical properties of microwave-assisted reflux-synthesized Mn3O4 nanoparticles in different electrolytes for supercapacitor applications. *J. Appl. Electrochem.* **2012**, *42*, 463–470. [CrossRef]
56. Fink, M.; Eckhardt, J.; Khadke, P.; Gerdes, T.; Roth, C. Bifunctional α—MnO 2 and Co 3 O 4 Catalyst for Oxygen Electrocatalysis in Alkaline Solution. *ChemElectroChem* **2020**, *7*, 4822–4836. [CrossRef]
57. Xia, H.; Zhu, D.; Luo, Z.; Yu, Y.; Shi, X.; Yuan, G.; Xie, J. Hierarchically Structured Co3O4@Pt@MnO2 Nanowire Arrays for High-Performance Supercapacitors. *Sci. Rep.* **2013**, *3*, 2978. [CrossRef]
58. Paulraj, A.R.; Kiros, Y. La0.1Ca0.9MnO3/Co3O4 for oxygen reduction and evolution reactions (ORER) in alkaline electrolyte. *J. Solid State Electrochem.* **2018**, *22*, 1697–1710. [CrossRef]
59. Xie, G.; Chen, B.; Jiang, Z.; Niu, X.; Cheng, S.; Zhen, Z.; Jiang, Y.; Rong, H.; Jiang, Z.-J. High catalytic activity of Co3O4 nanoparticles encapsulated in a graphene supported carbon matrix for oxygen reduction reaction. *RSC Adv.* **2016**, *6*, 50349–50357. [CrossRef]
60. Zhao, Y.; Xu, L.; Mai, L.; Han, C.; An, Q.; Xu, X.; Liu, X.; Zhang, Q. Hierarchical mesoporous perovskite La0.5Sr0.5CoO2.91 nanowires with ultrahigh capacity for Li-air batteries. *Proc. Natl. Acad. Sci. USA* **2012**, *109*, 19569–19574. [CrossRef]
61. Kim, G.-P.; Sun, H.-H.; Manthiram, A. Design of a sectionalized MnO2-Co3O4 electrode via selective electrodeposition of metal ions in hydrogel for enhanced electrocatalytic activity in metal-air batteries. *Nano Energy* **2016**, *30*, 130–137. [CrossRef]

Article

Mixed Oxides NiO/ZnO/Al₂O₃ Synthesized in a Single Step via Ultrasonic Spray Pyrolysis (USP) Method

Duygu Yeşiltepe Özcelik [1], Burçak Ebin [2], Srecko Stopic [3,*], Sebahattin Gürmen [1] and Bernd Friedrich [3]

[1] Department of Metallurgical and Materials Engineering, Faculty of Chemistry and Metallurgy, Ayazağa Campus, Istanbul Technical University, İstanbul 34469, Turkey; yesiltepe15@itu.edu.tr (D.Y.Ö.); gurmen@itu.edu.tr (S.G.)
[2] Nuclear Chemistry and Industrial Material Recycling, Department of Chemistry and Chemical Engineering, Chalmers University of Technology, S-412 96 Gothenburg, Sweden; burcak@chalmers.se
[3] Department of Process Metallurgy and Metal Recycling, RWTH Aachen University, 52056 Aachen, Germany; bfriedrich@ime-aachen.de
* Correspondence: sstopic@ime-aachen.de; Tel.: +49-176-7826-1674

Abstract: Mixed oxides have received remarkable attention due to the many opportunities to adjust their interesting structural, electrical, catalytic properties, leading to a better, more useful performance compared to the basic metal oxides. In this study, mixed oxides NiO/ZnO/Al₂O₃ were synthesized in a single step via the ultrasonic spray pyrolysis method using nitrate salts, and the temperature effects of the process were investigated (400, 600, 800 °C). The synthesized samples were characterized by means of scanning electron microscopy, energy-dispersive spectroscopy, X-ray diffraction and Raman spectroscopy analyses. The results showed Al₂O₃, NiO–Al₂O₃ and ZnO–Al₂O₃ systems with spinel phases. Furthermore, the Raman peaks supported the coexistence of oxide phases, which strongly impact the overall properties of nanocomposite.

Keywords: nanocomposite; ultrasonic spray pyrolysis; mixed oxides; NiAl₂O₄; ZnAl₂O₄

Citation: Yeşiltepe Özcelik, D.; Ebin, B.; Stopic, S.; Gürmen, S.; Friedrich, B. Mixed Oxides NiO/ZnO/Al₂O₃ Synthesized in a Single Step via Ultrasonic Spray Pyrolysis (USP) Method. Metals 2022, 12, 73. https://doi.org/10.3390/met12010073

Academic Editor: Chang Woo Lee

Received: 20 November 2021
Accepted: 29 December 2021
Published: 2 January 2022

Publisher's Note: MDPI stays neutral with regard to jurisdictional claims in published maps and institutional affiliations.

Copyright: © 2022 by the authors. Licensee MDPI, Basel, Switzerland. This article is an open access article distributed under the terms and conditions of the Creative Commons Attribution (CC BY) license (https://creativecommons.org/licenses/by/4.0/).

1. Introduction

Nanocomposites are a research hotspot at present, with various applications in day-to-day technologies. A further improvement in properties is achieved when one of the components in the composite is reduced to the nanoscale (~1–100 nm). With an increased surface area, and the quantum effects that arise at this scale, this nanocomposite offers better electrical, mechanical, chemical, optical, and magnetic properties. Their melting point and dielectric constant can change when particles reach nanometre sizes [1,2]. Mixed-oxide nanocomposites are studied due to their potential for an enhanced functional performance in photocatalysis, sensors and other optoelectronic device applications [3]. The combination of two or more metals in an oxide matrix can produce materials with novel physical and chemical properties, leading to an increased performance in various technological applications [4]. Among the various mixed oxides, nickel oxide (NiO), zinc oxide (ZnO) and alumina (Al₂O₃) have been a focus in the semiconductor and chemical and petrochemical industry due to their distinguished electronic, magnetic and chemical properties. These mixed metal oxides are widely used in the field of adsorption and catalysis. They are used in many catalytic reactions in chemical and petrochemical industries, including cracking, hydrogenation dehydrogenation, reforming, and dehydration [5,6].

NiO is an eco-friendly, stable, low-cost and wide-bandgap material [7,8]. NiO nanoparticles have been a great candidate for ferroelectric p-type semiconductors with a wide band gap (3.6–4.0 eV). Recently, NiO materials have been used in technological fields such as electrochromic test equipment, supercapacitors, rechargeable lithium ion batteries as electrodes, magnetic recorders, photocatalysts, adsorbents, etc. [9]. ZnO nanoparticles, which have a large band gap (3.37 eV) and large exciton binding energy of 60 meV at room temperature, have a wide range of uses [10]. ZnO is of great interest due to its potential applications in various fields, such as gas sensors, biosensors, catalysis, solar batteries and as electronic, piezoelectric and optical devices, as well as in ultraviolet (UV) protection, cosmetics and paints [9,11,12]. Al_2O_3 has unique properties that made it one of the most important engineering materials of the late twentieth century, including chemical stability, high hardness, and a high melting temperature, which allowed it to be used in many areas, especially in the manufacture of ceramics, refinement, and optics. It is found in several patterns that differ from each other in terms of their crystalline structure and physical and chemical properties, in addition to its various applications [13]. Al_2O_3 is one of the most common ceramic materials, used as a catalyst, adsorbent and abrasion-resistant coating [14]. It is a very important adsorbent, with surface activity species (Al^{3+}, O^{2-}, OH^- group decomposed and proton defects), and acts as an adsorption center for different gases. In this regard, in the gas-sensing process, Al_2O_3 increases the adsorption amount of oxygen and the tested gases. Moreover, transition metal oxides deposited on the Al_2O_3 surface have a high dispersion form [15]. The ZnO/Al_2O_3 nanocomposite is used for UV emission [16]. Composites from Al_2O_3–NiO systems with spinel-phase nickel aluminate ($NiAl_2O_4$) are used as catalysts or precatalysts for steam reformation or as electrode materials in high-temperature fuel cells due to their unusual conductivity [17]. Zinc aluminate ($ZnAl_2O_4$), is a mixed oxide with spinel structure that is currently used as high-temperature material, sensors, electronic and optical materials, and catalyst support [18]. $ZnAl_2O_4$ has a much higher photocatalytic activity than a single oxide [19]. $NiO/ZnO/Al_2O_3$ nanocomposites are used as catalysts in industrial processes such as hydrogenation and dehydrogenation reactions, petroleum refining, deoxygenation, CO_2 reduction and fuel cells [20,21]. Other applications, such as protective barriers, electrochromic material and sensors, are also available [22].

Two strategies are used in the synthesis of nanomaterials: top-down and bottom-up. In the top-down approach, bulk particulate materials are broken down into smaller and smaller particles. This approach is mostly applicable for solids and dispersed solids. In the bottom-up approach, nanoparticles are built up one atom or molecule at a time. This is applied mostly in the gas or liquid phases. Usually, the nanoparticles obtained by the bottom-up approach are purer and have a better control of particle size and surface chemistry [23]. Different methods, such as sol-gel, hydrothermal, homogeneous precipitation [3], solid-state reaction, sonochemical method and the ultrasonic spray pyrolysis (USP) method, are used to produce nanocomposites [2,14]. The method selection depends on the ease of the method, the type and properties of the nanocomposite, etc., in its preparation. The growth of controlled-size nanoparticles is an intricate task. The reported methods have many drawbacks, since they need complex equipment, higher processing temperature and a longer reaction time [12]. The spray pyrolysis method consists of sequential and continuous processes of nebulization, precipitation, pyrolysis, and sintering to construct particles with homogeneous compositions, allowing for the precise control of solid-state reaction output and chemical composition [24,25]. Compared with the traditional nozzle, an ultrasonicator can nebulize the coating solution into ultra-tiny and foggy droplets between 1 µm and 5 µm with more homogeneous nano- and submicron particle sizes [25]. The USP technique is used for its low cost and its simplicity to implement, to fabricate oxide with good qualities [26].

Lu et al. [27] synthesized a flower-like NiO/ZnO composite by a two-step hydrothermal process, where the NiO nanosheets grew on the surface of the ZnO hexagonal nanorods. Kaur et al. [28] developed a gas sensor based on branch-like NiO/ZnO heterostructures. The synthesis process contained the growth of NiO nanowires on a substrate via the vapour–liquid–solid mechanism, and then the formation of ZnO nanowires directly on the former NiO nanowires using the vapour–solid technique. Zhu et al. [29] synthesized a hierarchical flower-like NiO/ZnO composite via a one-step hydrothermal approach. The gas-sensing properties of the NiO/ZnO composite were investigated via exposure to different ethanol concentrations at various operating temperatures. Li et al. [30] synthesized nanostructured ZnO/NiO microspheres with a nanorods-composed shell and a microsphere yolk via the controlled calcination treatment of bimetallic organic frameworks in air. Kim et al. [31] prepared a three-dimensional (3D) sphere-like structured ZnO–NiO nanocomposites via a simple, one-pot solution process. They investigated the effects of annealing temperatures on the morphological properties of sphere-like structured ZnO–NiO nanocomposites. Mahajan et al. [32] produced NiO/ZnO composite powder using a solid-state reaction method. The dielectric constant of the composite powder was measured using impedance spectroscopy. From room-temperature dielectric measurements, it was observed that, at 1 kHz frequency, the dielectric constant for ZnO, NiO and NiO/ZnO composite powder was 8.3, 43.9, and 14.9, respectively. Li et al. [15] combined NiO and Al_2O_3 into one system and investigated the gas-sensing properties of the composite. The activity species on its surface are centers of adsorption for different gases. Lei et al. [33] present a facile and low-cost method to synthesize hierarchical porous ZnO–Al_2O_3 microspheres through a hydrothermal route. The hierarchical porous ZnO–Al_2O_3 composite has a higher adsorption ability compared with pure ZnO and Al_2O_3. Ullah et al. [34] produced ZnO–Al_2O_3 composite oxides with an improved structure, synthesized by the freeze-drying modified cation–anion couple hydrolysis (CADH) technique and supported by Ni, and it was determined that the desulfurization capacity is high. Li et al. [35] compared NiO/γ–Al_2O_3 nanofibers with TiO_2 nanoparticles, one of the most commonly used photocatalysts. Considering their recyclability and structural integrity, it is understood that NiO/γ–Al_2O_3 may have practical photocatalyst applications in environmental controls such as air/water pollution.

In this study, mixed oxides NiO/ZnO/Al_2O_3 were synthesized via the USP method and a series of tests were conducted to characterize the nanocomposite particles. To the best of our knowledge, it is still a great challenge to utilize a simple and facile route to synthesize the mixed oxide nanocomposite. This study aims to introduce a single-step, facile process route for the production of new, mixed-oxide nanocomposite particles.

2. Materials and Methods

Mixed oxides NiO/ZnO/Al_2O_3 were synthesized via the USP method using an aqueous solution of nitrate salts under a 1 L min^{-1} air flow rate at different temperatures (400, 600, 800 °C). The chemicals used in the preparation of the mixed oxides were high-purity from Sigma Aldrich. The salts used were nickel nitrate hexahydrate ($Ni(NO_3)_2 \cdot 6H_2O$), zinc nitrate hexahydrate ($Zn(NO_3)_2 \cdot 6H_2O$), and aluminium nitrate nonahydrate ($Al(NO_3)_3 \cdot 9H_2O$). The nitrate salts were dissolved in distilled water with a concentration of 0.2 M. Then, the solution was magnetically stirred at room temperature for 15 min at 500 rpm. Experimental parameters of syntheses are summarized in Table 1.

Table 1. Experimental parameters.

Process Temperature (°C)	$Ni(NO_3)_2$ $6H_2O$ (M)	$Zn(NO_3)_2$ $6H_2O$ (M)	$Al(NO_3)_3$ $9H_2O$ (M)	Ultrasonic Frequency (MHz)	Air Flow Rate (L min^{-1})
400	0.2	0.2	0.2	1.3	1
600	0.2	0.2	0.2	1.3	1
800	0.2	0.2	0.2	1.3	1

The experimental setup is illustrated in Figure 1. The set up consists of an ultrasonic atomizer, quartz tube, furnace, and collection chamber. The quartz tube was placed inside the furnace (Nabertherm, R50/500/12–B) with a temperature control of ±1 °C. An ultrasonic nebulizer with a frequency of 1.3 MHz (Ramine Baghai Instrumentation, Pyrosol 7901, Meylan, France) was used for aerosol generation from the precursor solution. The aerosols were carried into the preheated furnace by air. The removal of water occurred in the first sections of the furnace. Continuous thermal decomposition reaction of nitrates to oxides took place in the quartz tube at 400, 600 and 800 °C. The particles were collected in collection chambers in distilled water.

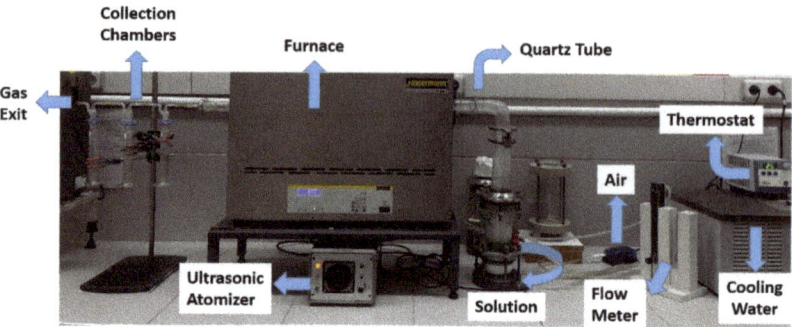

Figure 1. General view of the experimental setup.

The thermodynamics of the decomposition reaction of metal nitrates were investigated by HSC Chemistry software (HSC Chemistry 10, Metso Outotec, Helsinki, Finland). The crystal structures of the synthesized $NiO/ZnO/Al_2O_3$ were analysed with an X-ray diffractometer (XRD: Rigaku MiniFlex, Tokyo, Japan), using Cu K_α radiation (λ = 1.5405 Å). The samples were scanned in the 2θ range of 5–90°. The step angle and scan speed were kept at 0.02° and 2° min^{-1}. Morphology investigations of the specimens were inspected with scanning electron microscopy (SEM: FESEM; JSM 7000F, JEOL Ltd., Tokyo, Japan). The working distance of the samples from the tip of the electron gun and the accelerating voltage was adjusted to 10–15 mm and 5 kV, respectively. Energy dispersive X-ray spectroscopy (EDS: Oxford INCA, Abingdon, UK) was also used to investigate the chemical composition of mixed oxides. Raman spectrums of mixed oxides $NiO/ZnO/Al_2O_3$ composite specimens were measured using Raman spectrometer (Horiba HR800UV, Kyoto, Japan) equipped with a 632 nm laser for sample excitation.

3. Results and Discussion

3.1. Thermodynamic Analysis

The possible reaction equations during the decomposition of metal nitrates to mixed metal oxide particles were assumed as shown in Equations (1)–(3). The Gibb's free energy ($\Delta G°$), the reaction equilibrium constant and reaction equilibrium amounts were computed by HSC Chemistry software for the temperature range of 0–1000 °C. In spray pyrolysis, aerosol droplets of the dissolved salts are carried into the furnace. When the droplets meet the hot zone, the salt precipitates due to the evaporation of solvent, which is water, and then pyrolysis reactions take place. Thus, the reaction thermodynamic was calculated using metal salts. Figure 2 shows the changes in the Gibbs free energies and the logarithmic values of reaction equilibrium constant (Kp; partial pressure of reaction products divided by partial pressure of reactants in equilibrium condition) as a function of temperature.

$$2Ni(NO_3)_2 \cdot 6H_2O \rightarrow 2NiO + 4NO_{2(g)} + 12H_2O + O_{2(g)} \quad (1)$$

$$2Zn(NO_3)_2 \cdot 6H_2O \rightarrow 2ZnO + 4NO_{2(g)} + 12H_2O + O_{2(g)} \quad (2)$$

$$2Al(NO_3)_3 \cdot 6H_2O \rightarrow Al_2O_3 + 6NO_{2(g)} + 12H_2O + 3/2O_{2(g)} \quad (3)$$

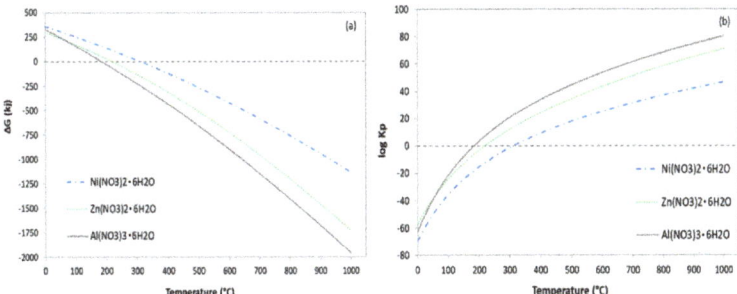

Figure 2. Thermodynamic investigation of the decomposition reactions for Ni, Zn and Al -nitrate salts to oxide forms; changes of (**a**) Gibbs free energy and (**b**) reaction equilibrium constant as a function of temperature.

At the chosen reaction temperatures, the Gibbs free energies for the decomposition of Al-, Zn- and Ni- nitrate salts to their oxides obtain negative values at temperatures above 185, 220 and 314 °C, respectively. Considering the continuous gas flow and their effect on removing the gas reaction products around the nucleated particles, the obtained reactions are spontaneous at the chosen process temperatures.

The changes in reaction equilibrium amounts caused by an elevation of temperature are given in Figure 3. According to the thermodynamic results, the formation of Al_2O_3 starts at a lower temperature than the others, and ZnO nucleation follows. In a tube furnace, as used in the experiments, there is a slight temperature gradient from the entry into the inner zone. Thus, Al_2O_3 should primarily nucleate, and the following decompositions of metal salts can lead to the formation of metal (zinc and nickel) aluminates.

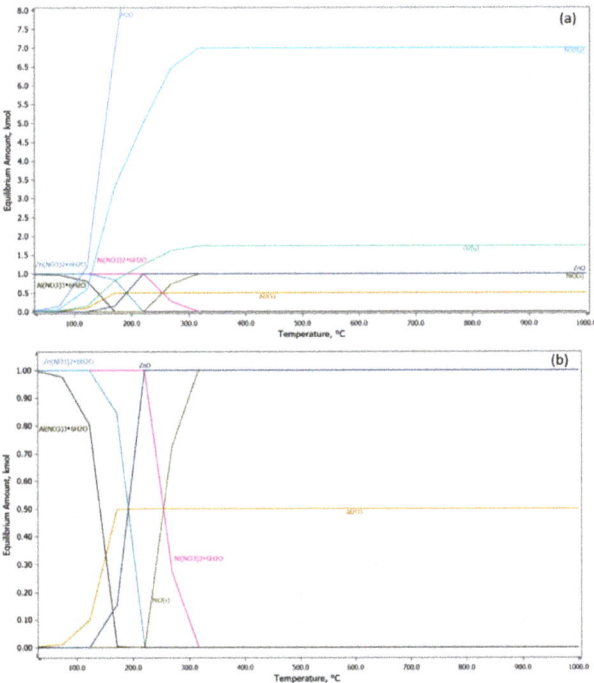

Figure 3. (**a**) Calculated reaction equilibrium amounts for decomposition reactions of metal nitrates. (**b**) Shows the magnified region of the graphic (**a**) (between Equilibrium amounts, kmol 0.00–1.00).

3.2. XRD Analysis

The XRD patterns of samples via produced at different temperatures are shown in Figure 4. X'Pert Highscore Plus was used to assign the phases. The diffraction peaks of mixed oxides were identified and assigned to crystalline alumina (Al_2O_3), a cubic $NiAl_2O_4$ (space group: Fd-3m) and a cubic $ZnAl_2O_4$ (space group: Fd-3m).

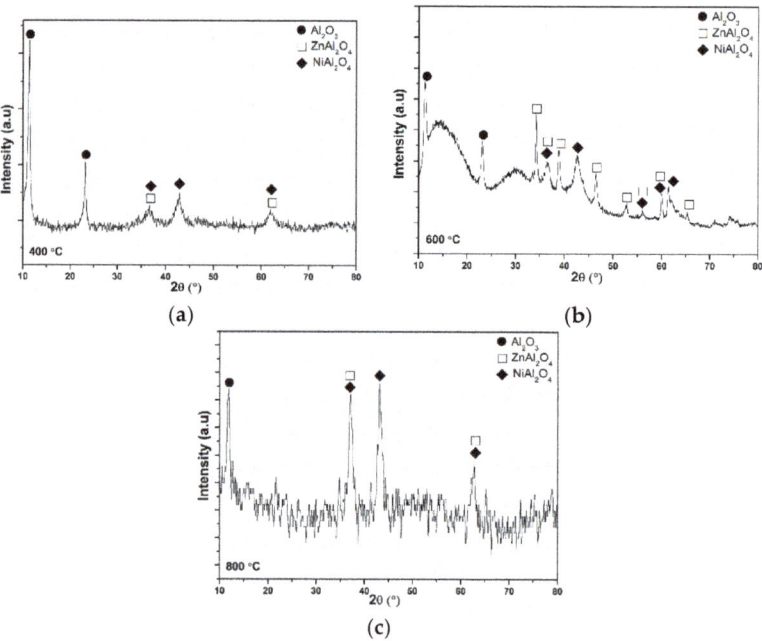

Figure 4. XRD patterns of mixed oxides $NiO/ZnO/Al_2O_3$ obtained with different reaction temperature: (**a**) 400 °C; (**b**) 600 °C; (**c**) 800 °C.

Al_2O_3 peaks' positions were identified by comparing with JCPDS file No. 00-046-1215. The peaks of the Al_2O_3 were the hkl reflection of 010 and 021 at 2θ of 11.09° and 23.68°, respectively. $ZnAl_2O_4$ peaks' positions were identified by comparison with JCPDS file No. 00-005-0669. The characteristic peaks at 2θ of 31.2°, 36.75°, 44.7°, 49.1°, 55.6°, 59.3°, and 65.3° are corresponding to (220), (311), (400), (331), (422), (511), and (440) diffraction planes [18,36,37]. The peaks of the $ZnAl_2O_4$ were the hkl reflection of 311 and 440. $NiAl_2O_4$ peaks' positions were identified by comparison with the diffraction data from the JCPDS file No. 10-0339. The major peaks for a cubic phase $NiAl_2O_4$ (space group Fd-3m) were the hkl reflections of 311, 400, 422 and 440 [17,38–40]. The peaks at 2θ of 36.28°, 42.70°, 55.9° and 62.4° represent $NiAl_2O_4$. It was noted that the peaks of $ZnAl_2O_4$ and $NiAl_2O_4$ were shifted to slightly lower 2θ values compared to those of stoichiometric $ZnAl_2O_4$ and $NiAl_2O_4$. This indicated that spinel in obtained nanocomposites was non-stoichiometric. When the reaction temperature was increasing, the diffraction peaks of the samples were slightly expanded (Figure 4). For $NiAl_2O_4$ and $ZnAl_2O_4$ samples, no peak characteristics of NiO and ZnO were seen, indicating the fine dispersion of these species on the $NiAl_2O_4$ and $ZnAl_2O_4$ supports, respectively, or a possible overlap with the supports' diffraction peaks [41].

3.3. SEM Analysis

SEM–EDS results of mixed oxides $NiO/ZnO/Al_2O_3$ particles, which were produced at different temperatures (400–800 °C) from initial solutions at 0.2 M concentration, are shown in Figure 5.

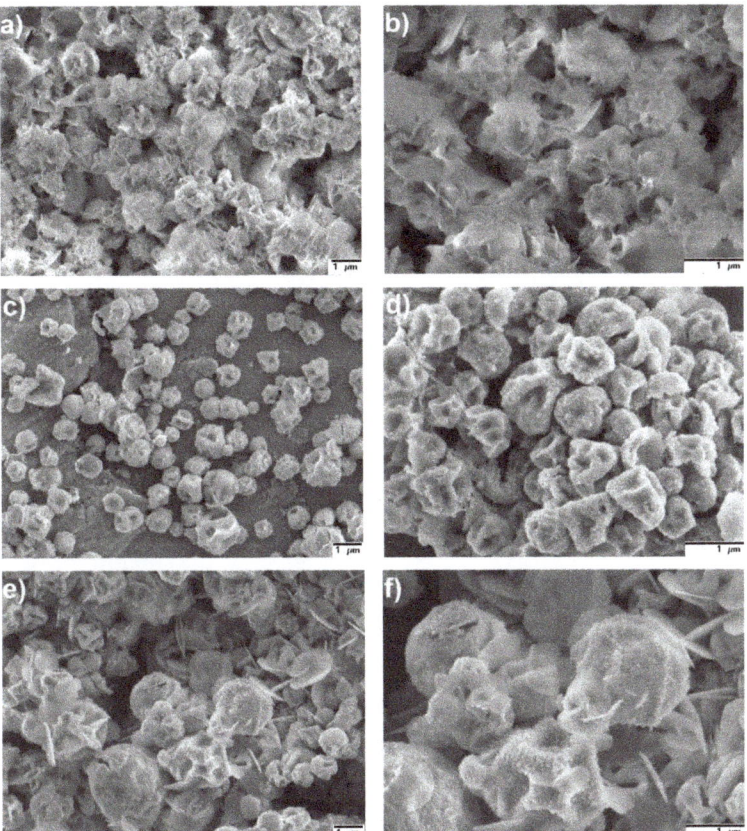

Figure 5. SEM results of the NiO/ZnO/Al$_2$O$_3$ particles: (**a**) ×10,000; (**b**) ×20,000 magnification for the samples of 400 °C; (**c**) ×10,000; (**d**) ×20,000 magnification for the samples of 600 °C; (**e**) ×10,000; (**f**) ×20,000 magnification for the samples of 800 °C.

The SEM images show the mixed oxide particles of nearly spherical and some foliated shapes. The size of the particles also varies: primary particles are in the nano range, and submicron size particles are obtained by the aggregation of primary particles. The significant reason for this difference is the temperature effects on aerosol droplets, decomposition reaction and sintering [42,43]. Additionally, particle formation starts with the nucleation of nano-size aerosol droplets, and then secondary particles form by aggregation due to the sintering effect of temperature. A calculation was carried out using the SEM analysis results with the ImageJ program to find the particle size. The results confirmed that the primary particles were nanometer size (Max. = 26.84 nm; Min. = 4.11 nm; Average = 8.65 nm) [44]. When the images were examined, it was determined that the particles have a foliated morphology at 400 °C (Figure 5a,b). Nearly spherical and hollow particles, which began to reshape through intraparticle sintering, replace the foliated morphology at 600 °C (Figure 5c,d). Starting with the appearance of a very irregular shape, both a spherical and foliated morphology were formed at 800 °C (Figure 5e,f). As the temperature increased, the morphology also changed.

EDS analysis was used to investigate the chemical composition of NiO/ZnO/Al$_2$O$_3$ nanoparticles obtained as a result of experimental studies at different temperatures using 0.2 M solution. The existence of nickel, zinc, aluminum and oxygen as elements in EDS results were determined. In addition, no impurities were detected in the synthesized samples. Table 2 shows the atomic ratios for these analyses.

Table 2. EDS results of NiO/ZnO/Al$_2$O$_3$ particles produced at different temperatures.

Molarity (M)/Temperature (°C)	Element (Atomic %)			
	O	Al	Ni	Zn
0.2 M/400 °C	63.5	7.8	18.0	10.7
0.2 M/600 °C	59.2	10.2	18.3	12.3
0.2 M/800 °C	58.7	9.1	17.1	15.1

3.4. Raman Analysis

Figure 6 shows the results of Raman analyses of NiO/ZnO/Al$_2$O$_3$ nanoparticles produced at 0.2 M from the initial solutions at different temperatures (400, 600, 800 °C).

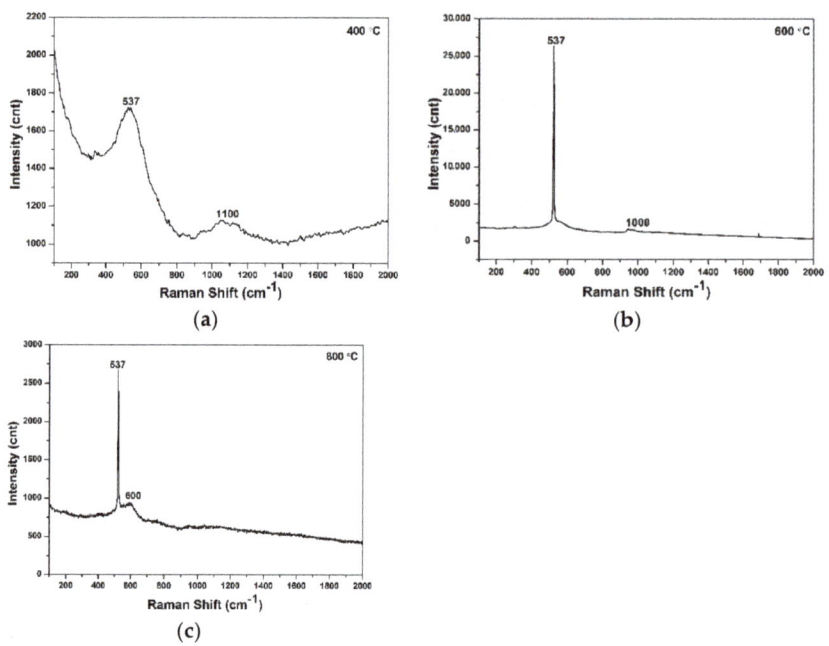

Figure 6. Raman spectra of mixed oxides NiO/ZnO/Al$_2$O$_3$ were obtained with different process temperatures: (**a**) 400 °C; (**b**) 600 °C; (**c**) 800 °C.

It is well known that the infrared spectra of spinels are characterized by absorption bands in the range 400–700 cm^{-1} [45]. A relatively strong absorption band corresponding to the stretching vibration of the atom in the tetrahedral oxygen environment was located at ∼537 cm^{-1} and ∼600 cm^{-1} (Figure 6). NiO has typical emission peaks at ∼1000–1100 cm^{-1} showing the vibrations occur among Ni–O [39]. Then, the Raman peak of spinels strongly increases at 600 °C and 800 °C, and NiO vibrations decrease and disappear. Increasing the process temperature explains the order of magnitude increase in the Raman signal. The Raman peaks support the observation of the coexistence of NiAl$_2$O$_4$ and ZnAl$_2$O$_4$ phases, as reported in the XRD analysis.

4. Conclusions

The present work primarily focuses on the production of mixed oxides NiO/ZnO/Al$_2$O$_3$ nanocomposite particles via the USP method. Secondly, the influence of temperature on both structural and morphology properties of particles was studied. NiO/ZnO/Al$_2$O$_3$

nanocomposite particles were synthesized in a single step and used a non-cost carrier gas of air in the system. The X-ray diffraction pattern indicated an Al_2O_3 phase, $NiO-Al_2O_3$ systems with spinel phase $NiAl_2O_4$, and $ZnO-Al_2O_3$ systems with spinel-phase $ZnAl_2O_4$ [13]. SEM and EDS analyses clearly showed the submicron sized nearly spherical and leafy morphology particles containing Ni, Zn, Al and O elements. It has been determined that the produced $NiO/ZnO/Al_2O_3$ nanocomposite particles are composed of primary and secondary particles. The primary particles are nanometer-size (Max. = 26.84 nm; Min. = 4.11 nm; Average = 8.65 nm). On the other hand, with increasing temperature, the sintering and agglomeration mechanism had the most important effect on particle size and morphology. The optimum operating parameters for $NiO/ZnO/Al_2O_3$ were determined and the process temperature was 600 °C. According to the results of the Raman analysis at 400 °C, a full conversion did not occur and the presence of NiO was detected. However, no phases other than $NiAl_2O_4$ and $ZnAl_2O_4$ were found at 600 °C and 800 °C. For this reason, since conversion is achieved at a lower temperature of 600 °C, it is more advantageous in terms of cost and provides optimum conditions. This study opens up a promising route for high-quality $NiO/ZnO/Al_2O_3$, as well as various other mixed-oxide nanocomposites.

Author Contributions: Conceptualization, D.Y.Ö. and S.G.; methodology, D.Y.Ö. and S.G.; investigation, D.Y.Ö.; writing—original draft preparation, D.Y.Ö., S.G., B.E. and S.S.; supervision, S.G., S.S. and B.F. All authors have read and agreed to the published version of the manuscript.

Funding: This research received no external funding.

Institutional Review Board Statement: Not applicable.

Informed Consent Statement: Not applicable.

Data Availability Statement: Data is contained within the article.

Acknowledgments: Authors would like to thank; Gültekin Göller, Kürşat Kazmanlı, Fatma Ünal and Hüseyin Sezer for XRD, SEM–EDS and Raman analyses.

Conflicts of Interest: The authors declare no conflict of interest.

References

1. Hussain, F.; Hojjati, M.; Okamoto, M.; Gorga, R.E. Review article: Polymer-matrix Nanocomposites, Processing, Manufacturing, and Application: An Overview. *J. Compos. Mater.* **2006**, *40*, 1511–1575. [CrossRef]
2. Ravichandran, K.; Praseetha, P.K.; Arun, T.; Gobalakrishnan, S. Chapter 6—Synthesis of Nanocomposites, Synthesis of Inorganic Nanomaterials Advances and Key Technologies. In *Micro and Nano Technologies*; Bhagyaraj, S.M., Oluwafemi, O.S., Kalarikkal, N., Thomas, S., Eds.; Woodhead Publishing: Swaston, UK, 2018; pp. 141–168. [CrossRef]
3. Juma, A.O.; Matibini, A. Synthesis and structural analysis of ZnO-NiO mixed oxide nanocomposite prepared by homogeneous precipitation. *Ceram. Int.* **2017**, *43*, 15424–15430. [CrossRef]
4. Juma, A.O.; Arbab, E.A.; Muiva, C.; Lepodise, L.M.; Mola, G.T. Synthesis and characterization of CuO-NiO-ZnO mixed metal oxide nanocomposite. *J. Alloy. Compd.* **2017**, *723*, 866–872. [CrossRef]
5. Loos, M. Nanoscience and Nanotechnology. In *Carbon Nanotube Reinforced Composites*; CNR Polymer Science and Technology; Elsevier BV: Amsterdam, The Netherlands, 2015; pp. 1–36.
6. El-Nabarawy, T.; Attia, A.; Alaya, M. Effect of thermal treatment on the structural, textural and catalytic properties of the $ZnO-Al_2O_3$ system. *Mater. Lett.* **1995**, *24*, 319–325. [CrossRef]
7. Sajid, S.; Elseman, A.M.; Huang, H.; Ji, J.; Dou, S.; Jiang, H.; Liu, X.; Wei, D.; Cui, P.; Li, M. Breakthroughs in NiO_x-HTMs towards stable, low-cost and efficient perovskite solar cells. *Nano Energy* **2018**, *51*, 408–424. [CrossRef]
8. Goel, R.; Jha, R.; Bhushan, M.; Bhardwaj, R.; Ravikant, C. Hydrothermally synthesized nickel oxide (NiO) nano petals. In *Materials Today: Proceedings*; Elsevier BV: Amsterdam, The Netherlands, 2021.
9. Sharma, R.K.; Kumar, D.; Ghose, R. Synthesis of nanocrystalline ZnO–NiO mixed metal oxide powder by homogeneous precipitation method. *Ceram. Int.* **2016**, *42*, 4090–4098. [CrossRef]
10. Khudiar, S.S.; Mutlak, F.A.-H.; Nayef, U.M. Synthesis of ZnO nanostructures by hydrothermal method deposited on porous silicon for photo-conversion application. *Optik* **2021**, *247*, 167903. [CrossRef]
11. Tari, O.; Aronne, A.; Addonizio, M.L.; Daliento, S.; Fanelli, E.; Pernice, P. Sol–gel synthesis of ZnO transparent and conductive films: A critical approach. *Sol. Energy Mater. Sol. Cells* **2012**, *105*, 179–186. [CrossRef]
12. Tawale, J.; Dey, K.; Pasricha, R.; Sood, K.; Srivastava, A. Synthesis and characterization of ZnO tetrapods for optical and antibacterial applications. *Thin Solid Films* **2010**, *519*, 1244–1247. [CrossRef]

13. Mohammed, A.A.; Khodair, Z.T.; Khadom, A.A. Preparation and investigation of the structural properties of α-Al$_2$O$_3$ nanoparticles using the sol-gel method. *Chem. Data Collect.* **2020**, *29*, 100531. [CrossRef]
14. Yadav, S.K.; Jeevanandam, P. Synthesis of NiO–Al$_2$O$_3$ nanocomposites by sol–gel process and their use as catalyst for the oxidation of styrene. *J. Alloy. Compd.* **2014**, *610*, 567–574. [CrossRef]
15. Li, X.; Mu, Z.; Hu, J.; Cui, Z. Gas sensing characteristics of composite NiO/Al$_2$O$_3$ for 2-chloroethanol at low temperature. *Sens. Actuators B Chem.* **2016**, *232*, 143–149. [CrossRef]
16. Saedy, S.; Haghighi, M.; Amirkhosrow, M. Hydrothermal synthesis and physicochemical characterization of CuO/ZnO/Al$_2$O$_3$ nanopowder. Part I: Effect of crystallization time. *Particuology* **2012**, *10*, 729–736. [CrossRef]
17. Zygmuntowicz, J.; Wiecińska, P.; Miazga, A.; Konopka, K. Characterization of composites containing NiAl$_2$O$_4$ spinel phase from Al$_2$O$_3$/NiO and Al$_2$O$_3$/Ni systems. *J. Therm. Anal. Calorim.* **2016**, *125*, 1079–1086. [CrossRef]
18. Macedo, H.P.D.; Medeiros, R.L.B.D.A.; Medeiros, A.L.D.; Oliveira, Â.A.S.D.; Figueredo, G.P.D.; Melo, M.A.D.F.; Melo, D.M.D. Characterization of ZnAl$_2$O$_4$ Spinel Obtained by Hydrothermal and Microwave Assisted Combustion Method: A Comparative Study. *Mater. Res.* **2017**, *20*, 29–33. [CrossRef]
19. Zhang, L.; Yan, J.; Zhou, M.; Yang, Y.; Liu, Y.-N. Fabrication and photocatalytic properties of spheres-in-spheres ZnO/ZnAl$_2$O$_4$ composite hollow microspheres. *Appl. Surf. Sci.* **2013**, *268*, 237–245. [CrossRef]
20. Crisan, M.; Zaharescu, M.; Kumari, V.D.; Subrahmanyam, M.; Crişan, D.; DrĂgan, N.; Răileanu, M.; Jitianu, M.; Rusu, A.; Sadanandam, G.; et al. Sol–gel based alumina powders with catalytic applications. *Appl. Surf. Sci.* **2011**, *258*, 448–455. [CrossRef]
21. Chen, L.; Zhang, F.; Li, G.; Li, X. Effect of Zn/Al ratio of Ni/ZnO-Al$_2$O$_3$ catalysts on the catalytic deoxygenation of oleic acid into alkane. *Appl. Catal. A Gen.* **2017**, *529*, 175–184. [CrossRef]
22. Lin, F.; Nordlund, D.; Weng, T.-C.; Moore, R.G.; Gillaspie, D.T.; Dillon, A.C.; Richards, R.M.; Engtrakul, C. Hole Doping in Al-Containing Nickel Oxide Materials to Improve Electrochromic Performance. *ACS Appl. Mater. Interfaces* **2013**, *5*, 301–309. [CrossRef]
23. Sonawane, G.H.; Patil, S.P.; Sonawane, S.H. Chapter 1-Nanocomposites and Its Applications, Applications of Nanomaterials. In *Micro and Nano Technologies*; Bhagyaraj, S.M., Oluwafemi, O.S., Kalarikkal, N., Thomas, S., Eds.; Woodhead Publishing: Sawston, UK, 2018; pp. 1–22.
24. Messing, G.L.; Zhang, S.-C.; Jayanthi, G.V. Ceramic Powder Synthesis by Spray Pyrolysis. *J. Am. Ceram. Soc.* **1993**, *76*, 2707–2726. [CrossRef]
25. Hsieh, H.-C.; Yu, J.; Rwei, S.-P.; Lin, K.-F.; Shih, Y.-C.; Wang, L. Ultra-compact titanium oxide prepared by ultrasonic spray pyrolysis method for planar heterojunction perovskite hybrid solar cells. *Thin Solid Films* **2018**, *659*, 41–47. [CrossRef]
26. Ouhaibi, A.; Ghamnia, M.; Dahamni, M.; Heresanu, V.; Fauquet, C.; Tonneau, D. The effect of strontium doping on structural and morphological properties of ZnO nanofilms synthesized by ultrasonic spray pyrolysis method. *J. Sci. Adv. Mater. Devices* **2018**, *3*, 29–36. [CrossRef]
27. Lu, Y.; Ma, Y.; Ma, S.; Yan, S. Hierarchical heterostructure of porous NiO nanosheets on flower-like ZnO assembled by hexagonal nanorods for high-performance gas sensor. *Ceram. Int.* **2017**, *43*, 7508–7515. [CrossRef]
28. Kaur, N.; Zappa, D.; Ferroni, M.; Poli, N.; Campanini, M.; Negrea, R.; Comini, E. Branch-like NiO/ZnO heterostructures for VOC sensing. *Sens. Actuators B Chem.* **2018**, *262*, 477–485. [CrossRef]
29. Zhu, L.; Zeng, W.; Yang, J.; Li, Y. One-step hydrothermal fabrication of nanosheet-assembled NiO/ZnO microflower and its ethanol sensing property. *Ceram. Int.* **2018**, *44*, 19825–19830. [CrossRef]
30. Li, J.; Yan, D.; Hou, S.; Lu, T.; Yao, Y.; Chua, D.H.; Pan, L. Metal-organic frameworks derived yolk-shell ZnO/NiO microspheres as high-performance anode materials for lithium-ion batteries. *Chem. Eng. J.* **2018**, *335*, 579–589. [CrossRef]
31. Kim, K.H.; Yoshihara, Y.; Abe, Y.; Kawamura, M.; Kiba, T. Morphological characterization of sphere-like structured ZnO-NiO nanocomposites with annealing temperatures. *Mater. Lett.* **2016**, *186*, 364–367. [CrossRef]
32. Mahajan, A.; Deshpande, P.; Butee, S. Synthesis and characterization of NiO/ZnO composite prepared by solid-state reaction method. In *Materials Today: Proceedings*; Elsevier BV: Amsterdam, The Netherlands, 2021. [CrossRef]
33. Lei, C.; Pi, M.; Xu, D.; Jiang, C.; Cheng, B. Fabrication of hierarchical porous ZnO-Al$_2$O$_3$ microspheres with enhanced adsorption performance. *Appl. Surf. Sci.* **2017**, *426*, 360–368. [CrossRef]
34. Ullah, R.; Bai, P.; Wu, P.; Etim, U.; Zhang, Z.; Han, D.; Subhan, F.; Ullah, S.; Rood, M.; Yan, Z. Superior performance of freeze-dried Ni/ZnO-Al$_2$O$_3$ adsorbent in the ultra-deep desulfurization of high sulfur model gasoline. *Fuel Process. Technol.* **2017**, *156*, 505–514. [CrossRef]
35. Li, B.; Yuan, H.; Yang, P.; Yi, B.; Zhang, Y. Fabrication of the composite nanofibers of NiO/γ-Al$_2$O$_3$ for potential application in photocatalysis. *Ceram. Int.* **2016**, *42*, 17405–17409. [CrossRef]
36. Battiston, S.; Rigo, C.; Severo, E.D.C.; Mazutti, M.A.; Kuhn, R.C.; Gundel, A.; Foletto, E.L. Synthesis of zinc aluminate (ZnAl$_2$O$_4$) spinel and its application as photocatalyst. *Mater. Res.* **2014**, *17*, 734–738. [CrossRef]
37. Du, X.; Li, L.; Zhang, W.; Chen, W.; Cui, Y. Morphology and structure features of ZnAl$_2$O$_4$ spinel nanoparticles prepared by matrix-isolation-assisted calcination. *Mater. Res. Bull.* **2015**, *61*, 64–69. [CrossRef]
38. Ragupathi, C.; Vijaya, J.J.; Kennedy, L.J. Preparation, characterization and catalytic properties of nickel aluminate nanoparticles: A comparison between conventional and microwave method. *J. Saudi Chem. Soc.* **2017**, *21*, S231–S239. [CrossRef]

39. Benrabaa, R.; Barama, A.; Boukhlouf, H.; Guerrero-Caballero, J.; Rubbens, A.; Bordes-Richard, E.; Löfberg, A.; Vannier, R.-N. Physico-chemical properties and syngas production via dry reforming of methane over NiAl$_2$O$_4$ catalyst. *Int. J. Hydrog. Energy* **2017**, *42*, 12989–12996. [CrossRef]
40. García-Gómez, N.; Valecillos, J.; Remiro, A.; Valle, B.; Bilbao, J.; Gayubo, A.G. Effect of reaction conditions on the deactivation by coke of a NiAl$_2$O$_4$ spinel derived catalyst in the steam reforming of bio-oil. *Appl. Catal. B Environ.* **2021**, *297*, 120445. [CrossRef]
41. Rajkumar, T.; Sápi, A.; Ábel, M.; Farkas, F.; Gómez-Pérez, J.F.; Kukovecz, Á.; Konya, Z. Ni–Zn–Al-Based Oxide/Spinel Nanostructures for High Performance, Methane-Selective CO$_2$ Hydrogenation Reactions. *Catal. Lett.* **2019**, *150*, 1527–1536. [CrossRef]
42. Schneider, C.A.; Rasband, W.S.; Eliceiri, K.W. NIH Image to ImageJ: 25 Years of image analysis. *Nat. Methods* **2012**, *9*, 671–675. [CrossRef] [PubMed]
43. Mohaček-Grošev, V.; Vrankić, M.; Maksimović, A.; Mandić, V. Influence of titanium doping on the Raman spectra of nanocrystalline ZnAl$_2$O$_4$. *J. Alloy. Compd.* **2017**, *697*, 90–95. [CrossRef]
44. Gurmen, S.; Ebin, B.; Stopić, S.; Friedrich, B. Nanocrystalline spherical iron–nickel (Fe–Ni) alloy particles prepared by ultrasonic spray pyrolysis and hydrogen reduction (USP-HR). *J. Alloy. Compd.* **2009**, *480*, 529–533. [CrossRef]
45. Ebin, B.; Gençer, Ö.; Gürmen, S. Simple preperation of CuO nanoparticles and submicron spheres via ultrasonic spray pyrolysis (USP). *Int. J. Mater. Res.* **2013**, *104*, 199–206. [CrossRef]

Article

Microstructural and Cavitation Erosion Behavior of the CuAlNi Shape Memory Alloy

Tatjana Volkov-Husović [1], Ivana Ivanić [2], Stjepan Kožuh [2], Sanja Stevanović [3], Milica Vlahović [3,*], Sanja Martinović [3], Srecko Stopic [4,*] and Mirko Gojić [2]

1. Department of Metallurgical Engineering, Faculty of Technology and Metallurgy, University of Belgrade, Karnegijeva 4, 11000 Belgrade, Serbia; tatjana@tmf.bg.ac.rs
2. Department for Physical Metallurgy, Faculty of Metallurgy, University of Zagreb, Aleja Narodnih Heroja 3, 44000 Sisak, Croatia; iivanic@simet.hr (I.I.); kozuh@simet.hr (S.K.); gojic@simet.hr (M.G.)
3. Department of Electrochemistry, Institute of Chemistry, Technology and Metallurgy-National Institute of the Republic of Serbia, University of Belgrade, Njegoševa 12, 11000 Belgrade, Serbia; sanjas@ihtm.bg.ac.rs (S.S.); s.martinovic@ihtm.bg.ac.rs (S.M.)
4. IME Process Metallurgy and Metal Recycling, RWTH Aachen University, 52056 Aachen, Germany
* Correspondence: m.vlahovic@ihtm.bg.ac.rs (M.V.); sstopic@ime-aachen.de (S.S.); Tel.: +49-381-11-3370-698 (M.V.); +49-176-7826-1674 (S.S.)

Abstract: Microstructural and cavitation erosion testing was carried out on Cu-12.8Al-4.1Ni (wt. %) shape memory alloy (SMA) samples produced by continuous casting followed by heat treatment consisting of solution annealing at 885 °C for 60 min and, later, water quenching. Cavitation resistance testing was applied using a standard ultrasonic vibratory cavitation set up with stationary specimen. Surface changes during the cavitation were monitored by metallographic analysis using an optical microscope (OM), atomic force microscope (AFM), and scanning electron microscope (SEM) as well as by weight measurements. The results revealed a martensite microstructure after both casting and quenching. Microhardness value was higher after water quenching than in the as-cast state. After 420 min of cavitation exposure, a negligible mass loss was noticed for both samples. Based on the obtained results, both samples showed excellent cavitation resistance. Mass loss and morphological analysis of the formed pits indicated better cavitation resistance for the as-cast state (L).

Keywords: cavitation erosion; optical microscopy; electron microscopy; atomic force microscopy

1. Introduction

Among the variety of advanced materials with exceptional properties and applications, shape memory alloys (SMAs) have a unique ability to return to previously defined shapes or sizes if subjected to the relevant thermal treatment. The memory effect can be reached only in the presence of specific phase transformation, reversible austenite to the martensite phase. The conditions necessary for such phase transformation include mechanical (loading) or thermal (cooling and heating) methods.

Based on the literature, there are several basic types of SMAs, such as Ni-Ti (nitinol), Cu-based, and Fe-based alloys [1,2]. All of the above types have advantages and disadvantages, while economical aspects such as the price can be very important for material selection and application. Precisely, the economic effect (low price) is the main advantage of Cu-based SMAs compared with other SMAs. Namely, these alloys (Cu-Al-Ni alloys) can be applied in various industrial fields, especially when high transformation temperatures are required (near 200 °C), thanks to their high thermal stability and high transformation temperatures.

The selection for application of this group of alloys is affected by their characteristics such as high transformation temperatures (high thermal stability at elevated temperatures, above 200 °C); high corrosion resistance; high resistance to degradation of functional properties during aging processes; and, last but not least, the reasonable cost. Some of the usual applications include the different types of engineering sectors, such as automotive, aerospace, medical and biomedical, and construction [3–5].

Processing routes for the synthesis of Cu-Al-Ni shape memory alloys can be as follows:

1. Casting route (conventional casting route with and without quaternary addition);
2. Powder metallurgy route (mechanical alloying followed by sintering process, mechanical alloying followed by hot pressing and extrusion process, mechanical alloying of pre-alloyed powders followed by hot isostatic pressing, mechanical alloying of elemental and pre-alloyed powders followed by sintering and hot rolling);
3. Rapid solidification processing route;
4. Spray casting route.

In recent years, the continuous casting technique has been one of the most used technologies for the production of SMAs. This method is commonly used thanks to the special competitive growth mechanism of crystal and formation of cast products, which allowed to produce a favorable texture [6–8]. Functional properties strongly depend on microstructural changes of SMAs.

Different types of martensite plates (β'_1, γ'_1) and phases can occur, depending on the chemical composition, production technology, heat treatment, and stress conditions of Cu-Al-Ni shape memory alloys [8–10].

Cavitation presents a complex phenomenon of formation, growth, and condensation of bubbles in fluid flow. When those formed bubbles are transported by fluid flow in the region of pressure higher than the evaporation pressure, they disappear very fast. Cavitation begins even in the presence of positive pressures that is equal or close to the pressure of saturated vapor of the fluid at a given temperature. Extremely large pulses of stress are generated during the collapse of the bubbles, and the rapid repetition of the stress on nearby materials causes severe erosion [11]. The shock waves and micro jets can erode the surfaces of materials under vaporous cavitation conditions. This phenomenon is additionally referred to as cavitation erosion, vaporous cavitation, cavitation pitting, cavitation fatigue, and liquid impact erosion. The result of cavitation is fatigue wear, while cavitation resistance is the ability of a material to confront the degradation caused by cavitation.

Cavitation phenomena were studied in many papers and well described for the materials regarding many applications [12–14]. Cavitation damage test is usually used for metallic materials [15–18] and coatings [17–19], with great interest on material properties' influence [20] such as cobalt alloys [21], WC-12Co coatings [22], and Monel K-500 alloy [19]. Among shape memory alloys, the cavitation erosion was tested primarily for nitinol alloy [23–25] and Fe-based shape memory alloy [26], while for CuAlNi alloy, references were not observed. Moreover, non-metallic materials were investigated for the cavitation resistance behavior, such as ceramics [27–30] and polymer materials [31–33].

Based on the lack of literature data related to the cavitation behavior/degradation of Cu-Al-Ni SMAs, this paper represents the attempt to investigate the cavitation behavior of Cu-Al-Ni SMAs exposed to cavitation.

2. Experimental

2.1. Synthesis

The polycrystalline Cu-12.8Al-4.1Ni (wt.%) shape memory alloy was prepared from pure raw materials of copper, aluminum, and nickel in a vacuum induction furnace. The heating temperature was 1240 °C. A solid bar of 8 mm was produced directly from the melt by means of a device for the vertical continuous casting connected with a vacuum induction furnace. Continuous casting of the bar was carried out with a speed of 320 mm/min (as-cast state, sample L). After the casting, the heat treatment procedure was performed by the solution annealing at 885 °C for 60 min followed by water quenching (quenched state, sample K-2).

2.2. Cavitation Test

Cavitation erosion test was performed using the ultrasonic vibration method (with stationary sample), according to ASTM G32 standard [32]. The experimental set-up is described in detail in previous papers [18,20,30,31]. The gap between the surface of the samples and the transformer probe was 0.5 mm. The test was carried out with distilled water as the medium of room temperature to avoid corrosion effects. Samples' dimensions were diameter of 8 mm and height of 6 mm. The diameter of the sample was the same as the diameter of the solid bar and the height of the sample was adjusted to the followed AFM testing and performance of the holder of the sample. Three replicate specimens were used for the measurements, while each presented result is the mean value of the obtained results.

2.3. Methods for Monitoring the Cavitation Testing

2.3.1. Mass Loss

Mass measurements of the test specimens during the experiment were performed on an analytical balance with an accuracy of ±0.1 mg. Before being weighted, the test specimens were dried in a dryer at 110 °C for an hour.

2.3.2. Optical Microscopy

The surface of specimens was tested by trinocular metallurgical microscope (EUME, EU Instruments, Gramma Libero, Belgrade) using different magnitudes to analyze the effect of the surface erosion.

2.3.3. SEM

Microstructural characterization was performed on prepared metallographic samples. Samples were abraded by different grid emery papers (400–1200); polished in an Al_2O_3 solution; and etched in a solution containing 2.5 g $FeCl_3$, 10 mL HCl, and 48 mL of methanol.

2.3.4. AFM

The surface morphology was investigated by atomic force microscopy (AFM) with NanoScope 3D (Veeco, Santa Barbara, CA, USA) operated in contact mode under ambient conditions. Silicon Nitride probes with a spring constant of 20–60 N/m were used.

2.3.5. Microhardness

Measurement of microhardness of samples before and after heat treatment was performed using the Vickers method (HV10).

3. Results and Discussion

Figure 1 shows optical micrographs of the CuAlNi alloy after both continuous casting and quenching. Solution annealing (followed by quenching in water) as a heat treatment procedure in Cu-based shape memory alloys must be performed in order to achieve order in the alloy's structure, stabilization of the phase transformation temperatures, as well as a fully martensitic microstructure. The grain size of solidified alloys was determined by the amount of undercooling prior to crystallization. In Figure 1, grain boundaries are clearly visualized before and after heat treatment. The results showed that the size of grains increases after solution annealing and quenching in water. As can be seen, the micrographs of specimens (Figure 1) show the typical martensite microstructure. The continuous casting at a cooling rate of 320 mm/min was satisfied with the formation of martensite microstructure. Martensite laths have different orientations into particular grains. It can be explained by the nucleation of groups of martensite plates in numerous places within the grain and the creation of local strain within the grain [33,34].

Figure 1. Optical micrographs of CuAlNi shape memory alloy in the as-cast state (**a**) and after solution annealing at 885 °C/60 min H$_2$O (**b**).

Martensite microstructure obtained by OM was confirmed by SEM micrographs (Figure 2). This microstructure is the result of the beta-phase of CuAlNi alloys transforming into a martensite phase by cooling below the M$_s$-temperature. Martensite appears primarily as needle-like martensite. This microstructure consists of self-accommodating needle-like shape martensite in as-cast state and after heat treatment, which is characteristic for the β'_1 martensite in the CuAlNi alloy [33].

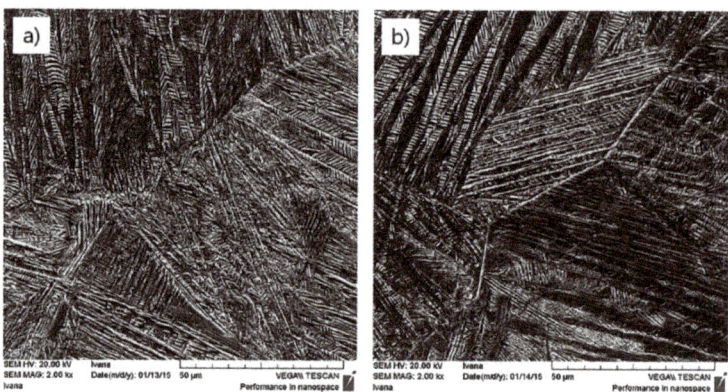

Figure 2. SEM micrographs of CuAlNi shape memory alloy in the as-cast state (**a**) and after solution annealing at 885 °C/60 min H$_2$O (**b**).

Average values of microhardness testing showed that, after quenching in water, microhardness is higher (480 HV10) than that in the as-cast state (344 HV10). Based on the obtained results for the mass loss (Figure 3), the sample after quenching exhibits higher values in comparison with the as-cast state. However, it is important to mention that both samples exhibited excellent resistance to cavitation erosion. After 420 min of exposure to cavitation testing, the mass loss was 0.0014 g for the specimen in a quenched state (Figure 3), while the mass loss was 0.0004 g for the sample after casting.

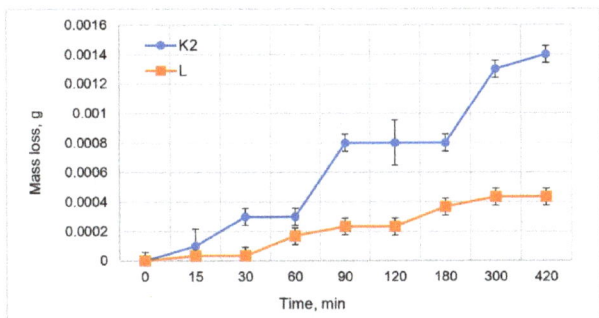

Figure 3. Mass loss during cavitation testing of specimens in the different states, as-cast (L) and quenched (K2), at 885 °C/60 min H$_2$O.

Cavitation erosion is a phenomenon that includes not only properties of liquid, but also the properties of material, for example, hardness, microstructure, grain size of material, and so on. It is known [20] that the material with a homogeneous and fine-grained structure has the highest cavitation erosion resistance, good mechanical properties, and high corrosion resistance. In the literature, there is no information about cavitation resistance testing of CuAlNi shape memory alloys. It was found that the average grain size of samples in as-cast condition was about 150 µm (Figures 1a and 2a), while after quenching, the average grain size was several times higher, up to about 1 mm (Figures 1b and 2b). The finer grain size after continuous casting of CuAlNi alloy resulted in better resistance to cavitation erosion than the sample in heat treated state (Figure 3). Moreover, it was observed that the CuAlNi alloy in as-cast condition is softer (344 HV10) than after quenching (480 HV10), which suggested that resistance to cavitation erosion is better after quenching than that in the as-casted state. This is in contrast to the behaviour of other materials, which show that higher hardness of the materials gives better resistance to cavitation erosion [20]. This area definitely requires further investigation.

In addition to grain size and hardness, resistance to cavitation erosion is related to the microstructure in as-casted and heat treated condition. It is known that CuAlNi shape memory alloys undergo a single transformation ($\beta \rightarrow \beta_1'$ or $\beta \rightarrow \gamma_1'$) or a mixed transformation ($\beta \rightarrow \beta_1' + \gamma_1'$), which depends on the alloy's chemical composition [35]. The previous works [36] confirmed that, after heat treatment in CuAlNi microstructure, along with β_1' martensite, γ_1' martensite also appears, while in the casted state, only β_1' martensite exists. This indicates that the different types of martensite affect alloys' cavitation erosion resistance. The difference in the obtained microstructures before and after heat treatment can be explained by possible changes in martensite morphology. Thus, the mixed martensite microstructure ($\beta_1' + \gamma_1'$) has better resistance to cavitation erosion than the single martensite β_1' microstructure. Self-accommodating zig-zag β_1' martensite benefits to alloys in terms of lower cavitation resistance, while the presence of γ_1' martensite benefits to alloys in terms of better cavitation resistance. This is confirmed by cavitation erosion tests (Figure 3) showing that mass loss of CuAlNi alloy after quenching was lower than for as-cast state. It can be assumed that appearance of γ_1' martensite during cooling has an effect on deformation processes of the CuAlNi alloy. This effect makes this alloy capable of obtaining bigger elastic deformations than for the as-casted condition, in this way delaying the plastic deformation process and fracture, and the consequent weight loss occurs [37]. J. Peña et al. observed that the critical stress for inducing martensite and the capacity of energy absorption related to the different deformation modes are the important parameters to justify the wear resistance of CuZnAl shape memory alloy.

Figure 4 shows the micrographs of microstructure in both as-cast and quenched states, before and after 420 min of exposure to cavitation testing obtained by an optical microscope.

According to those results, changes in microstructure are small and, for better detection, need larger magnification.

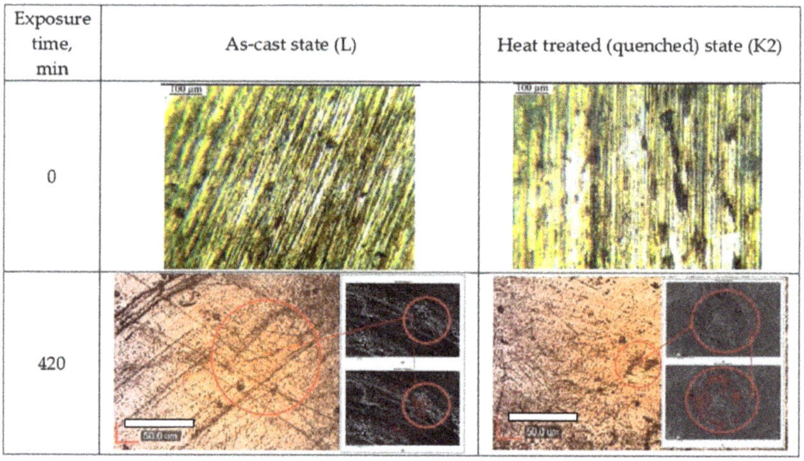

Exposure time, min	As-cast state (L)	Heat treated (quenched) state (K2)
0		
420		

Figure 4. Micrographs of samples before and after 420 min of cavitation exposure.

In Figure 4, the microstructure of the as cast sample (L) after 420 min of cavitation resistance testing is given. The first part of figure was taken by optical microscope, and some parts with pits are observed. Marked typical areas with pits are also given as SEM microphotographs. This marked area is taken for better visibility of the formed pits and is presented separately in Figure 5. The same approach was performed for the sample after heat treatment (K2), and is presented in Figure 4.

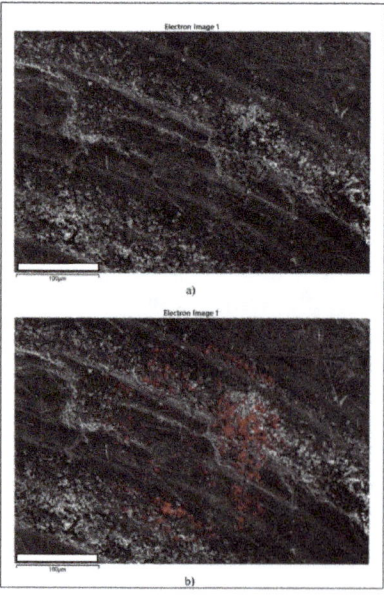

Figure 5. SEM of the as-cast state sample (L) after 420 min of cavitation exposure (the scale bar: 100 μm).

Figures 5a and 6a present SEM micrographs of the sample after 420 min of exposure to cavitation. In Figures 5b and 6b, the damaged area is colored for further image analysis in order to perform the analysis of morphological parameters, which describes the formed pits. The results are given in Table 1.

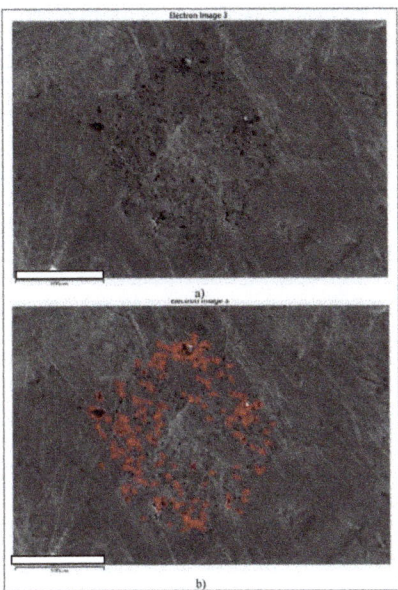

Figure 6. SEM of the sample after quenching and 420 min of cavitation exposure (K2) (the scale bar: 100 μm).

Table 1. Analysis of morphological parameters for the samples after exposure to cavitation testing (420 min).

Parameter	As-Cast State (L)	Heat-Treated (Quenched) State (K2)
	Average Values	
Area, μm	1.235312	36.13774
Diameter (max), μm	1.334329	9.553893
Diameter (min), μm	0.773391	3.257734
Diameter (mean), μm	1.051157	6.288539
Radius (min), μm	0.337882	0.724588
Perimeter, μm	1.858238	34.40823
Perimeter 2, μm	2.086906	47.0361
Fractal dimension	0.009532	1.25286
Perimeter 3, μm	2.147028	38.70619

These results are consistent with the results of mass loss (Figure 3). It has to be taken into account that the obtained results for both samples suggested very good cavitation resistance; however, there are differences in samples' behavior. Moreover, the formation of the typical cavitation ring was detected on the sample after heat treatment (K2), unlike on the sample in the as-cast state (L), which also indicates better cavitation resistance of the sample in the as-cast state (L).

Based on the obtained results for morphological analysis and parameters listed in Table 1, differences between the formed pits after 420 min can be observed. The results for the sample as cast (L) suggested better resistance to the cavitation erosion testing. Average values for observed parameters, such as average values for area, diameter, and perimeter of the formed pits, are smaller compared with values for the heat-treated state.

SEM micrographs of the investigated samples are given in Figure 7, while Tables 2 and 3 provide the results of EDS analysis.

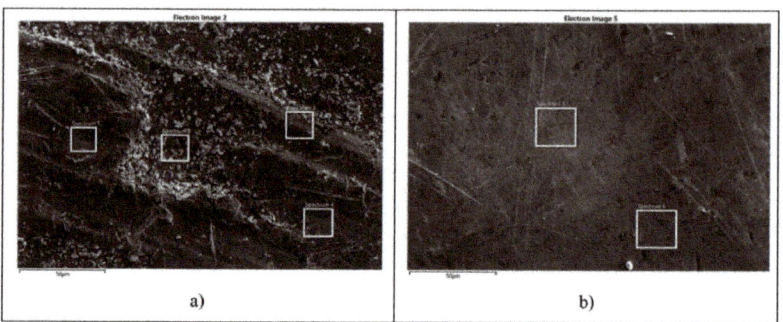

Figure 7. SEM for EDS of the sample after (**a**) 420 min of cavitation exposure of as-cast state (L) and (**b**) quenching and 420 min of cavitation exposure (K2).

Table 2. EDS analysis of as-cast state (L).

Result Type	Weight %			
Spectrum Label	Spectrum 1	Spectrum 2	Spectrum 3	Spectrum 4
L	12.55	12.41	12.93	12.9
Ni	4.73	5.65	4.8	4.62
Cu	82.72	81.94	82.26	82.48
Total	100	100	100	100
Statistics	Al		Ni	Cu
Max	12.93		5.65	82.72
Min	12.41		4.62	81.94
Average	12.7		4.95	82.35
Standard deviation	0.26		0.47	0.33

Table 3. EDS analysis after quenching (K2).

Result Type		Weight %	Weight %
Spectrum Label		Spectrum 5	Spectrum 6
Al		12.38	12.88
Ni		5.38	4.38
Cu		82.24	82.74
Total		100	100
Statistics	Al	Ni	Cu
Max	12.88	5.38	82.74
Min	12.38	4.38	82.24
Average	12.63	4.88	82.49
Standard deviation	0.35	0.71	0.35

According to Figures 4–6, where microstructures with different magnifications are presented, it can be concluded that both samples exhibited very good cavitation resistance. However, some differences between the samples were observed in OM and SEM images, indicating that the samples in the as-cast state exhibited better resistance to cavitation.

An additional test for the surface damage determination was conducted using an atomic force microscope (AFM). Typical two-dimensional (2D) images of the sample in the as-cast state and after cavitation testing are presented in Figure 8.

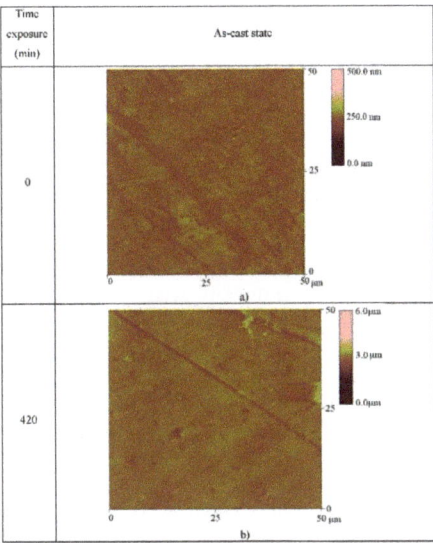

Figure 8. 2D AFM images of the sample in the as-cast state (**a**) before (50 × 150 × 0.5 μm^3) and (**b**) after (50 × 50 × 6 μm^3) cavitation testing for 420 min.

The AFM images of the quenched sample before and after cavitation testing for 420 min are shown in Figure 9. Before cavitation testing, the sample after quenching exhibited roughness of 13.11 nm, but after 420 min exposure to cavitation testing, roughness increased to 66.84 nm.

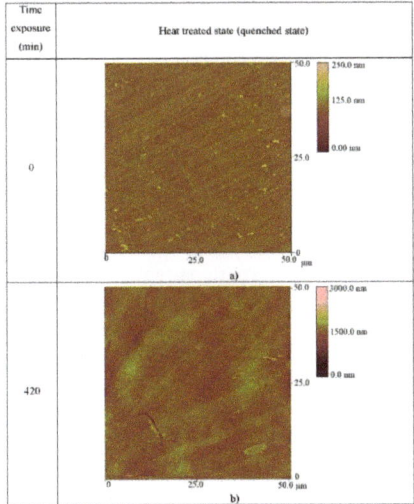

Figure 9. 2D AFM images of the sample after quenching: (**a**) before (50 × 50 × 0.25 μm^3) and (**b**) after (50 × 50 × 3 μm^3) cavitation testing for 420 min.

The results for surface roughness are listed in Table 4.

Table 4. Values of surface roughness of the samples in the different state and at different times of exposure to cavitation testing.

State of Specimens	Surface Roughness (nm) at Time of Exposure of 0 min	Surface Roughness (nm) at Time of Exposure of 420 min
As-cast state	9.33	84.185
Heat-treated (quenched) state	13.11	66.846

This method allows comparing these two samples, and the results for surface roughness pointed out the difference in their microstructure. Although both samples exhibited negligible mass loss difference during cavitation testing, according to the mass loss (Table 3), better cavitation resistance was observed for the sample in the as-cast state (L). Further analysis of morphological parameters given in Table 1 showed that all parameters that describe formed pits have grater values for the heat-treated state using quenching (K2), such as diameter, radius, and perimeter average area. Roughness measurements presented at Table 4 accompanied by the results of AFM (Figure 8) and previous results (mass loss, morphological parameters) before heat treatment showed better cavitation resistance for the sample in the as-cast state (L). For samples after 420 min of cavitation testing, roughness was lower for the heat-treated state. The influence of heat treatment on cavitation behavior of the samples can be related to different grain size, as it was about 150 μm in the as-cast sample (Figures 1a and 2a), while after quenching, the average grain size was several times higher, up to about 1 mm (Figures 1b and 2b). It is well known that grain size has a great influence on mechanical properties, as well as hardness, which can also be related to the cavitation resistance. The hardness values that were higher for the as-cast state confirmed this explanation for the behavior of the samples.

Before cavitation testing, the sample after quenching exhibited a similar value of surface roughness to the sample after casting.

However, after 420 min exposure to cavitation testing, better behavior was observed for the quenched sample (surface roughness of 66.846 nm) compared with the sample after casting (surface roughness of 84.185 nm).

The results of AFM monitoring presented in Table 4 and Figures 8 and 9 confirmed the excellent behavior of the samples during cavitation erosion testing. Differences in roughness were observed in samples before testing, as a higher value of roughness was observed for the heat-treated (quenched) state. After cavitation testing, roughness changed in a way such that the heat-treated sample showed lower roughness values, indicating better cavitation resistance.

The results of morphological analysis and surface roughness, which suggested better cavitation resistance for the as-cast state (L), can be correlated to the smaller grain size and better hardness of the sample.

The difference in the obtained microstructures before and after heat treatment can be explained by possible changes in martensite morphology. It is well known that functional properties of shape memory alloys depend on diffusionless martensite transformation between high temperature β-phase and low temperature martensite phase [35]. The previous work confirmed that, after heat treatment in CuAlNi shape memory alloy, along with β_1' martensite, γ_1' martensite also appears in the microstructure. CuAlNi shape memory alloys undergo a single transformation ($\beta \rightarrow \beta_1'$ or $\beta \rightarrow \gamma_1'$) or a mixed transformation ($\beta \rightarrow \beta_1' + \gamma_1'$), which depends on alloys' chemical composition [36]. It can be concluded that the morphology of different types of martensite affects alloys' cavitation erosion resistance. Self-accommodating zig-zag β_1' martensite provides benefits to alloys in terms of lower cavitation resistance, while the presence of γ_1' martensite provides benefits to alloys in terms of better cavitation resistance.

The obtained results for cavitation erosion can be related to other investigated materials. Compared with ceramic materials [27,29,30,38], it is usual that cavitation exposure time does not exceed 240 min. However, for this time of exposure, the damage level was significantly higher, with more and larger pits created. A similar situation can be observed when metallic materials [39] or coatings were investigated [18,40]. For samples based on carbon steel [39], the exposure time was also 240 min, as for coatings based on nickel and Cr3Si Film [18,40], where the exposure time was 240 and 180 min. For these lower exposure intervals, samples had a higher level of degradation, as well as with the greater number and area of formed pits compared with the results obtained for the investigated samples (for as-cast state (L) and heat-treated (K2)). This comparison suggests that of the SMA materials, the polycrystalline Cu-12.8Al-4.1Ni (wt.%) shape memory alloy exhibited excellent cavitation resistance.

4. Conclusions

Examination of the microstructure of Cu-12.8Al-4.1Ni (wt. %) shape memory alloy reveals martensite microstructure after casting and heat treatment (quenching in water). The grain size of samples is higher after solution annealing and quenching in water than in the as-cast state. Martensite appears primarily as needle-like martensite. This microstructure consists of self-accommodating needle-like shape $\beta'1$ martensite in as-cast state and after heat treatment. Measurements of microhardness showed that, after quenching in water, hardness was higher (480 HV10) than that in the as-cast state (344 HV10). Samples in as-cast and quenched state were investigated in order to measure the cavitation erosion behavior. After an exposure time of 420 min to cavitation erosion testing, very low values of mass loss were measured for both samples (as-cast and quenching state). Based on the obtained results, both samples showed excellent cavitation erosion resistance. Mass loss and morphological analysis of the formed pits pointed out differences between the samples and, based on the obtained results, better cavitation resistance was observed for the as-cast state (L).

Author Contributions: Conceptualization, T.V.-H. and M.G.; Synthesis, S.K. and I.I.; Methodology; S.K., I.I., S.S. (Sanja Stevanović). M.V. and S.M.; Formal Analysis, S.S. (Sanja Stevanović), S.S. (Srecko Stopic). and M.V.; Writing—Review & Editing, M.V. and S.M.; Supervision, M.G.; Data Curation, S.M. and M.V. All authors have read and agreed to the published version of the manuscript.

Funding: This research was supported by the Ministry of Education, Science, and Technological Development of the Republic of Serbia (Contract Nos. 451-03-9/2021-14/200135 and 451-03-9/2021-14/200026) and by the Croatian Science Foundation (Project IP-2014-09-3405).

Conflicts of Interest: The authors declare that they have no conflict of interest.

References

1. Otsuka, K.; Wayman, C.M. *Shape Memory Materials*; Cambridge University Press: Cambridge, UK, 1998.
2. Lexcellent, C. *Shape-Memory Alloys Handbook*; John Wiley & Sons: New York, NY, USA, 2013.
3. Jani, J.M.; Leary, M.; Subic, A.; Gibson, M.A. A review of shape memory alloy research, applications and opportunities. *Mater. Des.* **2014**, *56*, 1078–1113. [CrossRef]
4. Dasgupta, R. A look into Cu-based shape memory alloys: Present scenario and future prospects. *J. Mater. Res.* **2014**, *29*, 1681–1698. [CrossRef]
5. Arlic, U.; Zak, H.; Weidenfeller, B.; Riehemann, W. Impact of Alloy Composition and Thermal Stabilization on Martensitic Phase Transformation Structures in CuAlMn Shape Memory Alloys. *Mater. Res.* **2018**, *21*, 20170897. [CrossRef]
6. Agrawal, A.; Dube, R.K. Methods of fabricating Cu-Al-Ni shape memory alloys. *J. Alloys Compd.* **2018**, *750*, 235–247. [CrossRef]
7. Gojić, M.; Vrsalović, L.; Kožuh, S.; Kneissl, A.; Anzel, I.; Gudić, S.; Kosec, B.; Kliškić, M. Electrochemical and microstructural study of Cu–Al–Ni shape memory alloy. *J. Alloys Compd.* **2011**, *509*, 9782–9790. [CrossRef]
8. Lojen, G.; Gojić, M.; Anžel, I. Continuously cast Cu–Al–Ni shape memory alloy—Properties in as-cast condition. *J. Alloys Compd.* **2013**, *580*, 497–505. [CrossRef]
9. Pereira, E.C.; Matlakhova, L.; Matlakhov, A.N.; De Araújo, C.J.; Shigue, C.; Monteiro, S.N. Reversible martensite transformations in thermal cycled polycrystalline Cu-13.7%Al-4.0%Ni alloy. *J. Alloys Compd.* **2016**, *688*, 436–446. [CrossRef]

10. Wang, Z.; Liu, X.-F.; Xie, J.-X. Effects of solidification parameters on microstructure and mechanical properties of continuous columnar-grained Cu–Al–Ni alloy. *Prog. Nat. Sci.* **2011**, *21*, 368–374. [CrossRef]
11. Hammit, F.G. *Cavitation and Multiphase Flow Phenomena*; McGraw-Hill: New York, NY, USA, 1980.
12. Franc, J.-P.; Michel, J.-M. *Fundamentals of Cavitation*; Springer Science and Business Media LLC: Berlin, Germany, 2005.
13. Okada, T.; Iwai, Y.; Hattory, S.; Tanimura, N. Relation between impact load and the damage produced by cavitation babble collapse. *Wear* **1995**, *184*, 231–239. [CrossRef]
14. Hattori, S.; Mori, H.; Okada, T. Quantitative Evaluation of Cavitation Erosion. *J. Fluids Eng.* **1998**, *120*, 179–185. [CrossRef]
15. Czyzniewski, K.A. Cavitation erosion resistance of Cr–N coating deposited on stainless steel. *Wear* **2006**, *260*, 1324–1332.
16. Zhao, J.; Jiang, Z.; Zhu, J.; Zhang, J.; Li, Y. Investigation on Ultrasonic Cavitation Erosion Behaviors of Al and Al-5Ti Alloys in the Distilled Water. *Metals* **2020**, *10*, 1631. [CrossRef]
17. Szala, M.; Łatka, L.; Walczak, M.; Winnicki, M. Comparative study on the cavitation erosion and sliding wear of cold-sprayed Al/Al2O3 and Cu/Al2O3 coatings, and stainless steel, aluminium alloy, copper and brass. *Metals* **2020**, *10*, 856. [CrossRef]
18. Kazasidis, M.; Yin, S.; Cassidy, J.; Volkov-Husović, T.; Vlahović, M.; Martinović, S.; Kyriakopoulou, E.; Lupoi, R. Microstructure and cavitation erosion performance of nickel-Inconel 718 composite coatings produced with cold spray. *Surf. Coatings Technol.* **2020**, *382*, 125195. [CrossRef]
19. Singh, N.K.; Ang, A.S.; Mahajan, D.K.; Singh, H. Cavitation erosion resistant nickel-based cermet coatings for monel K-500. *Tribol. Int.* **2021**, *159*, 106954. [CrossRef]
20. Zakrzewska, D.E.; Krella, A.K. Cavitation Erosion Resistance Influence of Material Properties. *Adv. Mater. Sci.* **2019**, *19*, 18–34. [CrossRef]
21. Szala, M.; Chocyk, D.; Skic, A.; Kaminski, M. Effect of nitrogen ion implatation on the cavitation erosion resistance and cobalt-based solid solution phase transformations of HIPed Stellite 6. *Metals* **2021**, *14*, 2324.
22. Ding, X.; Ke, D.; Yuan, C.; Ding, Z.; Cheng, X. Microstructure and Cavitation Erosion Resistance of HVOF Deposited WC-Co Coatings with Different Sized WC. *Coatings* **2018**, *8*, 307. [CrossRef]
23. Cheng, F.; Shi, P.; Man, H. Cavitation erosion resistance of heat-treated NiTi. *Mater. Sci. Eng. A* **2003**, *339*, 312–317. [CrossRef]
24. Hattori, S.; Tainaka, A. Cavitation erosion of Ti–Ni base shape memory alloys. *Wear* **2007**, *262*, 191–197. [CrossRef]
25. Yang, L.; Tieu, A.; Dunne, D.; Huang, S.; Li, H.; Wexler, D.; Jiang, Z. Cavitation erosion resistance of NiTi thin films produced by Filtered Arc Deposition. *Wear* **2009**, *267*, 233–243. [CrossRef]
26. Wang, Z.; Zhu, J. Cavitation erosion of Fe-Mn-Si-Cr shape memory alloys. *Wear* **2004**, *256*, 66–72. [CrossRef]
27. Matovic, B.; Maksimovic, V.; Bucevac, D.; Pantic, J.; Lukovic, J.; Volkov-Husović, T.; Gautam, D. Oxidation and erosion behaviour of SiC-HfC multilayered composite. *Process. Appl. Ceram.* **2014**, *8*, 31–38. [CrossRef]
28. Cheng, F.; Jiang, S. Cavitation erosion resistance of diamond-like carbon coating on stainless steel. *Appl. Surf. Sci.* **2014**, *292*, 16–26. [CrossRef]
29. Fatjo, G.G.-A.; Hadfield, M.; Tabeshfar, K. Pseudoplastic deformation pits on polished ceramics due to cavitation erosion. *Ceram. Int.* **2011**, *37*, 1919–1927. [CrossRef]
30. Martinović, S.; Vlahović, M.; Boljanac, T.; Dojčinović, M.; Volkov-Husović, T. Cavitation resistance of refractory concrete: Influence of sintering temperature. *J. Eur. Ceram. Soc.* **2013**, *33*, 7–14. [CrossRef]
31. Algellai, A.A.; Tomić, N.; Vuksanović, M.M.; Dojčinović, M.; Volkov-Husović, T.; Radojević, V.; Heinemann, R.J. Adhesion testing of composites based on Bis-GMA/TEGDMA monomers reinforced with alumina based fillers on brass substrate. *Compos. Part B Eng.* **2018**, *140*, 164–173. [CrossRef]
32. Ashor, A.A.; Vuksanović, M.M.; Tomic, N.; Petrović, M.; Dojčinović, M.; Husović, T.V.; Radojević, V.; Heinemann, R.J. Optimization of modifier deposition on the alumina surface to enhance mechanical properties and cavitation resistance. *Polym. Bull.* **2019**, *77*, 3603–3620. [CrossRef]
33. Obradovic, V.; Vuksanovic, M.; Tomic, N.; Stojanovic, D.; Volkov-Husovic, T.; Uskokovic, P. Improvement in cavitation resistance of poly (vinyl butyral) composite films with silica nanoparticles: A technical note. *Polym. Compos.* **2021**. [CrossRef]
34. ASTM. ASTM Standard G32-98 Standard, Test Method for Cavitation Erosion Using Vibratory Apparatus. In *Annual Book of ASTM Standards*; ASTM: West Conshohocken, PA, USA, 2000.
35. Recarte, V.; Pérez-Landazábal, J.; Rodríguez, P.; Bocanegra, E.; Nó, M.; Juan, J.S. Thermodynamics of thermally induced martensitic transformations in Cu–Al–Ni shape memory alloys. *Acta Mater.* **2004**, *52*, 3941–3948. [CrossRef]
36. Ivanić, I.; Kožuh, S.; Kurajica, S.; Kosec, B.; Anžel, I.; Gojić, M. XRD analysis of CuAlNi shape memory alloy before and after heat treatment. In Proceedings of the Mechanical Technologies and Structural Materials, Split, Croatia, 22–23 September 2016; Croatian Science Foundation: Opatija, Croatia, 2016; pp. 55–60.
37. Peña, J.; Gil, F.; Guilemany, J. Effect of microstructure on dry sliding wear behaviour in CuZnAl shape memory alloys. *Acta Mater.* **2002**, *50*, 3117–3126. [CrossRef]
38. Pošarac-Marković, M.; Veljovic, D.; Devečerski, A.; Matović, B.; Volkov-Husovic, T. Nondestructive evaluation of surface degradation of silicon carbide–cordierite ceramics subjected to the erosive wear. *Mater. Des.* **2013**, *52*, 295–299. [CrossRef]
39. Dojčinović, M.; Volkov-Husović, T. Cavitation damage of the medium carbon steel: Implementation of image analysis. *Mater. Lett.* **2008**, *62*, 953–956. [CrossRef]
40. Jiang, S.; Ding, H.; Xu, J. Cavitation Erosion Resistance of Sputter-Deposited Cr3Si Film on Stainless Steel. *J. Tribol.* **2016**, *139*, 014501. [CrossRef]

Review

Advances in Understanding of the Application of Unit Operations in Metallurgy of Rare Earth Elements

Srecko Stopic * and Bernd Friedrich

IME Process Metallurgy and Metal Recycling, RWTH Aachen University, 52056 Aachen, Germany; bfriedrich@ime-aachen.de
* Correspondence: sstopic@ime-aachen.de; Tel.: +49-176-7826-1674

Abstract: Unit operations (UO) are mostly used in non-ferrous extractive metallurgy (NFEM) and usually separated into three categories: (1) hydrometallurgy (leaching under atmospheric and high pressure conditions, mixing of solution with gas and mechanical parts, neutralization of solution, precipitation and cementation of metals from solution aiming purification, and compound productions during crystallization), (2) pyrometallurgy (roasting, smelting, refining), and (3) electrometallurgy (aqueous electrolysis and molten salt electrolysis). The high demand for critical metals, such as rare earth elements (REE), indium, scandium, and gallium raises the need for an advance in understanding of the UO in NFEM. The aimed metal is first transferred from ores and concentrates to a solution using a selective dissolution (leaching or dry digestion) under an atmospheric pressure below 1 bar at 100 °C in an agitating glass reactor and under a high pressure (40–50 bar) at high temperatures (below 270 °C) in an autoclave and tubular reactor. The purification of the obtained solution was performed using neutralization agents such as sodium hydroxide and calcium carbonate or more selective precipitation agents such as sodium carbonate and oxalic acid. The separation of metals is possible using liquid (water solution)/liquid (organic phase) extraction (solvent extraction (SX) in mixer-settler) and solid-liquid filtration in chamber filter-press under pressure until 5 bar. Crystallization is the process by which a metallic compound is converted from a liquid into a crystalline state via a supersaturated solution. The final step is metal production using different methods (aqueous electrolysis for basic metals such as copper, zinc, silver, and molten salt electrolysis for REE and aluminum). Advanced processes, such as ultrasonic spray pyrolysis, microwave assisted leaching, and can be combined with reduction processes in order to produce metallic powders. Some preparation for the leaching process is performed via a roasting process in a rotary furnace, where the sulfidic ore was first oxidized in an oxidic form which is a suitable for the metal transfer to water solution. UO in extractive metallurgy of REE can be successfully used not only for the metal wining from primary materials, but also for its recovery from secondary materials.

Keywords: rare earth elements; hydrometallurgy; recycling; non-ferrous metals; ultrasonic spray pyrolysis

Citation: Stopic, S.; Friedrich, B. Advances in Understanding of the Application of Unit Operations in Metallurgy of Rare Earth Elements. *Metals* **2021**, *11*, 978. https://doi.org/10.3390/met11060978

Academic Editor: Fernando Castro

Received: 23 May 2021
Accepted: 15 June 2021
Published: 18 June 2021

Publisher's Note: MDPI stays neutral with regard to jurisdictional claims in published maps and institutional affiliations.

Copyright: © 2021 by the authors. Licensee MDPI, Basel, Switzerland. This article is an open access article distributed under the terms and conditions of the Creative Commons Attribution (CC BY) license (https:// creativecommons.org/licenses/by/ 4.0/).

1. Introduction

Hydrometallurgy contains technologies derived from science of geochemistry where metals are extracted into an aqueous solution and subsequently recovered by a variety of methods. The hydrometallurgical unit operations are mostly used for metal recovery from ores, concentrates, and secondary materials: leaching under atmospheric pressure [1], treatment under high pressure in an autoclave [2], bioleaching [3], microwave dissolution [4], dry digestion [5], acid baking [6], filtration [7], neutralization [8], solvent extraction [9], purification of solution using anionic exchange resin [10], cementation [11], precipitation [12–17], crystallization [18], electrocoagulation [19–21], reduction in aqueous phase [22], aqueous electrolysis [23], electrochemical deposition with molten salt electrolysis [24], and ultrasonic spray pyrolysis [25–32].

One simplified combined hydrometallurgical process is shown at Figure 1:

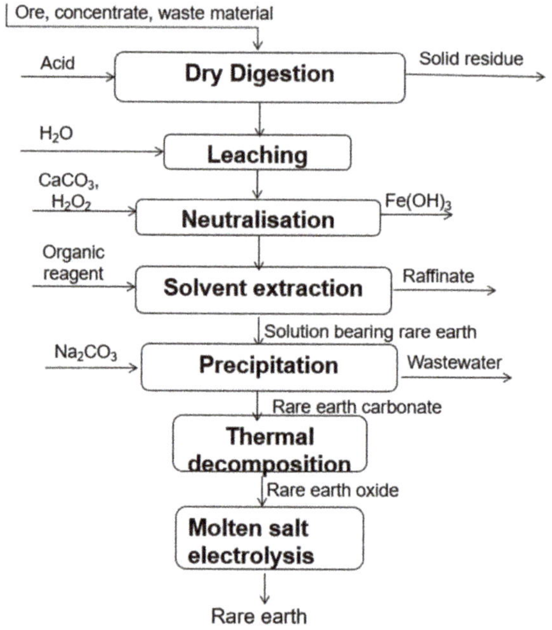

Figure 1. One combined hydrometallurgical and pyrometallurgical process for the production of rare earth elements.

One combined treatment of raw material (ore, concentrate, and waste bearing rare earth) contains a combination of pyrometallurgical operations (thermal decomposition of the obtained precipitate) and hydrometallurgical operations such as dry digestion, leaching under an atmospheric pressure conditions, solvent extraction, precipitation of impurities, and final obtention of rare earth via molten salt electrolysis. As shown at Figure 1, REE are firstly transferred from ores, concentrates and waste materials to the solution. A removal of impurities, such as iron, from the solution was performed via neutralization. The production of rare earth carbonate was reached through an addition of sodium carbonate. The rare earth oxides (REOs) were produced through thermal decomposition of rare earth carbonate above 850 °C. Final step is molten salt electrolysis in a special reactor for production of mixture of REE. Especially using of hydrometallurgical operations has to be adjusted according to strict environmental regulations in order to reach zero waste concept enabling sustainable green metallurgy and circular economy strategy.

Advantages of using of hydrometallurgical operations include: (a) highly selective reactions using leaching agents, (b) reduced energy consumption, (c) no off gas and flue dusts formations, and (d) economical recycling of waste materials. Disadvantages of using of hydrometallurgical operations represent: (a) formation of waste solutions and sludge, (b) small-grained powders are suitable for leaching, (c) formation of wet residue, (d) low reaction velocity, and (e) loss of basic metals.

Special attention is mentioned regarding to a hydrometallurgical treatment of raw materials such as ore, concentrates, and bauxite residues, which contain high amount of silica in order to prevent formation of silica gel, what blocks an extraction of rare earth elements. Alkan et al. [33] developed novel approach for enhanced scandium and titanium leaching efficiency from bauxite residue (BR) with suppressed silica gel formation. After treatment of bauxite residue, new step was a scandium extraction from iron–depleted red mud slags by dry digestion [34].

Borra et al. [35] reported that alkali roasting-smelting-leaching processes allow the recovery of aluminum, iron, titanium, and REEs from bauxite residue. Generally, recovery of critical metals such as REE from primary and secondary materials using new hydrometallurgical operations is open field for new research [36].

Potential-controlled selective recovery and separation of metals from secondary materials was considered as new metallurgical route in order to decrease number of metallurgical operations, but not yet applied in metallurgy of REE. Tian et al. [37] reported manganese and cobalt selective separation from zinc in a leaching solution of cobalt slag by potential-control oxidation with ozone. The separation mechanism of metals was discussed based on the electrochemical consideration. It is concluded that the manganese and cobalt could be oxidized and precipitated from the solution by potential-control respectively, and the manganese could be separated from solution prior to cobalt. Because of similar chemical characteristics of lanthanides, it can be very interesting subject to solve the problem of selective separations of REE without using solvent reactions.

Garg et al. [38] have investigated the chemical leaching of upgraded pyrrhotite tailings from the Sudbury over a wide temperature spectrum (30–80 °C), and in the presence and absence of oxidants such as Fe (III) and oxygen in order to obtain selectively nickel powder. Leaching tests were performed at 30 °C showed that the Ni recovery can be increased during an oxic acid leach in the presence of oxygen, wherein the oxidant Fe (III) is regenerated in-situ as a result of oxidation of Fe (II). However, commencement of leaching with an initial addition of Fe (III) was shown to negatively influence the kinetics of Ni dissolution due to the formation of a protective sulfur coating around unreacted pyrrhotite grain. New experimental setup is presented for this methodology in order to measure and regulate of the redox potential.

Kücher et al. [39] studied the potential of controlling the ORP (oxidation reduction potential) in order to selectively leach the binder metal (commonly cobalt) from a hard metal (WO_3/Co) substrate in a 2.0 mol/L HCl. The obtained results show how the experimental parameters control the nature of reaction mechanism in form of an empirical kinetic model equation. It is concluded that a very small feed rate of H_2O_2 enhances the leaching rate over a longer period of time. The idea of an ORP controlled acidic leaching of cobalt from a hard metal substrate presents a chance to enable semi-direct recycling of other metals from different secondary materials such as spent NdFeB-magnets and active materials from electrodes in Li-ion batteries as a feasible practical option.

Optimization of UO using artificial neural network and regression analysis is a new challenge in order to establish optimal parameters for the maximal leaching efficiency. Especially, due to the complex chemical composition of lateritic ores, as well as the need for decreased production costs and increased of nickel extraction in the existing resources, computer modeling of nickel ore leaching process has seen increased demand. Milivojevic et al. [40] applied the design of experiments (DOE) theory determining the optimal experimental design plan matrix based on the D optimality criterion. In the high-pressure sulfuric acid leaching (HPSAL) process for nickel laterite in "Rudjinci" ore in Serbia, the temperature (T), sulfuric acid to ore ratio (c), stirring speed (v), and leaching time (t) as the predictor variables (X_0, X_1, X_2, X_3, X_4) and the degree of nickel extraction as the response have been considered, as shown at Figure 2.

To model the process, the multiple linear regression (MLR) and response surface method (RSM), together with the two-level and four-factor full factorial central composite design (CCD) plan, were successful used to predict nickel leaching efficiency. Hernandes [41] developed a statistical model for the recovery of rare earth elements in a leaching process, from Chilean ores, using the neural networks technique. The complexity for the elaboration of predictive and mathematical models for the conventional leaching process, besides the difficulty of obtaining REE from low-grade minerals; represent a big challenge for the development of theoretical studies. According to results shown in his work and previously by Milivojevic et al. [40], they proposed that this methodology of artificial neural networks can be used to determine the degree of recovery of a species of interest, according to the most important variables of the process. Milivojevic et al. [42] reported forward stepwise regression in determining dimensions of forming and sizing tools for self-lubricated bearings using the same mathematical tools. These methods can be always used for an optimization of different UO.

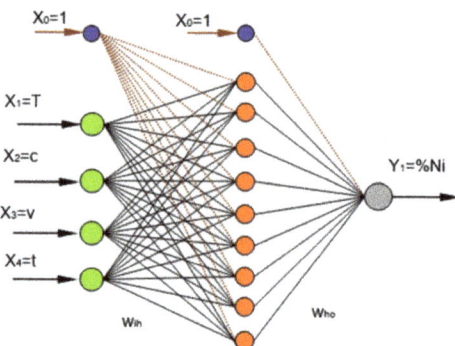

Figure 2. Signal flow in BP neural networks with four input neurons, nine hidden neurons, and one output neuron.

REE represent series of elements with increasing atomic numbers that begins with lanthanum or cerium and ends with lutetium together with yttrium. They have a wide range of uses in technological products and applications. Due to the increased demand and supply risk, most REEs have been added to the list of critical metals. The production of REEs from primary resource causes environmental problems [43]. The recovery of REEs from waste materials is the most suitable strategy to find the solution of environmental problems and ensure the sustainability for production of REE raw materials in the future, according to an increased demand in industrial application. Most developed countries are importing REEs from China; 95% of REEs are supplied from China and in addition to this situation, export quotas of REEs applied by China have increased the export prices of REEs [44].

In order to produce rare earth oxides (RE-oxides), most researchers have studied different hydrometallurgical and pyrometallurgical strategies in recycling processes of REE aiming at its higher extractions as new sustainable metallurgical route [45,46]. Önal et al. [47] studied recycling of NdFeB magnets using sulfation, selective roasting, and water leaching, enabling the production of a liquid with at least 98% rare earth purity. Furthermore, 98% extraction efficiency of REEs from NdFeB magnets was obtained by the acid-baking process with nitric acid [48]. After the acid baking process and subsequent water leaching of the treated concentrate, the produced suspension was filtrated in order to separate a pregnant leaching solution. To produce the REE oxides from leach liquor, all the proposed methods in the literature are completely based on precipitation methods by using various precipitation agents such as sodium carbonate and oxalic acid [49,50].

It is known that RE-carbonate or RE-oxalate can be produced from impurities present in sulfuric liquors using oxalic acid and sodium carbonate by a precipitation method [51]. It was reported that high purity RE-oxide (99.2%) was achieved using oxalic acid as a precipitation agent. Relatively lower purity RE-oxide was produced using sodium carbonate during precipitation [52]. The precipitation behavior of REEs with precipitation agents—including oxalate, sulfate, fluoride, phosphate, and carbonate—was examined using thermodynamic principles and calculations [53]. Construction of Eh-pH diagrams was enabled using different software such as FactSage. It was found that the pH of the system, types of the precipitation agent, and present anions in the leach liquor have a noteworthy impact on the purity of the REE precipitants. In contrast to the precipitation method, the production of nanosized RE-oxide such as Y_2O_3, La_2O_3 Gd_2O_3, and CeO_2 is mostly performed from synthetic solutions via an ultrasonic spray pyrolysis method [54–57]. Use of a real waste solution for production of nanosized REO is missing in literature.

Our aim in this review is to show different developed processes for the treatment of primary and secondary raw materials using mostly hydrometallurgical operations in order to obtain rare earth oxide, rare earth carbonate and mixture of rare earth in metallic form. Sometimes, a combination of hydrometallurgical and pyrometallurgical processes is required to obtain final product.

This review paper summarizes the application of the UO in metallurgy of REE starting from ores (Lovozero, Russia), concentrates (Eudialyte, Greenland and Bastnasite, Norway) and secondary materials (spent NdFeB-magnets, Ti-Al ceramics, acid mine drainage solution (AMD), bauxite residue (BR), phosphorgypsum, and coal flying ash (CFA)) in order to present the novelty in using of the combined UO.

2. Application of the Unit Operations in Primary Metallurgy of REE

2.1. Treatment of an Eudialyte Concentrate with Precipitation and Solvent Extraction

Eudialyte is a Na-rich zirconosilicate with varying amounts of Ca, Fe, Mn, REE, Sr, Nb, K, Y, and Ti (chemical formula: $Na_4(Ca,Ce)_2(Fe^{2+},Mn,Y)ZrSi_8O_{22}(OH,Cl_2)$). It was first described in 1819 by German chemist Stromeyer, who studied the samples from the Ilimaussaq complex in Greenland. The term "eudialyte" introduced by Stromeyer comes from the Greek language and refers to the good solubility of this mineral in acids.

Intensive studies of possible way of eudialyte processing have been conducted in several universities in Russia [58–64]. Most of them used eudialyte concentrate from the Lovozero massif of Kola Peninsula. Because of low REE-content in ore, in these investigations REE are considered mainly as secondary important products while recovery of zirconium was the main aim. The ore from Lovozero contains around 0.6–0.7 wt. % REO, 1.7–2.2 wt. % ZrO_2, and 51–52.5% SiO_2, after beneficiation, which includes flotation and electromagnetic separation, eudialyte concentrate is produced which contains (wt. %): 1.8–2.5 REO, 9.8–17 ZrO_2, and 45.7–52 SiO_2. According to literature [58–64] the mostly hydrometallurgical treatment was performed using sulfuric acid, as shown at Figure 3.

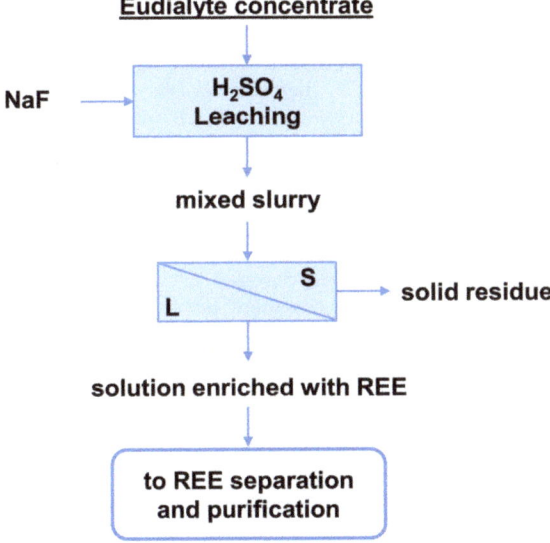

Figure 3. Treatment of eudialyte concentrate using sulfuric acid.

According to this scheme, products of the leaching are solution, where all REE are transferred, and solid residue of gangue material. An addition of NaF was aiming an avoiding of silica gel formation. By adjustment of the pH value of the solution containing REE impurity metals such as Al, Fe, and Zr can be precipitated from the solution. According to Litvinova [62] reaction of dissolution of eudialyte in sulfuric acid, as shown with Equation (1):

$$Na_{16}Ca_6Fe_2Zr_3Si_{26}O_{73}Cl_2 + 21H_2SO_4 = 7Na_2SO_4 + 6\,CaSO_4 + 2FeSO_4 + 3Zr(SO_4)_2 + 26SiO_2 + 21H_2O + 2NaCl \quad (1)$$

Experimental results confirmed ability to provide control of the silica in the solution during leaching with concentrated acid at 100 °C obtaining good filterable slurry. Davris et al. [65] dissolved eudialyte structure forming soluble metal salts and a secondary siliceous precipitate, leaving the remaining gangue minerals such as aegirine and feldspars intact during the two-stage treatment. Upon fuming pretreatment silica precipitates in an insoluble form generating a filterable sludge in the subsequent water leaching step. The most efficient process parameters during the proposed two stage treatment were found to be: 2M HCl solution addition to a heated concentrate at 100 °C, S/L ratio of 1/5 followed by water leaching of the treated concentrate at 30 °C, solid/liquid (S/L) ratio 1/10 for 1 h, resulting to 97% REE recovery. Using 2M H_2SO_4 solution addition to a heated concentrate at 110 °C for fuming, S/L ratio of 1/4 followed by water leaching of the treated concentrate at ambient temperature, S/L ratio 1/20 for 30 min, resulting in 91% REE recovery.

In contrast to previous work, Ma et al. [66] developed one environmentally friendly hydrometallurgical treatment of eudialyte in laboratory and scale up conditions using acid digestion with hydrochloric acid at room temperature, water leaching, and precipitation of impurities with calcium carbonate, as shown at Figure 4.

Mixing of the eudialyte concentrate with the concentrated hydrochloric step ('acid digestion') represents the first step. This strategy was firstly performed in laboratory conditions in 1 L reactor. After that this strategy was validated in scale up conditions using acid digestion in reactor of 40 L, as shown at Figure 5a and system of two digestion reactors, as shown in Figure 5b.

After the acid digestion process, an additional injection of water leads to further leaching of REE. An innovative step is very fast transport of formed suspension using an injected tube for a transport of suspension supported by a double membrane pump (approx. 8 L/min). The final products of the leaching process are the solution that is enriched with REE and other impurities, and solid residue with zirconium, and small amount of remained REE. Therefore, a washing step of the solid residue is required in order to increase the leaching efficiency and it is adopted in this leaching strategy. The obtained suspension was treated at the demonstration plant for the unit operations (neutralization, filtration, precipitation, and washing of final product of the REE-carbonate) developed by MEAB Chemie Technik GmbH Aachen and by the Institute of Process Metallurgy and Metal Recycling of the RWTH Aachen University (IME, RWTH Aachen) through work at the EURARE Project between 2013 and 2017, as shown at Figure 6.

Project designs for acid digestion of eudialyte concentrate contained different experiments three trials at the beginning according to the experience with different reaction parameters including the next optimal parameters: the ratio of HCl: eudialyte concentrate, weight of water and concentrate, leaching temperature, and reaction time during acid digestion process [66].

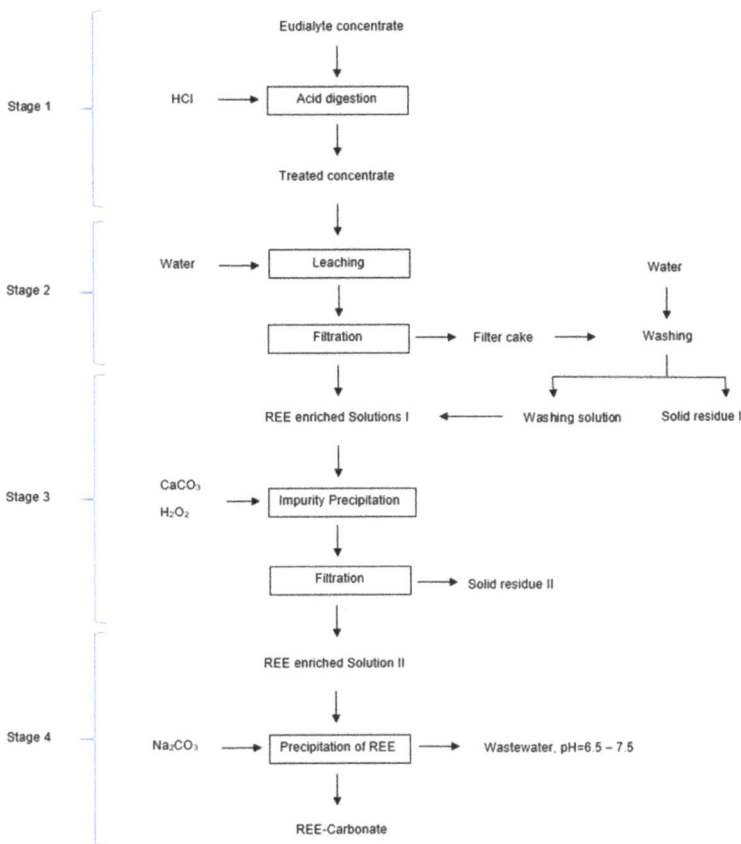

Figure 4. Treatment of eudialyte concentrate with hydrochloric acid at room temperature.

Figure 5. (**a**) Acid digestion reactor and (**b**) system of two digestion reactor during operations (key elements: anchor stirrer, glass reactor (40 L), engine, an injected tube for a transport of suspension supported by a double membrane pump).

(a) (b)

Figure 6. (a) Two reactors for leaching (2 × 100 L), (b) System for leaching, neutralization, and filtration. Main components: neutralization unit (3 reactors × 10 L and 1 × 8.5 L), stirred collecting tank (250 L) and separation unit (chamber filter press).

The hydrochloric acid for acid digestion (cHCl) to eudialyte concentrate ratio, mass of water for leaching (cH$_2$O) to mass of eudialyte concentrate ratio, leaching temperature (T) and leaching time (t) as the predictor variables, and the total rare earth elements (TREE) extraction efficiency as the response were in laboratory and scale up conditions by Ma et al. [67] considered in detail. After experimental work in laboratory conditions, according to design of experiment theory (DoE), the modeling process was performed using multiple linear regression (MLR), stepwise regression (SWR), and artificial neural network (ANN). The ANN model of REE extraction was adopted. Additional tests showed that values predicted by the neural network model were in very good agreement with the experimental results. Developed model shown with Equation (2) confirms the results of correlation analysis, that the leaching temperature (T) is not significant. Its changes throughout the range from 20 °C to 80 °C, cause a decrease of TREE extraction percentage only in a very small range from 0.20% to 0.81%, respectively.

$$\text{TREEeff.} = 59.9058 + 13.0437 \cdot c\text{HCl} + 1.5291 \cdot c\text{H}_2\text{O} - 0.010197 \cdot T + 0.0602969 \cdot t \quad (2)$$

After an optimization of acid digestion process new step at the IME, RWTH Aachen was a construction of new acid digestion reactor for an eudialyte concentrate for treatment in industrial conditions [68]. The development of a 100 L modular reactor for the dry digestion of highly silicone-rich ores and concentrates to prevent gel formation using the example of eudialyte concentrates is jointly realized by the IME, RWTH Aachen and Konzept GmbH, Engineering Services Düren, as shown in Figure 7.

The innovation of this reactor compared to process alternatives is the complete avoidance of external and cost-intensive heating energy, as well as the previously impossible implementation of the dry digestion process in large solution volumes [68]. In this process, it is possible to convert the silicon components into crystals prior to silica gel formation, which are filterable and stable. In order to separate heavy and light rare earth elements, the combination of dry digestion, precipitation, filtration, and solvent extraction is shown in Figure 8.

As shown at Figure 8, the solvent extraction ("SX") is applied for the enrichment of REE in poor solutions and the separation and refinement of metals. This is achieved by mixing an aqueous phase with an organic phase. The driving force of the extraction process is the different metals concentration between the aqueous and the organic solution. In contrast to eudialyte, other REE- minerals such as monazite [69], steenstrupine [70], xenotime [71], and bastnasite [72–75] contain radioactive elements such as thorium and uranium. Therefore, an initial treatment is needed, as shown for bastnasite in Figure 9.

Figure 7. New developed dry digestion reactor (main components: reaction vessel, system with nozzles for injection of water and acid; cover; special mixing system, engine, discharging system, and electronics).

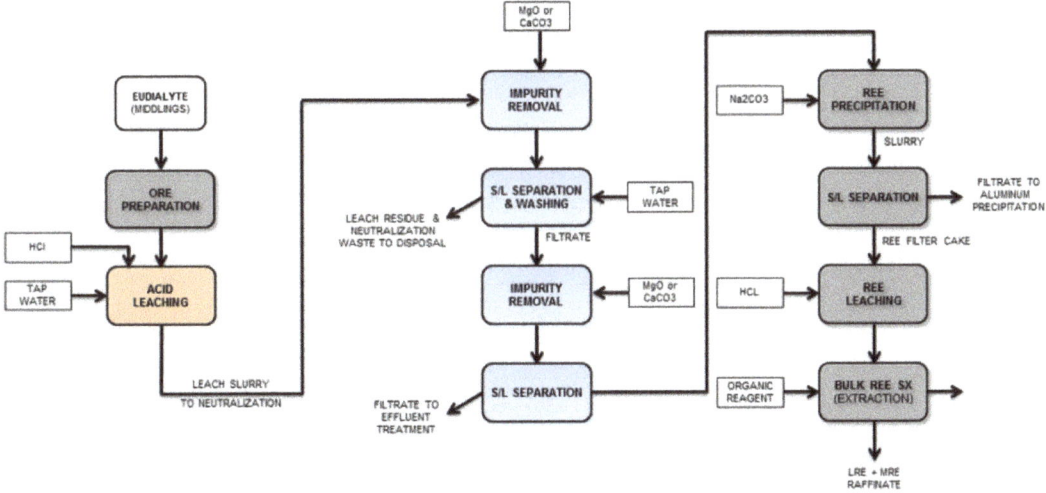

Figure 8. Treatment of eudialyte concentrate by MEAB Chemie Technik GmbH and IME, RWTH Aachen.

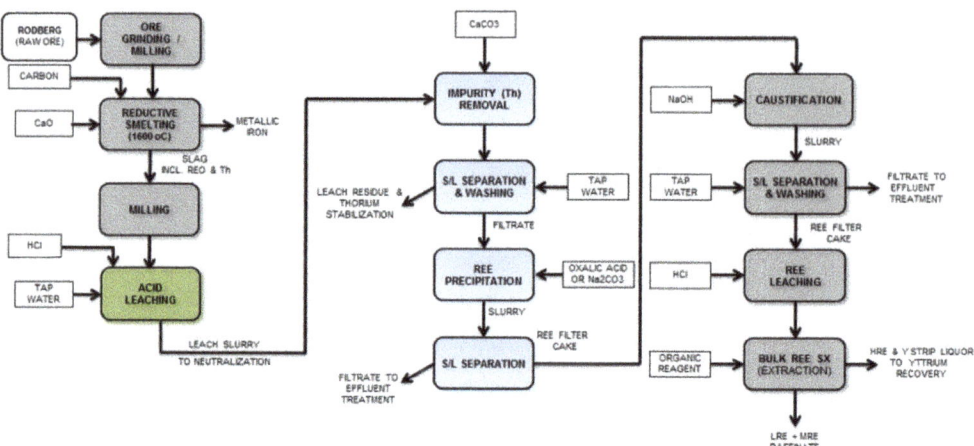

Figure 9. Treatment of bastnasite concentrate by MEAB Chemie Technik GmbH and IME, RWTH Aachen.

After grinding of ore, the smelting of bastnasite ore, Norway [74] was carried out in a small electric arc furnace and carbon was added as reducing agent. The furnace was kept at 1600 °C for 45 min. After this time slag formed at the top of the furnace was poured onto a ladle and left to cool. The leaching of slag (approx. 8.1% rare earth oxides) was performed using hydrochloric acid. Thorium was precipitated from solution using sodium hydrogen phosphate. The fully purification of obtained pregnant solution was performed using solvent extraction process producing raffinate of light rare elements (LRE) with middle rare earth (MRE). The second product is heavy rare earth (HRE) and yttrium strip liquor.

2.2. Molten Salt Electrolysis

In order to find one of most effective alternatives to the carbon reductant, molten salt electrolysis has enormous potential for the REE production. Oxide-fluoride electrolysis at 1050 °C is one of the dominant technologies to produce REE or alloys. Neodymium (Nd) and praseodymium (Pr) are two REE that have significant applications as magnets, such as those used in wind turbines. These elements have very similar properties, making them very difficult to separate using traditional methods, such as solvent extraction. Therefore, alternative methodologies were investigated under the EURARE-Project. The synthesis of Nd and Pr as a metal alloy ('didymium') was achieved at RWTH Aachen [76]. Electrolysis trials revealed that alloy composition is directly influenced by electrolyte composition (i.e., the activity concentration is transferred to the composition). An increase in Pr content in the electrolyte leads to an increased content in the alloy, where the obtained alloy had high purity (approx. >99%). Vogel et al. [77] studied reducing greenhouse gas emission from the Nd_2O_3. Electrolysis starting with an analysis of the anodic gas formation and calculating the theoretical voltages of formation of the relevant anodic gas products. The concluded the behavior of the electrochemical system at linear voltammetries of NdF_3-LiF-Nd_2O_3 shows a first passivation, termed the partial anode effect, where the current density falls turbulent with the increasing voltage.

In order improve an environmental impact, Vogel et al. [77] reported basics of a process control avoiding perfluorocarbons (PFC) emission, because they are potent greenhouse gases and are not filtered or destroyed in the off-gas. Because of the neodymium electrolysis produces unnecessary high emission of CF_4 and C_2F_6, a required process control was considered in analogy to the aluminum electrolysis reducing the PFC emission to a great extend and keep the process in a green process window. The feeding of 1% Nd_2O_3 has an immediate effect on the cell voltage and gas composition. Numerical simulation was performed by Haas et al. [78] for a comparison between two cell designs for electrochemical neodymium reduction. The numerical model uses the Eulerian volume of fluid approach to track phase boundaries between the continuous phases, while the Lagrangian discrete phase model is applied to compute the rising trajectories of emitted off-gas bubbles. Aiming an understanding of reduction process, Cvetkovic et al. [79] considered mechanism of Nd deposition onto W- and Mo-cathodes from molten oxide-fluoride electrolyte. The reported results indicated that the Nd(III) ions in the melts were reduced in two steps: Nd(III) → Nd(II) and Nd(II) → Nd(0). These consecutive processes are predominantly mass transfer controlled.

3. Application of the Unit Operations in Secondary Metallurgy

3.1. Recovery of REE from Waste of Electrical and Electronic Equipment (WEEE or e-Waste)

The production of REEs from primary resource causes environmental problems [80]. The recovery of REEs from waste materials is the most suitable strategy to find the solution of the growing environmental problems and ensure the sustainability for production of REE raw materials in the future, according to an increased demand in industrial application. Most developed countries are importing REEs from China (approx. 95%) and in addition to this situation, export quotas of REEs applied by China have increased the export prices of REEs [81]. Our aim is to give review about recovery of REE from secondary resources such as spent NdFeB-magnets, Ti-Al ceramic waste, acid mine drainage (AMD) solution, bauxite residue (BR), and coal flying ash (CFA).

Waste of electrical and electronic equipment (WEEE or e-waste) is dramatically increased in last time (approx. 50 million tons in 2018). Unfortunately, unsafe WEEE disposal leads to many health implications due to their hazardous nature, being composed of substances such as chlorofluorocarbons. Ambay et al. [82] analyzed challenges in the future perspective for the recovery of REEs from WEEE wastes which is increasing at an alarming rate all over the world. The research progress for REE recovery was found through bioleaching, biosorption, hydrometallurgy, pyrometallurgy, and reduction with carbon-based material. Because of growing industrial application of NdFeB magnets, highly efficient and selective recycling is mostly analyzed as WEEE. Yang et al. [83] written a critical review for REE recovery from end-of-life (EOL) NdFeB permanent magnet scrap. Most of the processing hydrometallurgical methods are still at various research and development stages. It was estimated that until 2030 the recycled REEs from EOL permanent magnets will play a significant role in the total REE supply in the magnet sector, providing that most efficient technologies will be developed and implanted in industrial practice. The selectivity of separation REE from Fe and B was high challenge in this study. Solvent extraction of Pr and Nd from chloride solution was performed by the mixtures of Cyanex 272 and amine extractants [84] and with organophosphorous acids and amine reagents [85] using MSU-0.5 mixer-settler units in four stage counter-current setup, MEAB Metallextraction AB, Göteborg, Sweden [86]. The used experimental setup reached complete scrubbing of Nd and Pr.

Kaya et al. [87] offered new process for treatment of spent NdFeB magnets without solvent extraction. Spherical particles of REE-oxides were produced from spent NdFeB magnets using a combined process containing following operations: nitric acid baking process at 200 °C, calcination, water leaching and ultrasonic spray pyrolysis between 700 °C and 1000 °C, as shown at Figure 10.

Iron was removed from water solution via a hydrolysis. XRD analysis of the obtained particles found a cubic and trigonal structure Nd_2O_3 with 20% Pr_2O_3, which is according to detected stoichiometry in solution after dissolution of spent NdFeB magnets. An increase in temperature from 700 °C to 1000 °C increases not only the crystallinity of the structure, but also the particle size. The minimal theoretical total particle size of prepared REE-oxides amounts to 215 nm. The differences between calculated and experimentally obtained values may be partially due to coalescence/agglomeration of aerosol droplets during transport to the furnace from an aerosol generator. This combined environmentally friendly process for recovery of nanosized powder mixture of Nd_2O_3 and Pr_2O_3 from spent magnets and re-use of nitric acid shall be reduced considered in scale up conditions. The final winning of the mixture of metallic Nd and Pr will be ensured using molten salt electrolysis [88–90].

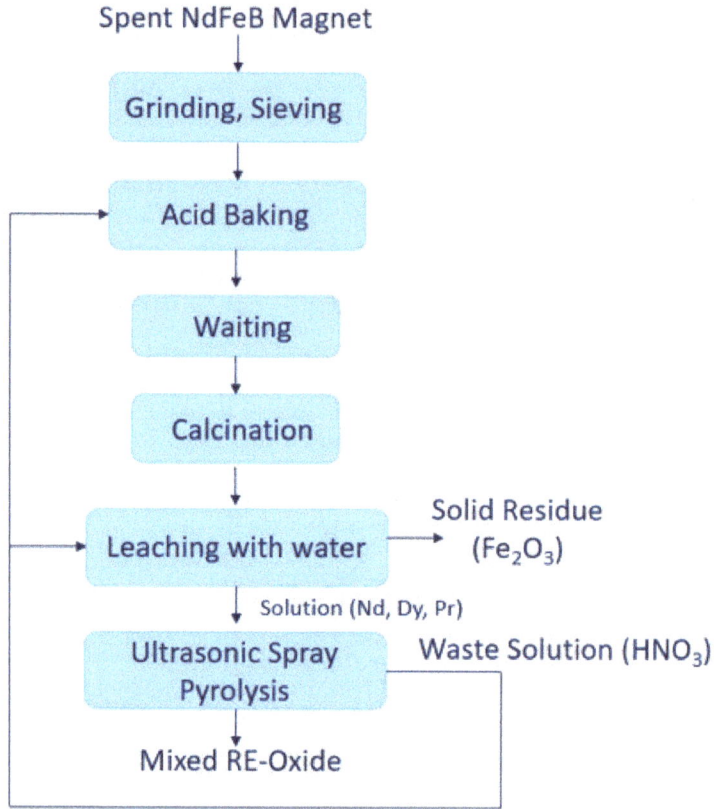

Figure 10. Combined process for treatment of spent NdFeB-Magnets.

Kruse et al. [90] proposed a pyrometallurgical treatment of the grinding slurries aiming recovering REE. The possibility of recycling the contained REE, which account for up to 30 wt. % in the alloy, through a pyrometallurgical process was investigated. In order to reach selective separation, the necessity of a thermal conditioning prior to the pyrometallurgical process is highly required. They concluded that the investigated recycling process must be compared to the pure hydrometallurgical treatment (leaching, precipitation, solvent extraction) of grinding slurries. In addition, a subsequent hydrometallurgical treatment must be considered as saleable concentrate with high purities are required. The future research on this subject will be focused on selective oxidation of REE and production of metallurgical slag with REO through smelting process in an induction furnace.

3.2. Recovery of Yttrium Oxide from Waste Materials

Because of similar characteristics to lanthanoides, yttrium belongs to REE. The importance of yttrium and REE is increasing in the transition to green economy because of their essential role in permanent magnets, lamp phosphors, catalysts, rechargeable batteries, and waste ceramics. In general, the recycling of yttrium can be distinguished in scrap metals which appear during the production or actual value-added chain and the scrap metals which result after the phase of utilization at the consumer and therefore at the end of their life cycle. Michelis et al. [91] studied recovery of yttrium from fluorescent powder starting from dismantling of spent fluorescent tubes. Yttrium and impurities are dissolved by using nitric, hydrochloric, and sulfuric acids and ammonia in different leaching tests. These tests show that ammonia is not suitable to recover yttrium, whereas HNO_3 produces toxic vapors. A full factorial design is carried out with HCl and H_2SO_4 to evaluate the influence of operating factors.

One of the major challenges in the processing of used phosphors for the extraction of REE lies in a large number of different compounds and their individual chemical properties, such as solubility in an aqueous phase. Poscher et al. [92] studied hydrochloric leaching of a screened luminophore powder followed by the precipitation of a contaminated RE oxalate, converting the mixture into their oxides and subsequent refining in order to dissolve most of the alkali metal oxides leads to a rare earth concentrate which could be applied as a new raw material for the subsequent process steps of refining.

Stopic et al. [93–95] studied a combination of hydrometallurgical and pyrometallurgical recovery of yttrium oxide from waste crucible materials, Ti-Al based wastes and ceramic dust. A target fraction rich in yttrium with a grain size < 250 μm was obtained after the grinding and sieving of the primary fraction between 8% and 14.25% of the target element yttrium. This target fraction became the input material for the following tests to the very selective leaching of yttrium involving dissolution under atmospheric pressure as well as pressure leaching in the autoclave. For a comprehensive investigation of the influence of various parameters on the leaching of waste materials, trials were performed under laboratory and scale up conditions in which parameters like leaching time, temperature, concentration of the hydrochloric acid, and the ratio of solid to leaching agent were varied. At process temperatures of 150 °C leaching efficiency was 97%. Selective precipitation with oxalic acid at fixed pH-Value after leaching and phase separation was crucial for a formation of yttrium oxalate. Yttrium oxide was finally formed after thermal decomposition of yttrium oxalate at 850 °C in 1 h in a muffle furnace, as shown at Figure 11.

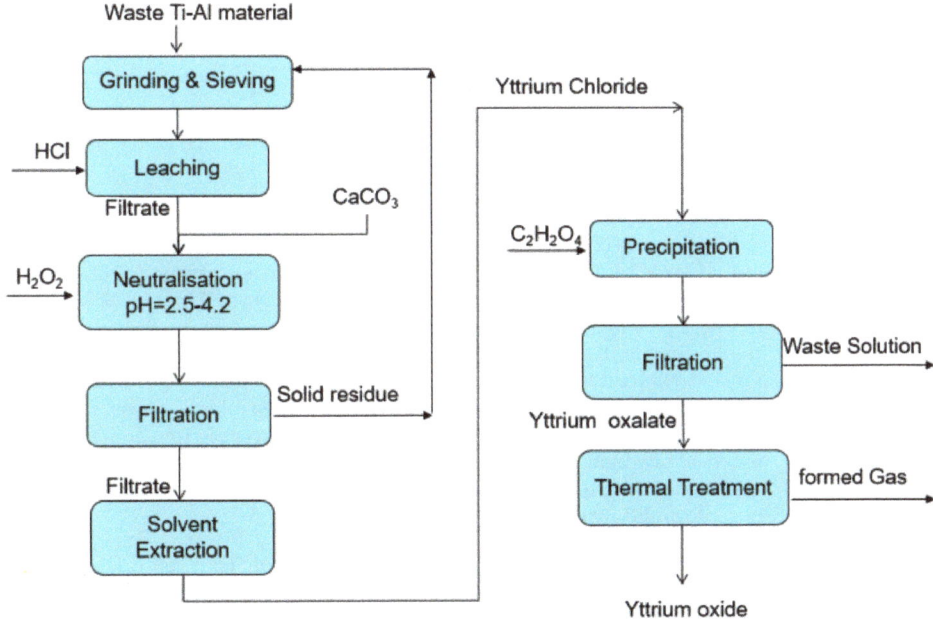

Figure 11. Combined hydrometallurgical and pyrometallurgical method for production of yttrium oxide.

3.3. Recovery of REE from Acid Mine Drainage (AMD) Solution and Bauxite Residue (BR)

AMD and BR are frequently available in the metallurgical and mining industry. The AMD sample was collected from Mpumalanga, South Africa [96] containing the initial concentration of cerium of 5.51 mg/L and yttrium of 2.05 mg/L. Treating AMD solutions requires the generation of enough alkalinity to neutralize the acidity excess. Recovery of REE was performed during precipitation after iron removal [97]. A near zero waste valorization vision for BR through experimental results was performed including a treatment of AMD [98]. BR were used for the recovery of iron, aluminum, and titanium and critical elements such vanadium and REE using a combination of pyrometallurgical and hydrometallurgical methods. The neutralization of AMD, as shown at Figure 12a and capture of the carbon dioxide from different processes, as shown at Figure 12b was performed using BR.

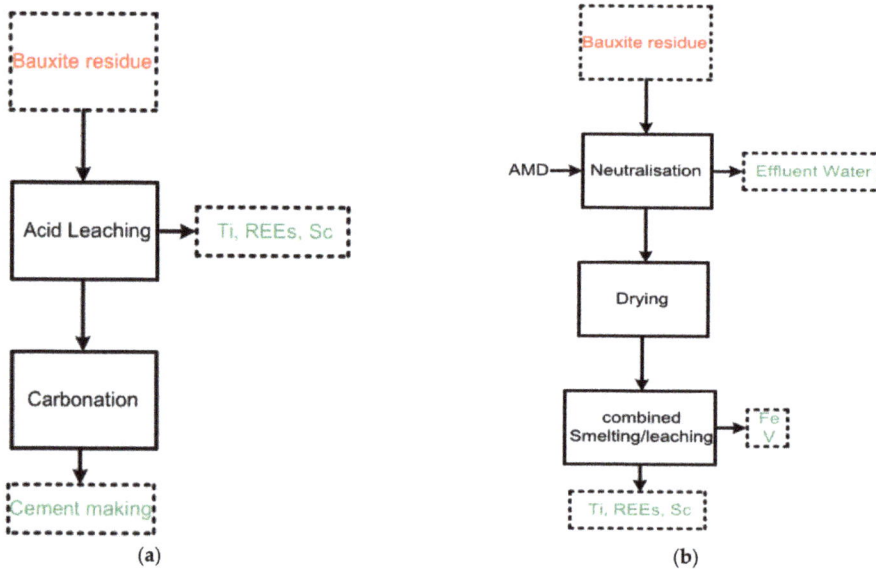

Figure 12. A zero-waste valorization vision for bauxite residue through experimental results: (**a**) for cement industry (left), and (**b**) treatment of AMD solution and recovery of Fe, V, Ti, REE, and Sc (right).

BR, recognized as a waste generating high alkalinity solution when it is in contact with water, was chosen to treat AMD from South Africa at room temperature. A German and a Greek BR ("red mud") have been evaluated as a potential low-cost material to neutralize and immobilize harmful chemical ions from AMD [97,98]. Results showed that heavy metals and other hazardous elements such as As, Se, Cd, and Zn had been immobilized in the mineral phase [99]. Valuable elements presented in BR such as La, Nd, and Nb tend to remain in the mineral phase. Elements such as Ce and Y present in AMD precipitates under the effect of RM enriching the mineral phase. Several authors had explored a pyrometallurgical treatment of BR to recover pig iron and enhance the content of critical raw material (CRM) in the final slag [100–102]. This approach can be beneficial to increase both pig iron and CRM from the filter cakes produce after coagulating AMD ions into an BR matrix.

3.4. Recovery of REE from Phosphogypsum (PG) and Coal Flying Ash (CFA)

PG is a waste by-product from the processing of phosphate rock in plants producing phosphoric acid and phosphate fertilizers, such as superphosphate. CFA is waste product formed by the burning of pulverized coal in a coal-fired boiler and collected from the flue gas by means of electrostatic precipitators and cyclones. PG and CFA became very interesting subject for the REE recovery using unit operations in extractive metallurgy in last decade. Rychkov et al. [103] reported recovery of REE from PG as a common large-tonnage waste of phosphoric fertilizers industry containing 0.43–0.52% of REE. Joint co-crystallization of REE with gypsum provided a low degree of REE leaching in the direct leaching process, but a combination of mechanical grinding treatment, ultrasonic impact and resin-in-pulp process provided significantly higher degrees of REE leaching from the PG (from 15% to 17%, to more than 70%). In addition, it was shown that the products after PG treatment can be successfully used instead of natural gypsum as an addition for cements in the production of building materials.

Mahandra et al. [104] studied an improvement in the developed hydrometallurgical methods and discovery of a new strategies with safe environment and cost effectiveness for the separation and recovery of REE from CFA. The REE recovered from CFA might find applications in clean energy technologies such as electric vehicles, phosphors, and wind turbines. The recycling of CFA using the unit hydrometallurgical operations can help to reduce the environmental burden of hazardous waste and will improve the economy. Pan et al. [105] studied recovery of REE from CFA by integrated physical separation and acid leaching. Using combined physical separation processes, the REE of CFA was enriched from 782 µg/g to 1025 µg/g. The acid leaching process was optimized for various parameters via the Taguchi three-level experimental design. Upon optimization, the physical separation product was leached at the optimum condition and 79.85% leaching efficiency was obtained. Transforming CFA to zeolites can increase the adsorption of dissolved metals [106]. Flying ash can be used for the neutralization of AMD [96] and carbonation process [107].

Vavlekas [108] studied microbial recovery of rare earth elements from metallic wastes and scrap. The microbial recovery of neodymium is being examined using biofilm of the bacterium *Serratia sp. N14*, which has been used previously for the removal of lanthanide elements from liquid solutions. The final aim was to establish the potential usefulness of the biofilm with respect to its tolerance to low pH values or to high salt concentrations (neutralized aqua regia), since the overall goal is to recover metals from solid scraps, which may have been leached in strong acid.

4. Conclusions

Better understanding of the application of UO in metallurgy of REE is very important and required step in order to develop a sustainable metallurgical treatment. REE belong to critical metals and their recovery from of primary and secondary materials is of high significance for the industrial development. The basic UO were presented in different combinations in order to maximize the REE recovery. The main conclusions were found:

- Acid digestion of eudialyte concentrate was successfully performed without heating and at 100 °C for fuming in order to prevent silica gel formation in laboratory conditions. Scale up of this process was tested in two reactors, each 40 L. The extraction efficiency reached a high level in a short time, increasing from 82.2% to 88.8% when the digestion time increased from 20 min to 40 min. The optimization of process was studied using regression analysis and artificial neural network determining one final equation with four reaction parameters (temperature, time, ratio between concentrate and hydrochloric acid, and ratio between water and used acid). The new reactor of 100 L volume was built for a digestion of an eudialyte concentrate. A REE carbonate containing 30.0% total REE was finally produced, with an overall REE recovery yield of 85.5%, what is an advantage in comparison to the previously existing solutions in hydrometallurgy.
- Purification of obtained solution was performed using calcium carbonate and hydrogen peroxide in order to remove iron after acid digestion and water leaching. The final product of a treatment of eudialyte concentrate is REE-carbonate. Heavy REE and light REE are separated in subsequent step using solvent reaction and precipitation. When adjusting the pH to ~4.0 using calcium carbonate, zirconium, aluminum, and iron were removed at 99.1%, 90.0%, and 53.1%, respectively, with a REE loss of 2.1%
- In comparison to an eudialyte concentrate, bastnasite ore was firstly reduced at high temperature and after that the obtained slag with 8% rare earth oxide (REO) was treated by combination of UO.

- REE were produced from secondary materials such as WEEE (spent NdFeB-magnets), Ti-Al waste materials, AMD, BR, PG, and CFA. The combination of hydrometallurgical and pyrometallurgical methods was successfully applied.
- The preparation of fine REO was performed using ultrasonic spray pyrolysis from water solution of rare earth nitrate between 700 °C and 1000 °C after treatment of spent NdFeB magnets.
- Thermal treatment of yttrium oxalate at 850 °C leads to formation of yttrium oxide starting from Ti-Al spent materials.
- A zero-waste valorization vision for BR through experimental results combines AMD and BR as two waste materials in order to absorb REE.
- Molten salt electrolysis was applied for the production of mixture of Nd and Pr.
- The future of an application of UO in extractive metallurgy of REE is depending on a construction of more efficient reactors and digitalization of the whole process.
- Controlled potential leaching and separation process shall be studied as one solution for a selective separation of the REE, which is a new research challenge.
- Optimization of metallurgical process is successfully performed using regression analysis and artificial neural network.

Author Contributions: Conceptualization, investigation, methodology, supervision, writing—original draft, S.S.; funding acquisition and supervision, B.F. All authors have read and agreed to the published version of the manuscript.

Funding: The research at the Institute for Process Metallurgy and Metal Recycling of the RWTH Aachen University regarding to the mentioned results has received funding from 1. Project: "Recycling of Nd, Dy, Pr from NdFeB-spent magnets" RECMAG, (1.1.2021–31.12.2022), AIF- German Federation of Industrial Research Associations, Germany and TÜBITAK-The Scientific of Technological Research Council of Turkey (Call identifier CORNET 29th Call) under grant agreement EN03193/20; 2. Project "Removing the waste streams from the primary Aluminium production and other metal sectors in Europe REMOVAL (1.5.2018–30.4.2022, Gr. No 776469), Executive Agency for Small and Medium-sized Enterprises H2020 Environment & Resources; 3. Project: "Development of Sustainable Exploitation Scheme for Europe's Rare Earth Ore Deposits (1.1.2013–31.12.2017) EURARE, from the European Community's Seventh Framework Programme (Call identifier FP7-NMP2012-LARGE-6) under grant agreement No. 309373; 4. Project "Acid mine drainage treatment technologies for maximum value extraction and near zero waste generation" (1.5.2017–30.4.2021) ADDWATER, with Witwatersrand University in Johannesburg, South Africa by the International Office of the BMBF in Germany, grant no. 01DG17024, and by NRF in South Africa (grant no. GERM160705176077); 5. Project: "European Training Network for Zero-waste Valorisation of Bauxite Residue" (1.1.2014–31.12.2017) REDMUD, EU-Commission, (grant no. 636876); 6: Project "Development of a modular reactor for dry digestion of high silicone ores and concentrates to avoid gel formation using the example of Eudialyte concentrates" MAREKO (1.4.2019–31.3.2021), ZIM AIF, (grant no.: ZF4098204SU8); and Project 7: "Increasing raw material efficiency in TiAl component production processes, (1.7.2014–31.12.2016) TiAl-2020, Federal Ministry for Economy (BMWI): (grant no. FKZ:20T1319D).

Institutional Review Board Statement: Not applicable.

Informed Consent Statement: Not applicable.

Data Availability Statement: Not applicable.

Acknowledgments: I would like to thank Carsten Dittrich, MEAB Chemie Technik GmbH in Aachen for his assistance in preparation of this paper, his permission to publish some results such as Figures 8 and 9, and especially for his continuous support of my work in metallurgy of rare earth elements.

Conflicts of Interest: The authors declare no conflict of interest.

References

1. Havlik, T.; Turzakova, M.; Friedrich, B.; Stopic, S. Atmospheric leaching of EAF dust with diluted sulphuric acid. *Hydrometallurgy* **2005**, *77*, 41–50. [CrossRef]
2. Havlik, T.; Friedrich, B.; Stopic, S. Pressure leaching of EAF dust with sulphuric acid. *Erzmetall World Metall.* **2004**, *57*, 2, 113–120.
3. Falagan, C.; Grail, B.; Johnson, D. New approaches for extracting and recovering metals from mine tailings. *Miner. Eng.* **2017**, *106*, 15, 71–78. [CrossRef]
4. Harahsheh, M.; Kingman, S. Microwave assisted leaching—A review. *Hydrometallurgy* **2004**, *73*, 189–203. [CrossRef]
5. Voßenkaul, D.; Birich, A.; Müller, N.; Stolz, N.; Friedrich, B. Hydrometallurgical Processing of Eudialyte Bearing Concentrates to Recover Rare Earth Elements Via Low-Temperature Dry Digestion to Prevent the Silica Gel Formation. *J. Sustain. Metall.* **2017**, *3*, 79–89. [CrossRef]
6. Demol, J.; Ho, E.; Senanayake, G. Sulfuric acid baking and leaching of rare earth elements, thorium and phosphate from a monazite concentrate: Effect of bake temperature from 200 to 800 °C. *Hydrometallurgy* **2018**, *179*, 254–267. [CrossRef]
7. Matus, C.; Stopic, S.; Etzold, S.; Kremer, D.; Wotruba, H.; Dertmann, C.; Telle, R.; Knops, P.; Friedrich, B. Mechanism of nickel, magnesium, and iron recovery from olivine bearing ore during leaching with hydrochloric acid including a carbonation pre-treatment. *Metals* **2020**, *10*, 811. [CrossRef]
8. Pavlović, J.; Stopic, S.; Friedrich, B.; Kamberovic, Z. Selective removal of heavy metals from metal-bearing wastewater in a cascade line reactor. *Environ. Sci. Pollut. Res.* **2007**, *14*, 7, 518–522. [CrossRef]
9. Xie, F.; Zhang, T.; Dreisinger, D.; Doyle, F. A critical review on solvent extraction of rare earths from aqueous solutions. *Miner. Eng.* **2014**, *56*, 10–28. [CrossRef]
10. Ma, Y.; Stopic, S.; Friedrich, B. Selective Recovery and separation of Zr and Hf from sulfuric acid leach solution using anionic exchange resin. *Hydrometallurgy* **2019**, *189*, 105–143. [CrossRef]
11. Wang, H.; Friedrich, B. Development of a highly efficient hydrometallurgical recycling process for automotive Li–Ion batteries. *J. Sustain. Metall.* **2015**, *1*, 168–178. [CrossRef]
12. Mwewa, B.; Stopic, S.; Ndlovu, S.; Simate, G.; Xakalashe, B.; Friedrich, B. Synthesis of Poly-Alumino-Ferric Sulphate Coagulant from Acid Mine Drainage by Precipitation. *Metals* **2019**, *9*, 1166. [CrossRef]
13. Kilicarslan, A.; Voßenkaul, D.; Stoltz, S.; Stopic, S.; Nezihi, M.; Friedrich, B. Selectivity Potential of Ionic Liquids for Metal Extraction from Rare Earth containing Slags—A QEMSCAN assisted approach. *Hydrometallurgy* **2017**, *169*, 59–67.
14. Stankovic, S.; Stopic, S.; Markovic, B.; Sokic, M.; Friedrich, B. Review of the past, present and future of the hydrometallurgical production of nickel and cobalt from lateritic ores. *Metall. Mater. Eng.* **2020**, 199–208. [CrossRef]
15. Petronijevic, N.; Stankovic, S.; Radovanovic, D.; Sokic, M.; Markovic, M.; Stopic, S.; Kamberovic, Z. Application of the flotation tailings as an alternative material for an acid mine drainage remediation: A case study of the extremely acidic Lake Robule (Serbia). *Metals* **2019**, *10*, 16. [CrossRef]
16. Silin, I.; Hahn, K.; Gürsel, D.; Kremer, D.; Gronen, L.; Stopic, S.; Friedrich, B.; Wotruba, H. Mineral Processing and Metallurgical Treatment of Lead Vanadate Ores. *Minerals* **2020**, *10*, 197. [CrossRef]
17. Stopic, S.; Friedrich, B. Deposition of silica in hydrometallurgical processes. *Mil. Tech. Cour.* **2020**, *68*, 65–78. [CrossRef]
18. Peters, E.M.; Kaya, Ş.; Dittrich, C.; Forsberg, K. Recovery of scandium by crystallization techniques. *J. Sustain. Metall.* **2019**, *5*, 48–56. [CrossRef]
19. Rodriguez, J.; Stopic, S.; Friedrich, B. Feasibility assessment of electrocoagulation towards a new sustainable wastewater treatment. *Environ. Sci. Pollut. Res.* **2007**, *14*, 7, 477–482. [CrossRef]
20. Rodriguez, J.; Schweda, M.; Stopic, S.; Friedrich, B. Techno-Economical Comparison between Chemical Precipitation and Electrocoagulation for Heavy Metal Removal in Industrial Wastewater. *Metals* **2007**, *2*, 208–214.
21. Rodriguez, J.; Stopic, S.; Friedrich, B. Continuous Electrocoagulation Treatment of Wastewater from Copper Production. *Erzmetall World Metall.* **2007**, *60*, 89–95.
22. Zhao, J.; Stopic, S.; Friedrich, B.; Rudolf, R. Mechanism of gold nanoparticle formation reduction from water solutions. In Proceedings of the European Metallurgical Conference 2013, Weimar, Germany, 23–26 June 2016; pp. 277–281.
23. Stopic, S.; Friedrich, B. Integrated treatment (electrolytic recovery and cascade line) of highly contaminated wastewaters. In Proceedings of the EMC 2007, Düsseldorf, Germany, 11–14 June 2007; pp. 433–444.
24. Cvetkovic, V.; Feldhaus, D.; Vukicevic, N.; Barudzija, T.; Friedrich, B.; Jovicevic, J. Investigation on the electrochemical behaviour and deposition mechanism of neodymium in NdF_3–LiF–Nd_2O_3 melt on Mo electrode. *Metals* **2020**, *10*, 576. [CrossRef]
25. Košević, M.; Stopic, S.; Cvetković, V.; Schroeder, M.; Stevanović, J.; Panic, V.; Friedrich, B. Mixed RuO_2/TiO_2 uniform microspheres synthesized by low-temperature ultrasonic spray pyrolysis and their advanced electrochemical performances. *Appl. Surf. Sci.* **2019**, *464*, 1–9. [CrossRef]
26. Majeric, P.; Rudolf, R. Advances in ultrasonic spray pyrolysis processing of noble metal nanoparticles—Review. *Materials* **2020**, *13*, 3485. [CrossRef]
27. Emil, E.; Alkan, G.; Gurmen, S.; Rudolf, R.; Jenko, D.; Friedrich, B. Tuning the morphology of ZnO nanostructures with the ultrasonic spray pyrolysis process. *Metals* **2018**, *8*, 569. [CrossRef]
28. Ardekani, S.R.; Aghdam, A.S.R.; Nazari, M.; Bayat, A.; Yazdani, E.; Saievar-Iranizad, E. A comprehensive review on ultrasonic spray pyrolysis technique: Mechanism, main parameters and applications in condensed matter. *J. Anal. Appl. Pyrolysis* **2019**, *141*, 104631. [CrossRef]

29. Kaya, E.E.; Kaya, O.; Alkan, G.; Gürmen, S.; Stopic, S.; Friedrich, B. New proposal for size and size-distribution evaluation of nanoparticles synthesized via ultrasonic spray pyrolysis using search algorithm based on image-processing technique. *Materials* **2020**, *13*, 38. [CrossRef]
30. Stopic, S.; Ilić, I.; Uskoković, D. Preparation and formation mechanism of submicrometer spherical NiO particles from water solution of NiCl$_2$ by ultrasonic spray pyrolysis. *J. Sci. Sinter.* **1994**, *26*, 145–156.
31. Stopic, S.; Ilić, I.; Uskoković, D. Structural and morphological transformations during NiO and Ni particles generations from chloride precursor by ultrasonic spray pyrolysis. *Mater. Lett.* **1995**, *24*, 369–376. [CrossRef]
32. Stopic, S.; Ilić, I.; Uskoković, D. Preparation of nickel submicrometer powders by ultrasonic spray pyrolysis. *Int. J. Powder Metall.* **1996**, *32*, 59–65.
33. Alkan, G.; Yagmurlu, B.; Cakmakoglu, S.; Hertel, T.; Kaya, S.; Gronen, L.; Stopic, S.; Friedrich, B. Novel approach for enhanced scandium and titanium leaching efficiency from bauxite residue with suppressed silica gel formation. *Nat. Sci. Rep.* **2018**, *8*, 5676. [CrossRef]
34. Alkan, G.; Yagmurlu, B.; Friedrich, B.; Dittrich, C.; Gronen, L.; Stopic, S.; Ma, Y. Selective silica gel scandium extraction from iron–depleted red mud slags by dry digestion. *Hydrometallurgy* **2019**, *185*, 266–272. [CrossRef]
35. Borra, C.R.; Blanpain, B.; Pontikes, Y.; Binnemans, K.; van Gerven, T. Recovery of rare earths and other valuable metals from bauxite residue (red mud): A review. *J. Sustain. Metall.* **2016**, *2*, 365–386. [CrossRef]
36. Binnemans, K.; Jones, P.T. Solvometallurgy: An Emerging Branch of Extractive Metallurgy. *J. Sustain. Metall.* **2017**, *3*, 570–600. [CrossRef]
37. Tian, Q.; Xin, Y.; Wang, H.; Guo, X. Potential-Control selective separation of manganese and cobalt from cobalt slag leaching solution. *Hydrometallurgy* **2017**, *169*, 201–206. [CrossRef]
38. Garg, S.; Judd, K.; Mahadevan, R.; Edwards, E.; Papangelakis, V. Leaching characteristics of nickeliferous pyrrhotite tailings from the Sudbury, Ontario area. *Can. Metall. Q.* **2017**, *8*, 1–9. [CrossRef]
39. Kücher, G.; Luidold, S.; Czettl, C.; Storf, C. Successful control of the reaction mechanism for semi-direct recycling of hard metals. *Int. J. Refract. Met. Hard Mater.* **2020**, *86*, 105131. [CrossRef]
40. Milivojevic, M.; Stopic, S.; Friedrich, B.; Stojanovic, B.; Drndarevic, D. Computer modeling of high pressure leaching process of nickel laterite by design of experiments and neural networks. *Int. J. Miner. Metall. Mater.* **2012**, *19*, 584–595. [CrossRef]
41. Hernandes, J. Development of a Predictive Model for the Recovery of Rare Earth Elements from the Leaching Process of Chilean Ores. *Appl. Math. Sci.* **2018**, *12*, 551–565. [CrossRef]
42. Milivojevic, M.; Stopic, S.; Stojanovic, B.; Drndarevic, D.; Friedrich, B. Forward stepwise regression in determining dimensions of forming and sizing tools for self-lubricated bearings. *Metall* **2013**, *67*, 147–153.
43. Ayres, R.U.; Peiró, L.T. Material efficiency: Rare and critical metals. *Phylosophical Trans. R. Soc. A* **2013**, *371*, 20110563. [CrossRef]
44. Tao, X.; Huiqing, P. Formation cause, composition analysis and comprehensive utilization of rare earth solid wastes. *J. Rare Earths* **2009**, *27*, 1096–1102.
45. Sprecher, B.; Xiao, Y.; Walton, A.; Speight, J.; Harris, R.; Kleijn, R.; Kramer, G.J. Life cycle inventory of the production of rare earths and the subsequent production of NdFeB rare earth permanent magnets. *Environ. Sci. Technol.* **2014**, *48*, 3951–3958. [CrossRef] [PubMed]
46. Binnemans, K.; Jones, P.T.; Blanpain, B.; van Gerven, T.; Yang, Y.; Walton, A.; Buchert, M. Recycling of rare earths: A critical review. *J. Clean. Prod.* **2013**, *51*, 1–22. [CrossRef]
47. Önal, M.; Borra, C.; Guo, M.; Blanpain, B.; van Gerven, T. Recycling of NdFeB Magnets using sulfation, selective roasting and water leaching. *J. Sustain. Metall.* **2015**, *1*, 199–215. [CrossRef]
48. Önal, M.A.R.; Aktan, E.; Borra, C.R.; Blanpain, B.; van Gerven, T.; Guo, M. Recycling of NdFeB magnets using nitration, calcination and water leaching for REE recovery. *Hydrometallurgy* **2017**, *167*, 115–123. [CrossRef]
49. Liu, Z.; Li, M.; Hu, Y.; Wang, M.; Shi, Z. Preparation of large particle rare earth oxides by precipitation with oxalic acid. *J. Rare Earths* **2008**, *26*, 158–162. [CrossRef]
50. Silva, R.G.; Morais, C.A.; Teixeira, L.V.; Oliveira, É.D. Selective precipitation of high-quality rare earth oxalates or carbonates from a purified sulfuric liquor containing soluble impurities. *Min. Metall. Explor.* **2019**, *36*, 967–977. [CrossRef]
51. Stopic, S.; Friedrich, B. Method of selective recovery of yttrium oxide from titanium-aluminium based wastes. In Proceedings of the 13th International Conference on Fundamental and Applied Aspects of Physical Chemistry, Belgrade, Serbia, 26 September 2016; pp. 510–516.
52. Ma, Y.; Stopic, S.; Wang, X.; Forsberg, K.; Friedrich, B. Basic sulfate precipitation of zirconium from sulfuric acid leach solution. *Metals* **2020**, *10*, 1099. [CrossRef]
53. Han, K.N. Characteristics of precipitation of rare earth elements with various precipitants. *Minerals* **2020**, *10*, 178. [CrossRef]
54. Yadav, A.A.; Lokhande, V.C.; Bulakhe, R.N.; Lokhande, C.D. Amperometric CO_2 gas sensor based on interconnected web-like nanoparticles of La_2O_3 synthesized by ultrasonic spray pyrolysis. *Microchim. Acta* **2017**, *184*, 3713–3720. [CrossRef]
55. Jung, D.S.; Hong, S.K.; Lee, H.J.; Kang, Y.C. Gd_2O_3: Eu phosphor particles prepared from spray solution containing boric acid flux and polymeric precursor by spray pyrolysis. *Opt. Mater.* **2006**, *28*, 530–535. [CrossRef]
56. Goulart, C.; Djurado, E. Synthesis and sintering of Gd-doped CeO_2 nanopowders prepared by ultrasonic spray pyrolysis. *J. Eur. Ceram. Soc.* **2013**, *33*, 769–778. [CrossRef]

57. Emil, E.; Gürmen, S. Estimation of yttrium oxide microstructural parameters using the Williamson–Hall analysis. *Mater. Sci. Technol.* **2018**, *34*, 1549–1557. [CrossRef]
58. Lebedev, V.N.; Schur, T.E.; Maiorov, D.V.; Popova, L.A.; Serkova, R.P. Specific Features of Acid Decomposition of Eudialyte and Certain Rare-Metal Concentrates from Kola Peninsula. *Russ. J. Appl. Chem.* **2003**, *76*, 1191–1196. [CrossRef]
59. Lebedev, V.N. Sulfuric Acid Technology for Processing of Eudialyte Concentrate. *Russ. J. Appl. Chem.* **2003**, *76*, 1559–1563. [CrossRef]
60. Zakharov, V.I.; Skiba, G.S.; Solovyov, A.V.; Lebedev, V.N.; Mayorov, D.V. Some Aspects of Eudialyte Processing. *Tsvetnye Met.* **2011**, *11*.
61. Zakharov, V.I.; Maiorov, D.V.; Alishkinm, A.R.; Matveev, V.A. Causes of Insufficient Recovery of Zirconium during Acidic Leaching of Lovozero Eudialyte Concentrate. *Izv. VUZ Tsvetnaya Metall.* **2011**, *5*, 26–31.
62. Litvinova, T.E.; Chirkist, D.E. Physic-Chemical foundations of the eudialyte processing. In Proceedings of the 2nd Russian Conference with International Participation "New Approaches to Chemical Engineering Minerals. The Use of Extraction and Sorption", St. Petersburg, Russia, 3–6 June 2013; pp. 84–87.
63. Dibrov, I.A.; Chirkist, D.E.; Litvinova, T.E. Distribution of the elements after sulfuric acid leaching of eudialyte concentrate. *Tsvetnye Met.* **2002**, *13*, 38–41.
64. Dibrov, I.A.; Chirkist, D.E. Processing technology of eudialyte concentrates. *Miner. Process. Extr. Metall. Rev.* **1995**, *15*, 141. [CrossRef]
65. Davris, P.; Stopic, S.; Balomenos, E.; Panias, D.; Paspaliaris, I.; Friedrich, B. Leaching of rare earth elements from Eudialyte concentrate by supressing silicon dissolution. *Miner. Eng.* **2017**, *108*, 115–122. [CrossRef]
66. Ma, Y.; Stopic, S.; Friedrich, B. Hydrometallurgical Treatment of a Eudialyte Concentrate for Preparation of Rare Earth Carbonate. *Johns. Matthey Technol. Rev.* **2019**, *63*, 2–13. [CrossRef]
67. Ma, Y.; Stopic, S.; Gronen, L.; Obradovic, S.; Milivojevic, M.; Friedrich, B. Neural network modeling for the extraction of rare earth elements from eudialyte concentrate by dry digestion and leaching. *Metals* **2018**, *8*, 267. [CrossRef]
68. Stopic, S.; Dertmann, C.; Gulgans, U.; Friedrich, B. New modular reactor for dry unearthing of silicate-rich ores and concentrates and for the extraction of critical metals. *World Metall. Erzmetall* **2019**, *72*, 180–181.
69. Calow, R.J. *The Industrial Chemistry of the Lanthanons, Yttrium, Thorium, and Uranium*; Pergamon Press: Oxford/London/Edinburgh, UK, 1967; Volume 7, p. 248.
70. Yun, Y.; Stopic, S.; Friedrich, B. Valorization of Rare Earth Elements from a Steenstrupine Concentrate via a Combined Hydrometallurgical and Pyrometallurgical Method. *Metals* **2020**, *10*, 248. [CrossRef]
71. Vijayalakshmi, R.; Mishra, S.; Singh, H.; Gupta, C. Processing of xenotime concentrate by sulfuric acid digestion and selective thorium precipitation for separation of rare earths. *Hydrometallurgy* **2001**, *61*, 75–80. [CrossRef]
72. Stopic, S.; Friedrich, B. Leaching of rare earth elements with sulfuric acid from bastnasite ores. *Mil. Tech. Cour.* **2018**, *66*, 757–770. [CrossRef]
73. Stopic, S.; Friedrich, B. Leaching of rare earth elements from bastnasite ore: Second part. *Mil. Tech. Cour.* **2019**, *67*, 241–254. [CrossRef]
74. Stopic, S.; Friedrich, B. Leaching of rare earth elements from bastnasite ore (third part). *Mil. Tech. Cour.* **2019**, *67*, 561–572. [CrossRef]
75. Milicevic, K.; Friedrich, B. Special Feature: Electrowinning of didymium. *Appl. Mineral.* **2017**, *2*, 2–4.
76. Vogel, H.; Flerus, B.; Stoffner, F. Reducing greenhouse gas emission from the neodymium oxide electrolysis. Part I: Analysis of the anodic gas formation. *J. Sustain. Metall.* **2016**, *3*, 99–107. [CrossRef]
77. Vogel, H.; Friedrich, B. Reducing greenhouse gas emission from the neodymium oxide electrolysis. Part II: Basics of a process control avoiding PFC emission. *Int. J. Nonferrous Metall.* **2017**, *6*, 27–46. [CrossRef]
78. Haas, T.; Hilgendorf, S.; Vogel, H.; Friedrich, B.; Pfeiffer, H. A Comparison between Two Cell Designs for Electrochemical Neodymium Reduction Using Numerical Simulation. *Metall. Mater. Trans. B* **2017**, *48*, 2187–2194. [CrossRef]
79. Cvetković, V.; Vukićević, N.; Feldhaus, D.; Barudžija, T.; Stevanović, J.; Friedrich, B.; Jovićević, J. Study of Nd Deposition onto W and Mo Cathodes from Molten Oxide-Fluoride Electrolyte. *Int. J. Electrochem. Sci.* **2020**, *15*, 7039–7052. [CrossRef]
80. Kuang-Taek, R. Effects of rare earth elements on the environment and human health: A literature review. *Toxicol. Environ. Health Sci.* **2016**, *8*, 189–200.
81. Mancheri, N.A.; Sprecher, B.; Bailey, G.; Ge, J.; Tukker, A. Effect of Chinese policies on rare earth supply chain resilience. *Resour. Conserv. Recycl.* **2019**, *142*, 101–112. [CrossRef]
82. Ambaye, T.; Vaccari Castro, F.; Prasad, S.; Rtimi, S. Emerging technologies for the recovery of rare earth elements (REEs) from the end-of-life electronic wastes: A review on progress, challenges, and perspectives. *Environ. Sci. Pollut. Res.* **2020**, *27*, 36052–36074. [CrossRef] [PubMed]
83. Yang, Y.; Walton, A.; Sheridan, R.; Gth, K.; Gauß, R.; Gutfleisch, O.; Buchertm, M.; Stenari, B.M.; van Gerven, T.; Jones, O.T.; et al. REE Recovery from End-of Life NdFeB permanent magnet scrap: A Critical review. *J. Sustain. Metall* **2017**, *3*, 122–149. [CrossRef]
84. Liu, Y.; Jeon, H.S.; Lee, M.S. Solvent extraction of Pr and Nd from chloride solution by the mixtures of Cyanex 272 and amine extractants. *Hydrometallurgy* **2014**, *150*, 61–67. [CrossRef]
85. Abreu, R.D.; Morais, C.A. Study on separation of heavy rare earth elements by solvent extraction with organophosphorous acids and amine reagents. *Miner. Eng.* **2014**, *61*, 82–87. [CrossRef]

86. Elwert, T.; Goldmann, D.; Römer, F. Separation of lanthanides from NdFeB magnets on a mixer-settler plant with PC-88A. *World Metall.* **2014**, *67*, 287–296.
87. Kaya, E.; Kaya, O.; Stopic, S.; Gürmen, S.; Friedrich, B. NdFeB Magnets recycling process: An alternative method to produce mixed rare rarth oxide from scrap NdFeB magnets. *Metals* **2021**, *11*, 716. [CrossRef]
88. Gussone, J.; Reddy, C.; Vijay, V.; Haubrich, J.; Milicevic, K.; Friedrich, B. Effect of vanadium ion valence state on the deposition behavior in molten salt electrolysis. *J. Appl. Chem.* **2018**. [CrossRef]
89. Milicevic, K.; Feldhaus, D.; Friedrich, B. Conditions and mechanism of gas emission from didymium electrolysis and its process control. *Light Met.* **2018**, 1435–1441. [CrossRef]
90. Kruse, S.; Raulf, K.; Trentmann, A.; Pretz, T.; Friedrich, B. Processing of grinding slurries arising from NdFeB Magnet production. *Chem. Ingen. Tech.* **2015**, *87*, 1589–1598. [CrossRef]
91. Michelis, I.; Ferrela, F.; Fioravante Varelli, E.; Veglio, F. Treatment of exhaust fluorescent lamps to recover yttrium: Experimental and process analyses. *Waste Manag.* **2011**, *31*, 2056–2068. [CrossRef]
92. Poscher, A.; Luidold, S.; Kaindl, M.; Antrekowitsch, H. Recycling of Rare Earth from Spent Phosphors. In Proceedings of the EMC Conference, Weimar, Germany, 23–26 June 2013; pp. 1217–1222.
93. Stopic, S.; Lerch, M.; Gerke-Cantow, R.; Friedrich, B. Hydrometallurgical Recovery of Yttrium from Waste Crucible Materials. In Proceedings of the European Metallurgical Conference in Weimar, Aachen, Germany, 23–26 June 2013; pp. 939–944.
94. Stopic, S.; Friedrich, B. Kinetics of yttrium dissolution from waste ceramic dust. *Mil. Tech. Cour.* **2016**, *64*, 383–395. [CrossRef]
95. Stopic, S.; Kalabis, S.; Friedrich, B. Recovery of yttrium oxide from titanium-aluminium based wastes. *J. Eng. Process. Manag.* **2018**, *10*, 9–20. [CrossRef]
96. Keller, V.; Stopic, S.; Xakalashe, B.; Ma, Y.; Ndlovu, S.; Simate, G.; Friedrich, B. Effectiveness of Fly Ash and Red Mud as Strategies for Sustainable Acid Mine Drainage Management. *Minerals* **2020**, *10*, 707. [CrossRef]
97. Stopic, S.; Ma, Y.; Mwewa, B.; Ndlovu, S.; Simate, G.; Xakalashe, B.; Friedrich, B. Hydrometallurgical treatment of acid mine drainage (AMD) solution. In Proceedings of the 20th Yucorr Conference, Tara, Serbia, 21–24 May 2018; pp. 44–48.
98. Stopic, S.; Dertmann, C.; Xakalashe, B.; Alkan, G.; Yagmurlu, B.; Lucas, H.; Friedrich, B. A near zero waste valorization vision for bauxite residue through experimental results. In Proceedings of the 21st YuCorr Conference, Tara, Serbia, 16–19 September 2019; pp. 120–125.
99. Lukas, H.; Stopic Xakalashe, B.; Ndlovu, S.; Friedrich, B. Synergism Red Mud-Acid Mine Drainage as a Sustainable Solution for Neutralizing and Immobilizing Hazardous Elements. *Metals* **2021**, *11*, 620. [CrossRef]
100. Yagmurlu, B.; Alkan, G.; Xakalashe, B.; Schier, C.; Gronen, L.; Koiwa, I.; Dittrich, C.; Friedrich, B. Synthesis of Scandium Phosphate after Peroxide Assisted Leaching of Iron Depleted Bauxite Residue (Red Mud) Slags. *Sci. Rep.* **2019**, *9*, 1–10. [CrossRef]
101. Kaußen, F.; Friedrich, B. Reductive smelting of red mud for iron recovery. *Chem. Ing. Tech.* **2015**, *87*, 1535–1542. [CrossRef]
102. Valeev, D.; Zinoveev, D.; Kondratiev, A.; Lubyanoi, D.; Pankratov, D. Reductive Smelting of Neutralized Red Mud for Iron Recovery and Produced Pig Iron for Heat-Resistant Castings. *Metals* **2019**, *10*, 32. [CrossRef]
103. Rychkov, V.N.; Kirillov, E.V.; Kirillov, S.V.; Semenishchev, V.S.; Bunkov, G.M.; Botalov, M.S.; Smyshlyaev, D.V.; Malyshev, A.S. Recovery of rare earth elements from phosphogypsum. *J. Clean. Prod.* **2018**, *196*, 674–681. [CrossRef]
104. Mahandra, H.; Hubert, B.; Ghahreman, A. Recovery of Rare Earth and Some Other Potential Elements from Coal Fly Ash for Sustainable Future. In *Clean Coal Technologies. Beneficiation*; Utilization, Transport Phenomena and Prospective; Rajesh, K.J., Pankaj, K.P., Eds.; Springer: Berlin/Heidelberg, Germany, 2021; pp. 339–380. [CrossRef]
105. Pan, J.; Nie, T.; Hassas, B.; Rezaee, M.; Wen, Z.; Zhou, C. Recovery of rare earth elements from coal fly ash by integrated physical separation and acid leaching. *Chemosphere* **2020**, *248*, 248–126112. [CrossRef]
106. Prasad, B.; Kumar, H. Treatment of Lignite Mine Water with Lignite Fly Ash and Its Zeolite. *Mine Water Environ.* **2019**, *38*, 24–29. [CrossRef]
107. Matus, C.; Stopic, S.; Friedrich, B. Carbonation of Minerals and Slags under High Pressure in an Autoclave. *Mil. Tech. Cour.* **2021**, *69*, 2. [CrossRef]
108. Vavlekas, D. *Microbial Recovery of Rare Earth Elements from Metallic Wastes and Scrap*; University of Birmingham, College of Life and Environmental Sciences: Birmingham, UK, 2014.

MDPI
St. Alban-Anlage 66
4052 Basel
Switzerland
Tel. +41 61 683 77 34
Fax +41 61 302 89 18
www.mdpi.com

Metals Editorial Office
E-mail: metals@mdpi.com
www.mdpi.com/journal/metals

www.ingramcontent.com/pod-product-compliance
Lightning Source LLC
LaVergne TN
LVHW070459100526
838202LV00014B/1752